Photocatalysis
Advanced Materials and Reaction Engineering

Edited by

Dr. Gaurav Sharma

Shoolini University

India

Dr. Amit Kumar

Shoolini University

India

Published by **Materials Research Forum LLC**
Millersville, PA 17551, USA

Published as part of the book series
Materials Research Foundations
Volume 100 (2021)
ISSN 2471-8890 (Print)
ISSN 2471-8904 (Online)

Print ISBN 978-1-64490-134-2
eBook ISBN 978-1-64490-135-9

This book contains information obtained from authentic and highly regarded sources. Reasonable efforts have been made to publish reliable data and information, but the author and publisher cannot assume responsibility for the validity of all materials or the consequences of their use. The authors and publishers have attempted to trace the copyright holders of all material reproduced in this publication and apologize to copyright holders if permission to publish in this form has not been obtained. If any copyright material has not been acknowledged please write and let us know so we may rectify this in any future reprints.

Distributed worldwide by

Materials Research Forum LLC
105 Springdale Lane
Millersville, PA 17551
USA
https://www.mrforum.com

Manufactured in the United States of America
10 9 8 7 6 5 4 3 2 1

Table of Contents

This Book is dedicated to my beloved father Late Sh. Inder Dev. Thanks for all the Love & Sacrifices. Your memories will be a blessing and motivation for life.......Amit Kumar

This book is tribute to my late uncle Sh. Inder Dev. His love, care and never ending support will guide me forever......miss you............Gaurav Sharma

Preface

Guaranteeing safe potable water and meeting the ever-increasing energy demands with the least impact on the environment has become a great challenge for statesmen, environmentalists and the research community worldwide. The sustainable chain of diversity has been disturbed and the energy crisis and environmental pollution issues have emerged as big issues to combat. Keeping in mind the limited resources, substantial efforts are focusing on the use of solar energy for high performance removal of pollutants and clean energy production. The scientific community and environmentalists are always amazed by nature inspired processes such as photocatalysis for sustainable solutions.

The phenomenon which is observed as "Honda-Fujishima" effect, "Photocatalysis" has been one of the pioneer techniques to achieve degradation and substantial mineralization of complex organic pollutants including pharmaceutical effluents, dyes, pesticides, endocrine disruptors and other emerging pollutants. The technique is most trusted after the secondary treatment because of high mineralization capability via hydroxyl and superoxide anion radicals, thus eliminating any secondary pollution. With rise in quantum and variety of emerging pollutants and contaminants various multi-functional materials are required with better photo-oxidative and photo-reduction potentials. This requires material engineering, reaction and optical engineering to harness visible & solar light and reduce the charge recombination issues. In addition, the challenge also lies in achieving cost effectiveness along with high quantum yield and reduced or eliminate the use of precious metals. Further, the catalysts with a multi-pronged approach for water detoxification, H_2 evolution, CO_2 conversion into fuels, N_2 fixation etc. are the demand of the current time.

With designing and development of hybrid advanced materials via selection of new photocatalysts, techniques for synthesis as chemical vapor deposition, electrospinning, membrane formation, immobilization, green synthesis, functionalization and sensitization are highly effective in a good photocatalytic response. In addition to popular photocatalysts i.e metal and metal oxides, various new age photocatalysts have appeared including nitrides, phosphides, perovskites, chalcogenides, organic semiconductors, conducting polymers with better and tuneable properties. Also, combining the materials with different and complimentary properties to form nano-composites, hetero-structures and heterojunctions is a popular strategy. This helps in the proper alignments of conduction and valence bonds for high production of reactive oxygen species. In addition, the use of adsorptional photocatalysis helps in mineralization and removal of micro-pollutants fast and effectively. Among organic materials, graphene oxide, carbon nanotubes, graphitic carbon nitride etc. have received huge attention.

The book focuses on some of the latest developments in the field of photocatalysis with special emphasis on new advanced materials and reaction engineering. The bottlenecks have been addressed in various research works focusing on heterojunction formations, composites, ion exchangers, photocatalytic membranes, etc. This book can provide useful insight for designing and fabrication of hybrid photocatalytic systems for photocatalytic water treatment and energy production.

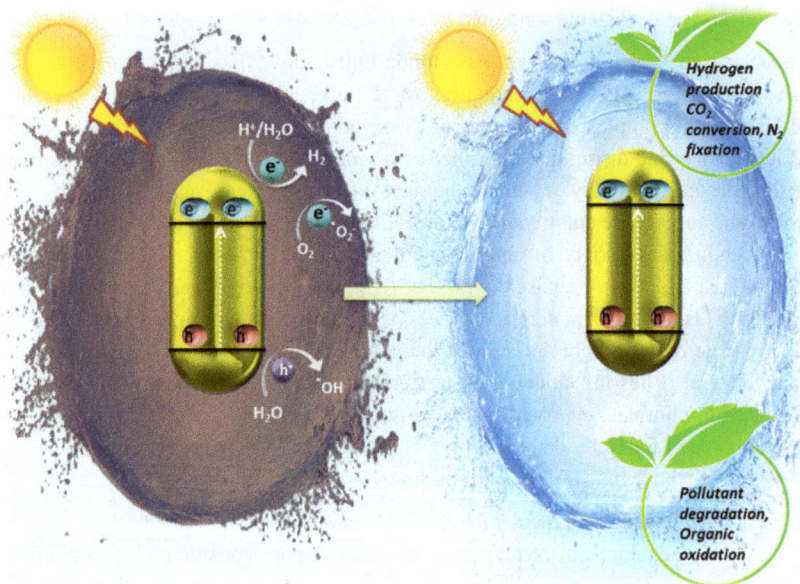

Dr. Gaurav Sharma

Shoolini University

India

Dr. Amit Kumar

Shoolini University

India

Photocatalysis
Materials Research Foundations **100** (2021) 1-56

Materials Research Forum LLC
https://doi.org/10.21741/9781644901359-1

Chapter 1

Photocatalytic Membranes in Degradation of Organic Molecules

Gisya Abdi[1]*, Mehdihasan I. Shekh[2], Jhaleh Amirian[3], Abdolhamid Alizadeh[4], Sirus Zinadini[5]

[1]Academic Centre for Materials and Nanotechnology, AGH University of Science and Technology, al. Adama Mickiewicza 30, 30-059 Krakow, Poland

[2]College of Materials Science and Engineering, Shenzhen University, Shenzhen 518060, P. R. China

[3]Institute of Physical Chemistry, Polish Academy of Sciences, Kasprzaka 44/52, 01-224 Warsaw, Poland

[4]Department of Organic Chemistry, Faculty of Physics Chemistry, Alzahra University, Tehran, 1993893973, Iran

[5]Environmental Research Center (ERC), Department of Applied Chemistry, Razi University, Kermanshah, Iran

* abdi_gisya@yahoo.com

Abstract

Heterogeneous photocatalysis is a technology widely applied to water purification and wastewater treatment under ultraviolet (UV) or even sunlight irradiation for the removal of a variety of environmental pollutants into harmless species. Application of membrane for immobilization of semiconductors and their suitability in photocatalytic degradation of dyes have recently been developed. Integration of photocatalysis with membrane processes significantly improve the membrane separation performance with reducing membrane fouling and improving permeate quality. This paper reviews recent progress in the photocatalytic membranes for wastewater treatment and water purification with an emphasis on the type of membranes, membrane fabrication, and applications in pollutant removal.

Keywords

Semiconductors, Organic Pollutants, Heterogeneous Photocatalysis, Ceramic Photocatalytic Membranes, Polymeric Photocatalytic Membranes

Photocatalysis
Materials Research Foundations **100** (2021) 1-56

Materials Research Forum LLC
https://doi.org/10.21741/9781644901359-1

Contents

1.   Introduction

The introduction of industrial, pharmaceutical, and agricultural chemicals into the ecosystems, as a result of human activities, are now a considerable subject of environmental remediation. The complex mixture of these compounds and their non-biodegradable character make conventional wastewater treatment methods (physical, chemical, and biological methods) unable to completely remove them. Thus, the interest in these environmental effects leads to an increase of research activities for elimination of such pollutants.

Advanced oxidation processes (AOPs) have been proposed as alternative methods for the elimination of persistent organic compounds in wastewaters, air, and soil. The principle of AOPs is to generate hydroxyl radical and other radical species in the water, the very powerful oxidants are capable of oxidizing a wide range of organic compounds with one or many double bonds. Nowadays, a photocatalytic process using a semiconductor as photocatalyst under light irradiation has been extensively applied for the oxidation of various organic pollutants.

Photocatalyst comes from the word "photo", which means light, and "catalyst", which refers to a substrate that accelerates chemical reactions by absorbing photons or light. It is noted that UV light works within the wavelength range of 280-400 nm (UVA, 315-400 nm, UVB, 280-315 nm, and UVC, 180-280 nm), while visible light works in the range between 400-700 nm. Since UV light activates most photocatalysts, a constant source is needed to ensure the continuity of the reaction. For doped photocatalyst that can function under visible light, direct exposure to sunlight is sufficient to initiate photocatalytic activity. The photocatalysis can be classified into homogeneous and heterogeneous categories [1]. In this review, we focus on heterogeneous photocatalysis using polymeric and ceramic membranes. A photocatalyst is a material that can induce reaction upon direct light absorption. Under an irradiation of light (hv) with equal or larger energy than the band-gap energy (the difference in energy between the valence band (VB) and the conduction band (CB)) of the semiconductor, electrons (e^-) migrate to the CB leaving holes (h^+) in the VB, thus the material is photoactivated due to the formation of e^-/h^+ pairs. The photogenerated e^- and h^+ are the basis of the heterogeneous photocatalysis using semiconductors (Figure 1).

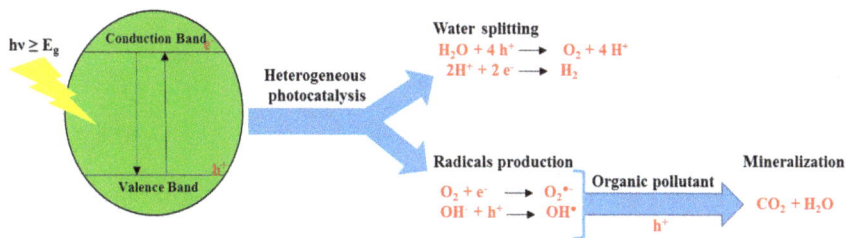

Figure 1: General mechanism of photocatalysis.

2. Heterogeneous photocatalytic reaction

In the photocatalysis process, the photocatalyst produces electron-hole pairs (e^-/h^+) by absorbing radiation energy ($\geq$ band-gap energy), excites the electrons from valence band (VB) to conduction band (CB) and leaves holes in the VB. Afterwards the photogenerated electrons reduce the surface adsorbed oxygen or oxygen dissolved in water to superoxide anions ($O_2^{\bullet-}$). Similarly, holes oxidize the surface adsorbed H_2O or hydroxyl (–OH) groups to hydroxyl radicals ($OH^{\bullet}$). Subsequently, these reactive radicals react with the pollutant molecules, and degrade them into inorganic ions. Moreover, the holes in the VB possess strong oxidizing power, thus these could directly oxidize the pollutants adsorbed on the photocatalyst surface. In addition, electrons in the CB possess strong reducing power, which indirectly degrade the pollutants using $OH^{\bullet}$ radicals formed by photo-cleavage of hydrogen peroxide (H_2O_2), which is produced by reaction of $O_2^{\bullet-}$ with the proton (H^+). Moreover, in the case of dye degradation, the surface adsorbed dye also excites under light irradiation, migrates an electron into the CB of the photocatalyst and reduces the surface adsorbed O_2 to $O_2^{\bullet-}$, which reacts with H^+ to form a H_2O_2 and further photo-cleaved into $OH^{\bullet}$ radicals under light irradiation. The reaction can be expressed as follows:

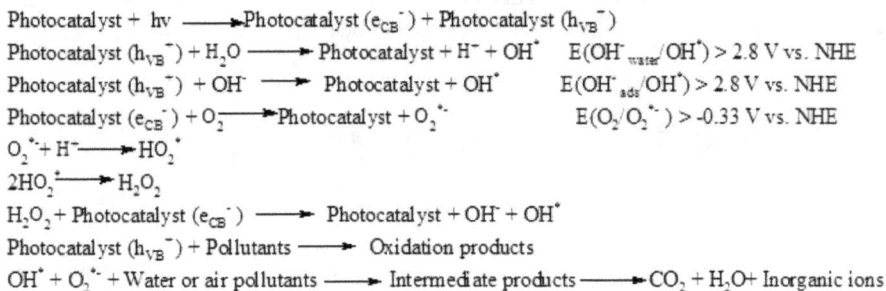

$$Photocatalyst + h\nu \longrightarrow Photocatalyst\ (e_{CB}^-) + Photocatalyst\ (h_{VB}^-)$$

$$Photocatalyst\ (h_{VB}^-) + H_2O \longrightarrow Photocatalyst + H^+ + OH^{\bullet} \quad E(OH^-_{water}/OH^{\bullet}) > 2.8\ V\ vs.\ NHE$$

$$Photocatalyst\ (h_{VB}^-) + OH^- \longrightarrow Photocatalyst + OH^{\bullet} \quad E(OH^-_{ads}/OH^{\bullet}) > 2.8\ V\ vs.\ NHE$$

$$Photocatalyst\ (e_{CB}^-) + O_2^- \longrightarrow Photocatalyst + O_2^{\bullet-} \quad E(O_2/O_2^{\bullet-}) > -0.33\ V\ vs.\ NHE$$

$$O_2^{\bullet-} + H^- \longrightarrow HO_2^{\bullet}$$

$$2HO_2^{\bullet} \longrightarrow H_2O_2$$

$$H_2O_2 + Photocatalyst\ (e_{CB}^-) \longrightarrow Photocatalyst + OH^- + OH^{\bullet}$$

$$Photocatalyst\ (h_{VB}^-) + Pollutants \longrightarrow Oxidation\ products$$

$$OH^{\bullet} + O_2^{\bullet-} + Water\ or\ air\ pollutants \longrightarrow Intermediate\ products \longrightarrow CO_2 + H_2O + Inorganic\ ions$$

The oxidizing agents can be semiconductor materials such as TiO_2, SnO_2, Sb_2O_3, Cu_2O, CeO_2, V_2O_5, ZrO_2, CdS, ZnO, MoS_2, In_2S_3, Bi_2WO_6, $BiVO_4$, WO_3, Semiconductors/Carbon Quantum Dots, and Semiconductors/g-C_3N_4 [2-5] (Table 1).

In the past years, a variety of semiconductors-based photocatalysts have been successfully prepared, such as TiO_2 and ZnO.

Table 1. Common semiconductors used in photocatalysis processes.

Name	Band gap	Valence band (V vs NHE)	Conduct band (V vs NHE)	Ref.
g-C_3N_4	2.68	1.56	-1.12	[6]
$Bi_{20}TiO_{32}$	2.36	2.59	0.23	[7]
CdS	2.4 5	1.5	-0.95	[8]
TiO_2 (rutile)	3.0	2.5	-0.5	[9]
TiO_2 (anatase)	3.2	2.5	-0.7	[9]
CdSe	1.749	1.520	-0.229	[10]
ZnS	3.6	1.8	-1.8	[9]
V_2O_5	2.6	2.9	0.3	[11]
ZrO_2	5	4.0	-1.0	[12]
ZnO	3.2	2.89	-0.31	[13]
CeO_2	2.86	2.626	-0.234	[14]
In_2S_3	1.96	1.18	-0.78	[15]
Bi_2WO_6	3.10	3.38	0.28	[16]
WO_3	2.8	3.2	0.4	[17]
$BiVO_4$	2.59	2.83	0.24	[18]
SnO_2	3.8	4.2	0.4	[19]
Sb_2O_3	3.0	3.32	0.32	[20]
Cu_2O	2	0.4	-1.6	[21]

TiO_2 seemed to be a desired photocatalyst owing to its stability and low cost. However, its practical application is limited due to quick recombination of the photogenerated electron–hole pairs (e^-/h^+) and absorbtion of UV light (solar spectrum contains only 4-5% of UV light). Several tactics have been used to solve these problems, such as manipulate its structure and morphology, doping with metal elements to form TiO_2/narrow-band-gap-conductor in order to enhance the charge separation, and sensitization using organic dyes [22]. Band gap and efficient surface area increase with reducing the size of nanostructures. Therefore, it is possible to create TiO_2 with specific properties by tuning its size and shape. TiO_2 has three main crystal phases, Rutile, Anatase and Brookite which their band gap increases, respectively. Decorating TiO_2 with metal oxide (e.g. ZnO, Cu_2O and WO_3) and noble metals (e.g. Ag, Au and Pt) shift the fermi level to the more negative potentials, therefore the electron-hole pair recombination time decreases. Developing a green and low-cost photocatalyst with visible-light response still remains a significant challenge. Recently, various structures of TiO_2 such as nanotube [23], nanorod [24], nanoparticle [25] and flower shape [26] nanostructures were investigated as a photocatalyst.

Due to the promising applications in photocatalytic degradation of organic pollutants from waters and disinfection, heterogeneous photocatalysis has attracted a lot of attention recently. An integrated water treatment process, known as a photocatalytic membrane reactors has been developed by combining photocatalytic oxidation and membrane filtration. Photocatalytic membrane reactors are prepared by the embedding of semiconductor particles such as titanium oxide into the membrane structure. Energy from light (i.e., visible or UV) is provided by photons and absorbed by catalytic particles to start the pollutant degradation reaction on the membrane surface. The high pollutant-degradation performance makes this technology a promising solution for the use in water- and wastewater-treatment processes. A variety of materials including organic, inorganic and metallic materials have been used as supports for the fabrication of photocatalytic membranes. Apart from immobilized semcondoctor membranes, some pure photocatalytic membranes have been fabricated using TiO_2 nanofibers, nanowires or nanotubes. The next sections focus on elaborating and expanding these topics.

3.   Photocatalytic membranes

Over the past few decades, membrane process has become one of the most effective technologies for water treatment due to its small footprint, superior separation efficiency and easy maintenance. However, in the conventional membrane processes, membrane fouling caused by the formation of a cake layer on the membrane surface usually results in pore blocking. This in turn results in a significant decrease in water flux and increased energy consumption and treatment costs. Moreover, membrane filtration can only concentrate pollutants into a high-concentration retentate, which needs further post-treatment before discharge. A hybrid photocatalysis-membrane system has been established to address these issues and shown to be a potentially effective treatment.

Currently, photocatalytic membrane reactors can be classified into four different configurations: (a) a slurry photocatalytic reactor followed by a membrane filtration unit, (b) inorganic or polymeric membrane submerged in a slurry photocatalytic reactor, (c) membrane placed inside a photoreactor whose internal walls are coated by a photocatalyst, and (d) photocatalytic membrane such as pure semiconductor or composite membrane (Figure 2). Among these four configurations, the photocatalytic membrane (d) has potential advantages over the other three configurations as it has the benefit having physical separation during membrane filtration as well as the organic degradation and anti-bacterial properties of the semiconductor achieved by photocatalysis in a single unit.

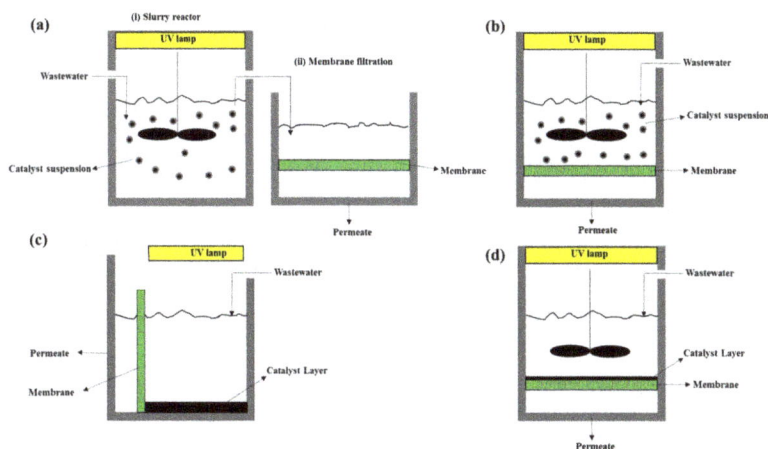

Figure 2. Different configurations of photocatalytic membrane reactors.

3.1 Polymeric photocatalytic membranes

Polymeric membranes are popular materials used as a support for photocatalysts in wastewater treatment. Various kinds of polymers such as polyamide (PA), polyvinylidenefluoride (PVDF), polyethersulfone (PES), polyurethane (PU), polyacrylonitrile (PAN), polytetrafluoroethylene (PTFE), cellulose acetate (CA), and sulfonated pentablock copolymer (s-PBC) have been modified by photocatalysts and applied in degradation of organic dyes. In preparation of photocatalytic membrane, photocatalyst particles can be either deposited onto the membrane surface or dispersed in the polymer solution for membrane casting.

A membrane process can be defined as splitting a feed stream by a membrane into a retentate (or concentrate) and a permeate fraction. Pressure-driven membrane processes use the pressure difference between the feed and permeate side as the driving force to transport the solvent (usually water) through the membrane. Particles and dissolved components are (partially) retained based on properties such as size, shape, and charge. Depending on their pore sizes, membranes can be classified as microfiltration (MF), ultrafiltration (UF), nanofiltration (NF) and reverse osmosis (RO) membranes Table 2.

Membranes applied in these processes can be divided, based on the material they are made of, into two types: polymeric and ceramic. Due to the lower price and simpler manufacturing, the polymeric membranes are more commonly used than the ceramic ones. However, polymeric membranes have some serious disadvantages when their application in photocatalytic membrane reactors (PMRs) is considered, namely low resistance to the action of UV light or oxidizing conditions [27].

Table 2. Different types of membranes.

Membrane process	Material type[a]	Pore size	Pressure range (bar)	Contaminant
Microfiltration (MF)	PTFE, PVDF, PP, PE, CE, PC, PS, PES), PI, PEI, PA, PEEK	0.1 μm to 10 μm	0.1-2	Bacteria
Ultrafiltration (UF)	PS, PES, S-PS, PVDF, PAN, CA, PI, PEI, PA, PEEK	2-100 nm	0.1-5	Macromolecules, bacteria, viruses
Nanofiltration (NF)	PA, PS, PES, S-PS, CA, P-PIP-A	~1 nm	5-20	Viruses, natural organic matter, multivalent ions
Reverse osmosis (RO)	CTA, Aramids ,PA, PEU	Non-porous	5-120	Viruses, natural organic matter, multivalent ions, monovalent ions (desalination,Ultr apure water)

[a]Refer to abbrivieations.

A wide variety of methods such as phase inversion, atomic layer deposition (ALD), chemical vapor deposition (CVD), sol–gel, electrospining, hydrothermal, electrophoretic deposition technique, dip-coating, vacuum filtration, solvent-casting method have been applied to modification and enhance separation and anti-fouling performance of polymers [28-33]. Following we briefly described some of these photocatalytic membranes and their preparation methods.

3.1.1 Development of polyethersulfone and polysulfone membranes

Polymeric membranes are widely applied in water purification and wastewater treatment. Polyethersulfone (PES) and polysulfone (PS) are popular polymeric membranes due to their low cost, excellent chemical resistance, mechanical strength, high rigidity, thermal and biological stability[34]. Nevertheless, the conspicuous drawbacks of PS are its hydrophobicity and bad ultraviolet (UV) resistance. To overcome aforementioned problems in the line of membrane fouling alleviation, various approaches such as grafting with hydrophilic nanoparticles especially inorganic nanoparticles (e.g., ZrO_2, Al_2O_3, SiO_2,

TiO_2, Fe_3O_4, carbon nanotubes (CNTs) and etc.), blending with hydrophilic polymers, surface-coating and embedding hydrophilic nanoparticles have been done [35-39].

TiO_2 nanoparticles could remarkably affect the hydrophilicity of PS membrane, enhance mineralization of pollutants under ultraviolet (UV) light, and improve self-cleaning/antifouling property of the modified membranes. Ma et al. reported that the removal rate of model pollutant, dye Reactive Red ED-2B (RR ED-2B), was improved obviously using the inorganic membrane under UV irradiation in comparison with photocatalysis or membrane separation alone [40].Yang et al. showed the water flux was nearly 30% improved and contact angle decreased from 84.7° to 41.4° with 2.0 wt.% TiO_2 blended [41]. Mozia et al. investigate the type of a TiO_2 photocatalyst (P25, ST-01, A700, and A800) on the stability of commercial polyethersulfone ultrafiltration membranes with different cut-off in a PMR. It was revealed that the shape and size of TiO_2 particles had a significant influence on the stability of the membranes during their operation in the PMR. It was concluded that the small and sharp-edged photocatalyst particles are especially detrimental to membrane stability [27].

The photocatalytic activity of fixed TiO_2 on the polymer structure decrease due to the loss of overall surface area associated with immobilizing the TiO_2. The design of smart TiO_2 morphologies (*e.g.*nanotubes, nanospheres, fibers) increases the photocatalytic activity. Schulze et al. showed that anodization of TiO_2 nanotubes on PES membrane has high photocatalytic activity in degrading methylene blue and diclofenac [42, 43].

Currently the stability of polymeric membranes under high energy UV irradiation is not fully documented, especially for extended exposure periods. Polymer degradation due to UV exposure or generation of free radicals during photocatalysis is in general an undesired process, often leading to a reduction in polymer molar mass and deterioration of the polymer properties.

One strategy of dealing with this disadvantage is using radical scavenger in the structure of polymeric membranes. It has been reported that PDA can act as free-radical scavenger as its chemical structure is similar to that of eumelanins, which are well known natural pigments for the protection of human body against UV by quenching reactive radicals generated by exposure to UV [44]. Wu. et al. and Feng et al. prepared doped polysulfone ultrafiltration membrane with TiO_2-polydopamine (PDA) nanohybrid for simultaneous self-cleaning and self-protection [45, 46]. In the TiO_2-PDA nanohybrid, PDA spheres acts as an adhesive substrate to hold densely covered photocatalytic TiO_2 nanoparticles, and also serves as a free-radical scavenger to protect PS membrane against the damage by free radicals produced by TiO_2 during UV exposure treatment. The spherical-shaped TiO_2-PDA can be easily doped into PS membrane via phase inversion method, which allows the hydrophilic TiO_2-PDA spheres migrating to the hydrophobic PS surface and benefiting the antifouling ability.

In addition, impurity doping is other typical approach to extend the spectral response of the titanium dioxide to visible light region and decrease the negative effect by UV irradiation. Some metal elements (such as Fe, Cr, Co, Mo, Pd, Ag and V), non-metals (such as B, C, N, S, and F) and attachment to graphite and carbon nanotubes have been employed to tune the electronic structure and enhanced the photocatalytic activity of the titanium dioxide [47-52]. Incorporating the co-doped TiO_2 into PS membranes will likely modify the optical properties of the membrane, potentially enabling the photocatalytic activation of the embedded co-doped TiO_2 under less energetic visible light irradiation [49, 53]. Different proportions of N,Pd co-doped TiO_2 were embedded in a PS polymeric membrane through the phase inversion method and the membrane evaluated for the degradation of a dye (Eosin yellow) under visible light irradiation by Kuvarega et al [54]. Up to 92% dye degradation was realized with the 7% N, Pd TiO_2/PS nanocomposite membrane after 180 min of visible light irradiation.

Recently, Hir et al. synthesized reduced TiO_2 (r-TiO_2) by a simple reduction method using $NaBH_4$ and immobilized into PES matrix via phase inversion [55]. The occurrence of surface defects induced by Ti^{3+} species or oxygen vacancies led to the extension of the light absorption in the visible region. Results indicate that a complete MO removal was obtained when one piece of PES-13 wt% of reduced TiO_2 (PrT-13) film was used in a very acidic medium.

Mahlambi et al. prepared PS membrane containing nanosized carbon-covered alumina supported TiO_2 (CCA/TiO_2) as an inorganic-organic nanomembranes to photodegrade Rhodamine B dye under visible light illumination [56]. Table 2 summarized recent progress in modification and application of PES and PE in removal of organic compounds (Table 3).

Table 3. Application of photocatalytic polyethersulfone and polysulfone membranes in degradation of various types of organic pollutants.

Photocatalyst	Preparation method	Dye/Condition/Light region (source)	Ref.
PES–reduced TiO_2	casting solution	-500 mL of MO (10 mg L^{-1}), pH = 5.8 -300-Watt halogen lamp (PHILIPS 13096 ELH) irradiation for 540 min	[55]
PES/TiO_2 film	Phase inversion	-1000 mL of MO (10 mg L^{-1}), pH = 5.8 -UV (6 W UV-A lamp (HITACHI F6T5/BL))	[57]
PS/CCA/TiO_2	Phase inversion	-100 mL of RhB (10 mg L^{-1}) -Visible light (Newport 9600 Full Spectrum Solar Simulator equipped with a 150 W ozone free xenon lamp)	[56]
Fe-doped TiO_2/PS	Hydrothermal, phase inversion	-250 mL BPA (10 mg·L^{-1}) -visible-light (500-W Xenon lamp)	[58]
PS/N,Pd co-doped TiO_2 composite	Phase inversion	-100 mL of EY(100 mg L^{-1}) -visible-light (solar simulator (Oriel, Newport), equipped with an Oriel 500W Xenon lamp)	[54]
TiO_2 nanotube-PES	Anodization	-4 mL of MB (20µmol L^{-1}), Diclofenac (25 mg L^{-1}) -UV-A sunlamp (Heraus Original Hanau Suncare tanning tube 21/25 Slim, radiant flux density 7.6 mW cm^{-2})	[42, 43]
PES/LSMM-OGCN	Phase inversion	-1 L of Phenol (10 ppm) -Ultraviolet lamp (Vilber Laurmat, $\lambda = 312$ nm, 30 W, light intensity 3.0 mW/cm^2) and a visible lamp (light-emitting diode (LED) lamp, $\lambda = 420$ nm, 30 W)	[59]

3.1.2 Fluoropolymeric membranes: PVDF and PTFE

Fluoropolymers like polyvinylidene fluoride (PVDF) has the advantages of high temperature resistance, chemical corrosion resistance, aging resistance and high strength due to stable –C–F bond in the main chain [60]. But the PVDF membrane application in the field of water treatment and separation is limited because of the strong hydrophobicity and easy pollution. An easy way to address this issue is applying casting solution method. This method entraps the hydrophilic nanoparticles within the membrane matrix with good stability. However, the aggregation of nanoparticles could affect membrane morphology and alter membrane porosity, roughness, and mechanical strength. Meanwhile, the photodegradation rate of pollutants is suppressed because the entrapped phtocatalysts are not easily accessible. Therefore, some researchers employed the self-assembly method, wherein semiconductor is bound on the surface of the membrane through non-covalent interactions [61]. The procedure is also straightforward, the fully formed membrane is dipped in a photocatalyst suspension for a certain period to allow coating [62-65]. In order to provide the necessary binding sites in the surface of PVDF, blending or grafting of hydrophilic materials like poly(styrene-alt-maleic anhydride) (SMA) [66] and PES [62, 67], poly(acrylic acid) (PAA) has been applied [68]. Plasma-induced graft polymerization is a simple technique that eliminates the need for chemical initiators or other components. The activated species of low energy plasma can trigger polymerization reaction on or near the surface of a wide variety of materials even with short bursts of plasma exposure [69]. You et al. grafted PAA on PVDF by plasma method to assembly of TiO_2 nanoparticles [68]. The membrane with 0.5% TiO_2 loading maintained the highest pure water flux and the best protein antifouling property. The TiO_2-modified membranes removed 30–42% of 50 mg L^{-1} aqueous Reactive Black 5 (RB5) dye.

Damodar et al. prepared composite PVDF/TiO_2 membranes using general phase inversion method [70]. The modified PVDF membranes were prepared by adding different amounts of TiO_2 particles (0–4 wt.%) into the casting solution. The TiO_2 entrapped PVDF membranes were tested for its antibacterial property using *Escherichia Coliform* (*E. Coli*), photoactive property using Reactive Black 5 (RB5) dye and self-cleaning (antifouling) properties by fouling using 1% Bovine Serum Albumin (BSA) solution. Results showed that TiO_2 addition significantly affects the pore size and hydrophilicity of the PVDF/TiO_2 membrane. This also improves the flux and permeability of prepared membrane. Antibacterial study showed that the composite PVDF/TiO_2 membrane removes *E. Coli* at a very faster rate than neat PVDF membrane and membrane with 4% TiO_2 possess highest antibacterial property. The RB5 dye removal using PVDF/TiO_2 occurs under UV by photolysis.

Alaoui et al. prepared porous PVDF-based membranes filled with anatase (TiO_2) nanocrystalline particles using phase inversion technique for entrapment of TiO_2 [71]. The best photocatalytic membrane had a pore size of 0.96 μm, with a maximum porosity of 86% at 0.5 TiO_2/PVDF weight ratio. The results indicate that after 7.5 h of irradiation for Brilliant Green (BG), the dye degradation ratio was about 81% for TiO_2/PVDF and 97%

for TiO_2, and after 5h of irradiation for Indigo Carmin (IC), the dye degradation ratio was about 89% for TiO_2/PVDF and 97% for TiO_2 (Degussa P25).

Li et al. investigated Ag/TiO_2 embedment on the surface property of PVDF membrane [72]. Ag/TiO_2 improved the surface hydrophilicity of PVDF membrane in contrast with virgin PVDF membrane. Besides, Ag/TiO_2 nanoparticle provided PVDF membrane with an enhanced visible-light response activity. The absorption in the visible light range was derived from the localized surface plasmon resonance (LSPR) of Ag nanoparticles. Ag/TiO_2/PVDF exhibits excellent activity on degradation of methylene blue (MB) and inactivation of bacteria [72, 73].

Nanocomposites of poly(vinylidene difluoride)-co-trifluoroethylene (PVDF-TrFE) with different concentrations of TiO_2 P25 nanoparticles (5, 10, and 15 wt.%) and ZnO nanoparticles (15 wt.%) were produced by Teixeira et al. and tested on the degradation of methylene blue (MB) [74]. It was found that increasing the photocatalyst concentration results in higher photocatalytic efficiencies; the degradation rates of 15% of TiO_2 and ZnO were similar; and the photocatalitic activity of composites in dye degradation with TiO_2 5%, TiO_2 10%, TiO_2 15%, and ZnO 15% decreased as 6%, 16%, 13%, and 11% after three utilizations [74].

Recently, many studies have shown that graphene-based membranes have great potential in filtration, separation and water desalination. Addition of GO (~ 1% rather than mass of polymer) can significantly improve the hydrophilicity, permeability and antifouling performance of the membranes [75-77].

One of the practical solutions for overcoming the photodegradation of the polymeric membrane is to protect the membrane surface with polydopamine (PDA) as a free-radical scavengers that is capable of absorbing and quenching the radicals produced by UV light [78]. Visible light photocatalysts have been highly developed owing to their efficient utilization of approximately 50% solar energy. In 2010, Ye team reported Ag_3PO_4, with the band gap of 2.36 eV, and a valence band potential of 2.43 eV, it can absorb the UV–vis wavelength in the solar light less than 530 nm [79]. Zhang et al. prepared PDA/RGO/Ag_3PO_4/PVDF membrane, which can efficiently degrade and separate dye molecules (99% removal of MB) under visible-light irradiation. Polydopamine has good absorption ability to ultraviolet and visible light, and its good photoconductivity can enhance the generation of photogenerated e^-/h^+ under the visible light irradiation [60].

Bi_2WO_6 with perovskite structure has attracted tremendous attention due to its superior intrinsic properties such as narrow band gap, high physicochemical stability, low cost and non-toxicity. However, bare Bi_2WO_6 exhibits a short visible-light absorption limiting its photocatalytic efficiency. To address this problem, various efforts have been made, such

Materials Research Forum LLC

https://doi.org/10.21741/9781644901359-1

as ions doping, noble-metal deposition, nanostructured control, crystal facet engineering and heterostructure construction [80-86].

Heterostructure construction is one of the most effective means to strengthen the visible-light photocatalytic activity of Bi_2WO_6. Recently, a few novel composites based on Bi_2WO_6 such as Bi_2WO_6/TiO_2 [87], $g-C_3N_4/Bi_2WO_6$ [84], WO_3/Bi_2WO_6 [88], $BiOI/Bi_2WO_6$ [89], $Bi_2O_2CO_3/Bi_2WO_6$ [90], $\alpha-Fe_2O_3/Bi_2WO_6$ [91], Ag_2O/Bi_2WO_6, Bi_2WO_6/GO [92], Bi_2WO_6/SiO_2 [93], Bi_2WO_6/Fe_3O_4 [94] and $Bi_2WO_6/BiPO_4$ [95] has been developed. Li et al. obtained PVDF membranes modified with RGO/ Bi_2WO_6 with a double-layer coating method through non-solvent-induced phase separation [96]. The water contact angle was improved by about 30° by RGO/Bi_2WO_6; this indicated the enhanced membrane hydrophobicity. The high desalination rate proved that all of the prepared membranes were appropriate for the membrane distillation (MD) process. The $RGO/Bi_2WO_6/PVDF$ achieve 26.26%–59.95% removal rates in 10 mg/L aqueous ciprofloxacin (CIP) under visible light for 7.5 h. The best photocatalytic performance (59.95% removal rate) was obtained in 300 mL of 10 mg/L CIP under the irradiation of visible light for 7.5 h; the amount of RGO/Bi_2WO_6 was influential in final result.

Metal-organic frameworks (MOFs), owing to their outstanding properties including large surface areas, well-ordered porosity and tunable molecular structure have been extensively researched in photocatalysis filed [97-99]. Guo et al. reported that by the hydrothermally prepared $g-C_3N_4/NH_2-MIL-88B(Fe)$ heterojunction (MIL-88B(Fe) = Iron-based MOFs); 100% of MB photodegradation was achieved in 120 min under visible light, much greater than the $g-C_3N_4$ and $NH_2-MIL-88B(Fe)$ individually [100]. Hu et al. fabricated composite film, $Bi_2WO_6/MIL-53(Al)/PVDF$, through a hydrothermal process combined with immersion phase inversion method [101]. The formation of heterojunction structure between Bi_2WO_6 and MIL-53(Al), increased photocatalytic activity (95.3% degradation of RhB). The photocatalytic degradation mechanism by quenching tests revealed that the predominant reactive species were h^+, $\cdot O_2^-$ and $\cdot OH$.

PVDF membrane was modified using atomic layer deposition (ALD) technique by coating three-dimensional (3D) TiO_2/ZnO photocatalyst on membrane surface and pore walls [102]. The hydrophilicity, permeability, photocatalysis and anti-fouling performances of modified membranes were investigated in detail. The composite modified membrane ($TiO_2:ZnO=1:3$) was more sensitive to visible light. The maximum flux of 0.017 cm/s was obtained for the composite modified membrane ($TiO_2:ZnO/1:3$ with ZnO content of 0.923 $\mu mol/cm^2$), due to enhanced hydrophilicity. Moreover, the 3D modified layer on membrane surface displayed excellent photocatalytic activity and stable reusability during methylene blue (MB) degradation.

Photocatalysis   Materials Research Forum LLC
Materials Research Foundations **100** (2021) 1-56   https://doi.org/10.21741/9781644901359-1

Compared with TiO_2, ZnO has rich micro-morphology, good biocompatibility and environmental safety, which make it quite popular in the field of photocatalysis. The electrons and holes generated on the surface of ZnO are easily recombined in the photocatalytic process, which reduces the photocatalytic efficiency. Therefore, considerable efforts including noble metal loading, semiconductor coupling, ion doping and special preparation process have been devoted to enhance the photocatalytic activity of ZnO [103].

Zhang et al. showed coupling ZnO with GO can effectively enhance the dispersibility and the photocatalytic efficiency [104]. They synthesized PVDF/GO/ZnO photocatalytic membranes for degradation of organic dyes. The photocatalytic degradation rate of methylene blue (MB) could reach 86.84% , under Xenon irradiation (300 W). The radical trapping experiments with different active radical scavengers showed that the oxidizing species ($\cdot O_2^-$) plays an important role in the decolorization process of MB. The photocatalytic efficiency of the composite membrane for MB increase by addition of H_2O_2, which may be attributed to the fact that the H_2O_2, which is a strong oxidant, could suppress the e^-/h^+ pair recombination on the surface of ZnO by trapping electrons generated on the surface of ZnO [105].

Zinc sulfide (ZnS) with strong oxidization and fast generation of charges widely used as photocatalyst for toxic organic pollutant degradation [102]. To alleviate the quick recombination rate of the photoinduced e^-/h^+ pairs, GO exhibits excellent electron transfer performance and can reduce the recombination rate, which is helpful to enhance the photocatalytic activity. Du et al. prepared zinc sulfide/graphene oxide/polyvinylidene fluoride (ZnS/GO/PVDF) composite membrane by immersed phase inversion method. Uniform surface, high hydrophilicity (water contact angle = 62.2°), and enhanced permeability to 326.1 $L/(m^2\ h)$ were characteristics of hybrid membrane [106].

$BiVO_4$/GO/PVDF system showed enhanced photocatalytic activity degradation of methylene blue (MB), Rhodamine B (RhB) and Safranin-O (SO) in water under visible light irradiation as compared to the pure $BiVO_4$ catalyst, $BiVO_4$ and PTFE (polytetrafluoroethylene) decorated on the graphene sheet [107]. Graphene sheets in this composite enhances photocatalytic performance under visible light. Photoluminescence spectra proved the effective quenching of the photogenerated e^-/h^+ pairs is the main reason of enhanced photocatalytic activity.

The electrospun nanofiber mats of fluoropolymers with micro-sized porous structure can offer high specific area and good enrichment ability for organic compounds, leading to the improvement of the photocatalytic efficiency, stability and recycle ability [108-116].

Materials Research Forum LLC

https://doi.org/10.21741/9781644901359-1

Following we have summarized recent development of PVDF photocatalytic membranes in degradation of organic pollutants (Table 4).

Table 4. Application of photocatalytic fluoropolymeric membranes in degradation of various types of organic pollutants.

Photocatalyst name	Preparation method	Condition/ Light source	Ref.
Anatase/PVDF	Phase inversion	-400 mL of BG, IC (20 mg L^{-1}) -UV Swig black lamp (Bioblock) at 365 nm	[71]
PDA/RGO/Ag$_3$PO$_4$/PVDF	Phase inversion	-100 mL of MB (20 mg L^{-1}) -Visible light (common incandescent lamp, 200 W	[60]
PVDF/GO/ZnO	Phase inversion	-60 mL of MB (10mg L^{-1}) -Xenon illumination (300 W)	[104]
Au-TiO$_2$/PDA/PVDF	Vacuum filtration	-20 mL Tetracycline (10 mg L^{-1}) -Visible light	[117]
TiO$_2$ entrapped PVDF membranes	Phase inversion	-25 mL of RB5 (100 mg L^{-1}) -UV light (UV-C, 15W)	[70]
Ag-TiO$_2$/PVDF film	Blending/photo reduction combined method	- 30 mL of MB (10 mg L^{-1}) -Visible light (18W fluorescent Lamp)	[72]
ZnS/GO/PVDF	Phase inversion	-100 mL of MB (10 mg L^{-1}) -photochemical reactor32 (Beijing NBET Technology Co., Ltd) in sunlight range (250-800 nm)	[106]
PVDF-TrFE/TiO$_2$	Solvent casting	-MB -17.6 µM /50 mL -UVA (365 nm ;6 Philips 8W mercurial fluorescent lamps– UMEX)	[74]
BiVO$_4$-GO-PVDF	Ultrasonication	-100 mL of MB, RhB, SO (20 mg L^{-1}) -visible light (500 W xenon lamp through UV cut-off filters)	[107]

3D-TiO_2/ZnO/ PVDF	ALD	-100 mL of MB (10^{-5} mol L^{-1}) -visible light (xenon light source, 200 W)	[102]
Bi_2WO_6/MIL-53(Al)/PVDF	Hydrothermal, immersion phase inversion	-250 mL of RhB (20 mg L^{-1}) -visible ligh 500 W Xe lamp ($\lambda > 420$ nm)	[101]
PVDF/ Bi_2WO_6/rGO	Double-layer coating method through non-solvent-induced phase separation	-300 mL CIP (10 mg L^{-1}) -visible light 500 W U-shaped xenon lamp	[96]
paper-like PVDF/TiO_2	Electrospinning and electrospraying	-45 mL of BPA, 4-CP, and CMT (10 μM) -UV (six 4-W blacklight blue lamps (λ= 350–400 nm, F10T8, Sankyo Denki, Japan)	[113]
ZnS/PVDF/MAA/TFA	Electrospinning	-15 mL of MB (5.3 mg L^{-1}) -UV (two 15 W shortwave UV lamps (254 nm) (Kexing Company, Changsha China), distance between two lamps =15.0 cm)	[114]
TiO_2/PVDF–TrFE membrane	Solvent casting	-Tartrazine -10 mg L^{-1} -sunlight irradiation	[118]
TiO_2-PVDF-steel mesh	Coating, electrospraying, and thermal fixation	-45 mL of MO, RB4, SMX (10 μM) - UV irradiation (six blacklight blue lamps (4 W, k = 350–400 nm, Sankyo Denki F10T8))	[115]
PVDF/Ag_2NCN	Phase inversion	-100 mL Acid Blue 1(10^{-5} M) -visible light (light irradiation density was 94.30 mW cm^{-2})	[119]
TiO_2-PVDF ultrafiltration membranes	Phase inversion	-80 mL of RhB (2×10^{-5}M), pH = 6 -UV (250 W, high-pressure Hg lamp at 365 nm)	[120]

TiO$_2$/GO/P(VDF-TrFE)	Electrospining	-13 mL MB (10^{-5} mol L^{-1}) -UVA (high-power LED source (Thorlabs, 700 mA) with an excitation peak at 365 nm)	[121]
PVDF/ meso-TiO$_2$	Dual-templated synthesis with solvent extraction	-20 mL of MO (10 mg L^{-1}) -UV (ZF-1 three UV analyzer, Shanghai Jihui scientific analysis instrument Co., Ltd., China) (25 W, 365 nm))	[122]
PVDF/TiO$_2$ nanofibers	Electrospining	-200 mL of BPA (10 ppm) -UV radiation (UV lamp, Vilber Laurmat, France, λ = 312 nm, 30 W).	[123]
Er-Doped and Er/Pr-Codoped TiO$_2$/ PVDF−TrFE	Solvent casting	-MB (1×10^{-5} M) pH 6.8 -UV (high-power light-emitting diode (LED) source (Thorlabs, 700 mA), at 365 nm (UV-A)	[124]
PVDF/anatase/silica nanocomposites	Thermal treatment	-MB (10 mg L^{-1}) -UV (UV lamp, λ_{max} = 365 nm)	[125]
PVDF/plasma-grafted poly(acrylic acid) membrane with self-assembled TiO$_2$	Plasma-induced graft polymerization	-RB5 (40 mg L^{-1}) -UV (UV lamp (254 nm, 15W)	[68]
Naked-TiO$_2$ capsulated in nanovoid microcapsule of PVDF	phase inversion	-100 mL of MO (15 mg L^{-1}) -UV (200 W UV-lamp at 365 nm)	[126]
PTFE/Co$_3$O$_4$	Dip coating	-Orange II (0.05 mM), pH = 7 -Suntest solar light simulator (90 mW/cm^2)	[127]
TiO$_2$/PTFE Membrane	electrospinning, immersion, and calcination	-100 mL MB solution (2.5 mg L^{-1}) -UV (lamp power 300 W)	[128]

Recently, there have been many reports of using modified poly(tetrafluoroethylene) (PTFE) fiber and membranes for the degradation of dyes [129, 130]. Modified PTFE fibers by acrylic acid and loaded Fe^{3+} were prepared and used as Fenton catalysts for the degradation of reactive Blue 222 in the pH range 3-9 under visible irradiation [131]. The study found that the loaded catalytic material had good performance in degradation of the azo dye, and the fibrous support could maintain good mechanical properties in the reaction process, which avoided the strength decreasing of PTFE fiber carrier during catalytic degradation. However, modification and loading operations are difficult , low efficiency and high cost. Besides, fibers with large diameters are difficult to achieve with better catalytic properties. Kang et al., prepared a TiO_2/PTFE membrane catalyst by electrospinning, immersion, and calcination, and the catalyst was applied to MB degradation [128].

3.1.3 Development of photocatalytic polyaniline (PANI) membranes

A number of processing technologies such as template synthesis, self-assembly and electrospinning have been proposed for the preparation of polymer nanofibres. Among these, the electrospinning process seems to be a promising method which can be further developed for mass production of one-by-one continuous nanofibres from various polymers [132]. Electrospun fibers of ceramic or polymeric materials possess many extraordinary properties, such as small diameters and large specific surface areas. Additionally, the non-woven fibrous mats made of electrospun polymer fibers offer a unique capability of readily controlling the pore sizes. Till now, many synthetic and natural polymers have been electrospun into fibrous mats for various applications.

Various strategies have been employed to make TiO_2 photocatalysts highly efficient under visible light, including dye sensitization, polymer modification, non-metals doping, semiconductor coupling, and transition metal doping [133]. Conjugated polymers, such as polyaniline (PANI), poly(fluorine-co-thiophene) (PFT), polythiophene, polypyrrole and their derivatives can function as sensitizers to extend photoresponse of TiO_2 into the visible region effectively [133, 134]. PANI is one of the most fascinating conductive polymers and has received much attention for easy polymerization, high yield, relative high conductivities, low cost, and high stability [135]. Under visible-light irradiation, PANI generates $\pi-\pi^*$ transition, injecting the excited electrons into the conduction band of TiO_2, and then the electrons transfer to an adsorbed electron acceptor to yield oxygenous radicals [136]. Meanwhile, synergetic effect of coupling photocatalysts with PANI proves to be effective to promote photoinduced charge separation and inhibit charge recombination [137, 138]. There are variously modified photocatalysts based on PANI such as PANI/TiO_2, PANI/$BiVO_4$, PANI/SnO_2, and PANI/MoO_3, PANI/PbS, and PANI/$CoFe_2O_4$/TiO_2 [139-142]. Lin et al. prepared a catalytic system of PANI/TiO_2 nanocomposites, which exhibited higher photocatalytic activity and stability for degradation of methyl orange than the bare TiO_2 under both UV and visible light irradiation [143]. Li et al. prepared PANI-doped TiO_2 and PLLA (poly(L-lactide)) fibers by a combination of coaxial-electrospinning and in-situ polymerization [144]. The aniline monomers were located in the core phase and in-situ polymerized by ammonium persulfate

(APS) after electrospinning. Photocatalytic degradation tests show that the membrane exhibit enhanced photocatalytic activity for degradation of methyl orange under visible light, likely due to the synergistic effect of PANI and TiO_2.

Li et al. applied electrospining method in preparation of bilayered composite fibrous membrane [145]. Polyaniline (PANI), Fe_3O_4 nanoparticles (NPs), and $Bi_2WO_6:Yb^{3+},Er^{3+}$ were incorporated into polyacrylonitrile (PAN) and electrospun into the obtained fibrous membrane with $[Bi_2WO_6:Yb^{3+}, Er^{3+}/PAN]$ nanofibers as one layer and $[PANI/Fe_3O_4/PAN]$ nanofibers as the other layer. The degradation of Rhodamine B (RhB) was used to characterize the photocatalytic activity of the photocatalyst.

Table 5. Application of photocatalytic polyaniline membranes in degradation of various types of organic pollutants.

Photocatalytic name	Preparation method	Condition & light source	Ref.
TiO_2/PANI/PEO	Electrospinning & electrospraying	-10 mL CEPS solution in heptane at concentration of 0.05(v/v) -UV irradiation	[146]
PANI-doped TiO_2/PLLA	Coaxial-electrospinning and in-situ polymerization	-25 mL of MO solution (0.5 mg L^{-1}) -Visible light (250 W xenon lamp with a 420 nm cut-off glass filter was used as a visible-light source)	[144]
PANI/Fe_3O_4/PAN/$[Bi_2WO_6:Yb^{3+},Er^{3+}/PAN]$ bilayered composite fibrous membrane	Electrospinning	-100 mL of RhB aqueous solution (0.05 mg L^{-1}) -UV-vis light irradiatio (A 500 W mercury lamp)	[145]
PANI/TiO_2/SiO_2 nanofiber membrane	Electrospinning	- 3 mL of MO solution (1.5 mg L^{-1}) -Visible light (500 W xenon lamp with a 420 nm cut-off glass filter)	[133]

Photocatalysis   Materials Research Forum LLC
Materials Research Foundations **100** (2021) 1-56   https://doi.org/10.21741/9781644901359-1

Neubert et al. prepared conductive electrospun fibrous membrane. by integration of electrospinning and electrospraying. The non-conductive polyethylene oxide (PEO) was blended with (±)-camphor-10-sulfonic acid (CSA) doped conductive polyaniline (PANI) for electrospinning. The conductive CSA/PANI/PEO composite fibers were produced upon electrospinning, then TiO_2 nanoparticles were sprayed and allowed to adsorb on the fibers. The photocatalytic activity of the prepared TiO_2/PANI/PEO was tested against the toxicant simulant 2-chloroethyl phenyl sulfide (CEPS) under UV irradiation. It was observed that the TiO_2 nanoparticles catalysts embedded PANI/PEO fibrous membrane decontaminated the toxicant CEPS significantly. Table 5 summarized reported photocatalytic polyamide membranes in degradation of organic pollutants.

3.1.4 Development of photocatalytic polyamide membranes

Kwak et al. prepared hybrid organic/inorganic reverse osmosis (RO) membranes composed of aromatic polyamide thin films underneath TiO_2 nano-particles by a self-assembly process, with photocatalytic destructive capability on microorganisms[64, 147]. Lombardi et al. prepared electrospun polyamide-6 membranes containing TiO_2 photocatalyst with self-cleaning, antibacterial and photochromic textiles, and high surface area [132]. By tailoring the electrospinning parameters, two different membranes (thicknesses 5 and 20 µm, with inorganic/organic ratios of 10 and 20 wt%.) were successfully produced, The photocatalytic activity of both hybrid systems was evaluated by degradation of methylene blue under UV irradiation.

Zhang et al. applied immobilized TiO_2 nanoparticle on polyamide fabric in degradation of methylene blue [148]. It was found that when polyamide-6 (PA-6) fabric was treated in titanium sulfate and urea aqueous solution, anatase nanocrystalline titanium dioxide was synthesized and simultaneously adhered onto the fiber surface. The average crystal size of titanium dioxide nanoparticles was about 13.2 nm.

Daels et al. prepared TiO_2 functionalized nanofiber membranes via electrospinning on a single nozzle electrospin set-up by adding two types of TiO_2 nanoparticles (commercial Degussa P25 ~21 nm and synthetic colloids ~ 6 nm) at different concentrations to the PA-6 prior to the spinning solution or by post-functionalizing the electrospun membranes [149]. Both methods improved the degradation of methylene blue under UV irradiation although the post-functionalization with colloidal TiO_2 nanoparticles showed the best photocatalytic activity. Table 6 represents the recent progress in degradation of organic pollutants by photocatalytic polyamide membranes.

Materials Research Forum LLC
https://doi.org/10.21741/9781644901359-1

Table 6. Application of photocatalytic polyamide membranes in degradation of various types of organic pollutants.

Photocatalyst name	Preparation method	Condition & light source	Ref.
PA6 fabric/TiO_2 nanoparticle	Hydrothermal method	-50 mL of MB solution (20 mg L^{-1}) at natural pH -UV(20 W, $\lambda = 254$ nm, quartz ultraviolet lamp)	[148]
Electrospun PA 6/ TiO_2 nanoparticles	Electrospinning	-MB solution (10 mg L^{-1}) -UV (300 W Osram Ultra-Vitalux lamp)	[149]
Electrospun PA6/TiO_2	Electrospinning	-MB (5 ppm in ethanol) -UV (medium-pressure mercury arc lamp (Hamamatsu), average light intensity was ca 50 mW cm^{-2})	[132]
TiO_2/PA 6 nanofibrous membranes	Dip coating	-MB (4.2 mg L^{-1}) and isoproturon (5 and 10 mg L^{-1}) solutions -UV-A, similar to sunlight (300 W Osram Ultra-Vitalux lamp, intensity 5 m W/cm^2)	[150]
TiO_2/PA 12 electrospun fiber mats	Electrospinning	-13 mL of MB solution (10^{-2} mM), pH = 6.8 -UV-A (high power LED source (Thorlabs, 700 mA, at 365 nm))	[151]

3.1.5 Development of cellulose-based photocatalytic membranes

Cellulose acetate (CA) is the most widely used cellulose ester, but its usefulness is limited by its low dimensional stability at high temperatures. Several strategies have been employed in an attempt to broaden the applications of this indirectly renewable polymer, since it is derived from cellulose, such as the formation of polymer blends, chemical modification through the introduction of functional groups, or grafting to the polymer chain. Preparation of organic/inorganic hybrid composites such as cellulose acetate with niobium, zirconium, titanium, aluminum, bismuth, and silicon oxides is a suitable route for its chemical modification [152].

Campos et al. prepared a cellulose acetate/TiO_2 composite by dissolving cellulose acetate in acetone:acetic acid (70:30 in volume) and adding quantities of titanium (IV) tert-butoxide to obtain varying oxide content after phase inversion [152]. TiO_2 in the composite was a crystalline anatase phase. The composite showed effective photocatalytic activity in the degradation of MB under solar radiation. Yang et al. [153] fabricated the CA/SiO_2-TiO_2 hybrid via water vapor-induced phase inversion of CA solution and simultaneous hydrolysis/condensation of 3-aminopropyltrimethoxysilane (APTMS) and tetrabutyl titanate (TBT) at room temperature. Micro-nano hierarchical structure was constructed on the surface of the film. The flux of the film for the emulsion separation was up to 667 L m^{-2} h^{-1}, while the separation efficiency was up to 99.99 wt%. Meanwhile, the film exhibited excellent stability during multiple cycles, performed excellent photo-degradation under UV light.

Bi_2O_3 is an attractive candidate semiconductor material because of its low cost, favorable direct bandgap of 2.5-2.8 eV, and good photostability in acidic conditions. Zhang et al. [154] prepared nanosized Bi_2O_3 to enhance the photocatalytic activity by a simple sonochemical route. In addition, Liu et al. [155] showed the crystalline phase evolution of Bi_2O_3 had an effect on photocatalytic performance. The effective utilization of visible light for photocatalytic membranes is main challenge. To extend the lifespan of photogenerated electrons and holes for higher photocatalytic activity, various strategies have been reported, such as doping with metal elements (Au-doped Bi_2O_3 [156]), nonmetal elements (S-doped Bi_2O_3 [157]) and coupling with other semiconductors to form heterojunction structures (Bi_2O_3/Bi_2MoO_6 [158], Bi_2O_3/g-C_3N_4 [158]). The heterojunction structure can not only combine at least two disparate functional materials in one system but also reduce the recombination rate of the electron-hole pairs. For instance, Juntrapirom et al. [159] found that the novel SnS/BiOI heterostructures presented higher photocatalytic performance and better stability than a single semiconductor. Liu et al. [160] synthesized another $BiVO_4/WO_3$ heterojunction with an excellent visible-light response, a good photostability and high photoelectrochemical activity. However, it is difficult to separate it from treated water, which limits its practical application. The challenge is to develop an effective method to improve the recovery and reuse of the catalyst. CA-based membranes have been widely utilized as support membranes due to the advantages of biodegradability nontoxicity and low cost. Li et al. synthesized the Bi_2O_3/ZnS heterojunction, by a one-step hydrothermal method and applied phase inversion method to immobilize Bi_2O_3/ZnS nanocomposite in CA membrane. The prepared photocatalytic membrane was applied to degradation of RhB [1].

$Bi_{12}O_{17}Cl_2$, a new type of promising layered materials, has been demonstrated with excellent visible light photocatalytic activity for pollutant degradation [161]. However, the practical application of pure $Bi_{12}O_{17}Cl_2$ is still limited by the fast recombination electron-hole pairs and many methods have been used to modify and adjust the catalyst to visible light region [162-164]. The visible light photoactivity of the obtained catalyst generally have a good effect on dye degradation. However, the visible light activity for the colorless organic contaminant still needs to be improved. Yu et al. [165] combined the $Bi_{12}O_{17}Cl_2$ with graphene oxide (GO) and dopamine together on surface of CA membrane. The $Bi_{12}O_{17}Cl_2$ can tune the interlayer spacing of GO to increase the flux [166]. During the reduction process, the hydrophilic interfaces of GO may change into hydrophobic because of the loss of oxygen functionalities [167]. Therefore, a proper reducing agent both achieving the reduction of graphene oxide and keeping the hydrophilicity of the GO interface is very important. Many researches indicated that dopamine with physicochemical properties of adhesion and self-polymerization is a milder and nontoxic reducing agent at weak alkaline pH. Therefore, dopamine has been introduced to reduce GO and obtain uniformed dispersion membrane solution. Results indicated that the as-prepared membrane can simultaneously degrade dye/colorless organic contaminant and separate oil/water emulsions. In addition, the use of dopamine has improved the adherence between $Bi_{12}O_{17}Cl_2$ with graphene matrix and made the $RGO/PDA/Bi_{12}O_{17}Cl_2$ firmly together with the CA membrane. The photodegradation of MB solution and 4-CP solution under visible light was used to estimate the photocatalytic performance of the prepared membrane [165].

It was reported that electrospun CA nanofibrous membranes with large specific surface area, high porosity, and high permeability could be an effective support for photocatalysts such as ZnO , ZnS, TiO_2 [168-170]. This could create a synergistic effect that enhanced the catalytic performance of the individual materials.

Bacterial cellulose (BC)-hydrated membranes with nanosize superfine structure are good templates in the preparation of new organic–inorganic hybrids with specific morphology and size [171-173]. Among many possible nanoparticles to be incorporated in BC matrixes, TiO_2 nanoparticles have attracted considerable attention because of the relative ease of synthesis and their strong photodegradation activity. Table 7 summarizes the application of cellulose-based photocatalyst nanostructured materials for photocatalytic degradation of various organic pollutants.

Table 7: Application of cellulose-based hybrid photocatalysts in photocatalytic degradation of various types of organic pollutants.

Photocatalyst name	Preparation Method	Condition & light source	Ref.
RGO/PDA/Bi$_{12}$O$_{17}$Cl$_2$-CA	Vacuum filtration	-100 mL MB solution (15 mg L^{-1})) and 4-CP solution (15 mg L^{-1}) -visible light (long-arc Xe lamp, 500 W)	[165]
Cellulose (core)-TiO$_2$ (sheath) fibers	Coaxial-lectrospinning	-Keyacid Blue dye (0.1 wt %) -A halogen light source (power density of ~27 mW/cm^2)	[174]
BC/titanium dioxide	Sol–Gel method	- MO solution (20 mg L^{-1}) -ultraviolet lamp.	[175]
CA/TiO$_2$ composite	Phase inversion	-25 mL of MB solution (12 × 10^{-6} mol L^{-1}) -Uv radiation. (150W Hg lamp)	[176]
ZnS/BC nanocomposites	In Situ Precipitation using BC as template	-100 mL of MO solution(10 mg L^{-1}) -UV light at a wavelength of 254 nm.	[177]
CA-TiO$_2$ P90 (ratio: 10:1)	Casting soltion	-30 mL of MB or RhB solution (2 × 10^{-5} M) -UV radiation (two 15 W lamps, 365 nm)	[178]
CA/TiO$_2$ composite ultrafine fiber	Electrospinning	-10 mL of MB solution (10 mg L^{-1}) - UV source. (300 W mercury lamp)	[179]
SrTiO$_3$/TiO$_2$ heterojunctioned nanofibers assembled on commercial CA	Electrospinning and hydrothermal treatment	-60 mL of AO 7 solution (50 mg L^{-1}) -Upland UVP lamp (254 nm, 40 mW/cm^2)	[180]
TiO$_2$ nanoparticles/cellulose fibers	Hydrothermal method	-4 mL of MO solution (0.25 mM) -UV-A lamp (50 cm long, 72 W, emission range 320–400 nm)	[181]
N-doped TiO$_2$ nanorods in regenerated cellulose thin films	Phase inversion	- 150 mL of MB solution (40 mg L^{-1}) - UV lamp (Vilber Laurmat, λ = 312 nm, 30 W) and white light-emitting diode (LED) flood light (CS-FL-30W, λ > 420 nm, 30 W).	[182]
C/SiO$_2$/TiO$_2$	Dip-Coating or spin-coating	-MV dye (0.1%) - UV light (Xe 150-W lamp, the range of emission wavelength between 290-400 nm (UVA and UVB))	[183]
Dispersed-TiO$_2$ on CA	Phase inversion	-25 mL of MB solution (12 × 10^{-6} M) -UV radiation (150W Hg lamp)	[152]

CA-polystyrene-ZnO	Solution dispersion blending method	-CR or RY-105 solution (100 mg L^{-1}, Neutral pH) -Sunlight irradation	[184]
Bi_2O_3/ZnS Heterojunction Functionalized Porous CA Membranes	Phase-inversion	-50 mL of RhB solution (5.0 mg L^{-1}) -Visible-Light (xenon light source with the power of 200 W)	[185]
RGO/PDA/$Bi_{12}O_{17}Cl_{2}$/CA membrane.	Vacuum filtration method	-100 mL MB solution (15 mg L^{-1}) and 4-CP solution (15 mg L^{-1}) -visible light (long-arc Xe lamp, 500 W).	[165]
RGO-Ag-TiO_2-CA membrane	Hydrothermal and vacuum filtration method	-MB (20 mg L^{-1}) and RhB (30 mg L^{-1}) -visible light (500 W Xe lamp coupled with a UV-cutoff filter ($\lambda > = 400$ nm))	[186]
CA-PEG-PU/ZnO	Casting material and solution dispersion blending method	-50 mL of RR 11 (or RO 84) sulution (100 mg L^{-1}) -Sunlight	[187]
Electrospun $H_4SiW_{12}O_{40}$/CA	Electrospinning	-100 mL of MO (or TC) solution (10 mg L^{-1}) ,pH = 2 -UV irradiation (300 W high pressure mercury lamp)	[188]
RGO/PDA/g-C_3N_4/CA composite	Vacuum Filtration	-100 mL of MB solution (5 mg·L^{-1}), 1 mL H_2O_2 (wt% 30) -visible light irradiation (long-arc Xe lamp, 500 W	[189]
CA-PU/ZnO	Casting material and solution dispersion blending method	-50 mL of RR 11 (or RO 84) (100 mg·L^{-1}) -UV-light	[190]
CA supported Ag@AgCl	Double diffusion technique	-10 mL of MO solution (10 mg·L^{-1}) -visible-light irradiation (500 W Xe arc lamp equipped with an ultraviolet cutoff filter ($\lambda > 420$ nm)).	[168]

Materials Research Forum LLC

https://doi.org/10.21741/9781644901359-1

4. Inorganic membranes

Some studies found polymeric membranes damaged by UV irradiation, hydroxyl radicals and the penetration of the hard inorganic semiconductors particles. Further, organic solvents, high temperatures acidic or caustic solutions affect applications these polymeric membranes and restrict their utilization [191].

Since the pioneering work of Anderson et al. [192], synergistic coupling of ceramic membrane filtration and UV-assisted photocatalytic degradation has been receiving much attention [193]. Recently, ceramic membranes preferred over the conventional polymeric membranes due to their excellent thermal, chemical and mechanical stability and biological inertia for wide pH range adaptability, long service life and good recoverability [194, 195]. The cost of ceramic membranes has been continuously reduced during the last years.

Ceramic membranes, which can be used in microfiltration and ultrafiltration, have superior chemical, thermal, and mechanical stability compared to polymeric membranes, and the pore size can be more easily controlled [196]. Sintering and the sol-gel process are the most common techniques to prepare ceramic microfiltration membranes. Sintering is limited to pore sizes around 0.1 μm, practically all UF membranes are prepared with the sol-gel technique. The base materials for the preparation of ceramic membranes are alumina (Al_2O_3), titania (TiO_2), silica (SiO_2), and zirconia (ZrO_2). Ceramic nanofiltration membranes can also be produced, but, to date, the pore size of most of these membranes is still relatively high.

4.1 TiO₂-based membranes

Since the pioneering work on the fabrication of TiO_2 ceramic membrane by Anderson and co-workers [192] many research studies have been carried out to fabricate TiO_2 membranes by coating TiO_2 films on various supports [197]. Titania membranes have received significant attention in recent years due to their unique characteristics such as high water flux, semiconducting properties, photocatalysis, and chemical resistances over other membrane materials such as alumina, silica, and zirconia [198]. Currently, TiO_2 membranes have been mainly prepared via conventional tape casting processes. Chemical vapor deposition (CVD), sputtering, and sol-gel methods are also applicable for the fabrication of homogeneous TiO_2 membranes. Ma and Quan modified the sol–gel dip-coating method to enhance the performance of TiO_2 membrane via doping with Ag and Si and HAP coupling [199]. Furthermore, a visible light responsive TiO_2 membrane was fabricated via co-doping C, N and Ce through a weak alkaline sol–gel process by Cao and co-workers [200]. Most recently, Liu deposited Ag nanoparticles on TiO_2 nanofiber and subsequently formed a Ag/ TiO_2 flat membrane on glass fiber substrate for disinfection under solar light irradiation [201].

In these methods, however, strict processing conditions are commonly required for high crystallinity of the ceramic membranes, which is very cumbersome and time consuming, and are often not economically viable. In addition, the TiO_2-based membranes studied so far are in the form of finite sized tubes with diameters of at least several millimeters or flat discs and consequently have low surface area to volume ratios. These low area to volume ratios compares unfavorably with polymeric hollow fiber modules with high values; this limits the application of current inorganic tubular and disc membranes. This limitation is most evident in catalytic membrane reactors, where it is desirable to maximize the area of the membrane to increase the permeation rate to remove the product species from the reaction zone. Recently various morphologies of one-dimensional (1D) nanostructured TiO_2, including nanowires, nanofibers, nanorods and nanotubes, have been synthesized by chemical or physical methods (Figure 3) [173, 202, 203].

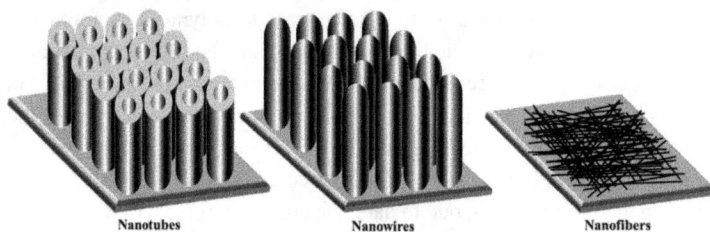

Nanotubes   Nanowires   Nanofibers

Figure 3. Various morphologies of one-dimensional (1D) nanostructured TiO_2.

Regarding the geometry of the membrane, hollow fibers have the highest surface area per volume ratio, which makes them very appealing because of the possibility of obtaining small membrane modules with large surface areas [203]. There are a few methods available for preparing inorganic hollow fibers, including dry spinning a system of inorganic material and binder, wet spinning a suitable inorganic material-containing solution, depositing fibers from the gas phase on to a substrate, or pyrolyzing the polymers[204, 205]. Recently, the well-known phase inversion method, commonly employed for spinning polymeric hollow fiber membranes, have been successfully adopted in preparing the inorganic aluminum oxide (Al_2O_3) hollow fibers [198, 206, 207]. Because of the phase inversion characteristics, the prepared inorganic hollow fibers possess an asymmetric structure, which provides a better permeability for a given thickness. Thus, they can be used not only in many separation processes, but also to be served as a porous support for composite membrane formation.

Typically, TiO_2 nanowires are obtainable by hydrothermal method, through treating TiO_2 powders within steel pressure vessels autoclave under controlled temperature and/or pressure in strong basic NaOH solution. It found during hydrothermal reaction, some of the Ti–O–Ti bonds were broken and Ti–O–Na and Ti–OH bonds were formed, which result in the formation of TiO_2 nanowires [208]. These nanowires exhibit superior photocatalytic efficiency relative to conventional bulk materials as a result of its larger surface area and presence of quantum size effect.

Imai, et al. [197] synthesized TiO_2 nanotube membrane, by deposition of anatase films on the inner walls of alumina porous membranes. After dissolving the alumina membrane using an aqueous solution of ammonia, the deposited titania films were obtained as nanotubes. The channel diameter of the tubes was controllable over the range 50–150 nm by changing the deposition time. In this method the morphology of the anatase nanotubes was highly controllable because no annealing for crystallization was required.

Albu et al. [209] reported formation of a nanotubular layer of TiO_2, a thickness of >100 μm, electrochemically by controlled anodization of a Ti foil in a fluoride-containing ethylene glycol. This leads to a dense array of aligned TiO_2 nanotubes, attached to the Ti substrate with a diameter of 160 ± 30 nm and a wall thickness of 20 ± 5 nm. The titanium foil was removed by selective dissolution in water-free CH_3OH/Br_2 solution, followed by exposure to hydrofluoric acid (HF) vapor for 30 min to open the bottom of the tube.

More recently, immense efforts are devoted to the study of organizing of nanowire/fiber/tube into 2D nanomaterials, including membrane and sheet. These 2D nanomaterials exhibit new properties while retaining the properties of 1D nanomaterials.

Many researchers have endeavored to introduce TiO_2 on/into high-surface-area substrate. Among them, TiO_2–SiO_2 composites have been widely used in industrial applications and most extensively studied [133]. SiO_2 is an excellent catalyst support because of its chemical inertia, thermal stability and adsorption of reactants. Particularly, the nanofibrous SiO_2 supports have both high surface-area-to-volume ratio and favorable recycling properties. Electrospinning in combination with sol–gel processes has been proved a simple and effective method to produce polymer/metal oxide composite nanofibers. In many cases, the metal oxide nanofibers can be obtained by subsequent pyrolysis. Ding et al. fabricated the flexible and amphiphobic SiO_2 nanofibrous mats via electrospinning the blend solutions of poly(vinyl alcohol) (PVA) and SiO_2 gel, followed by calcination to remove the organic component [210]. Wang et al. prepared anatase mesoporous titanium nanofibers from calcination of the electrospun tetrabutyl titanate/poly(vinylpyrrolidone) (PVP)/pluronic123 (P123) composite nanofibers [211].

Table 8. Ceramic photocatalytic membranes for degradation of organic pollutants.

Photocatalyst name	Preparation method	Condition & light source	Ref.
TiO$_2$ nanostructured hollow fiber membrane	Spinning-sintering technique	-20 mL of AO7 at a concentration of 20 mg L^{-1} -UV light (Four UV-A lamps (SYLVANIA Blacklite F8 W/BL350, emit at 330–370 nm))	[203]
TiO$_2$/Al$_2$O$_3$ composite membranes	Dip coating	-8 mLof 30 µM MB solution, pH = 3 - UV radiation Two 15 W low pressure mercury UV tubes (Spectronics), at 365 nm ,light intensity of 3.48 mW/cm^2	[212]
ZnO quantum dots on hollow SiO$_2$ nanofibers	Electrospinning and calcination	-UV (50 W high-pressure mercury lamp with main emission at 313 nm)) -100 mL of the RB solution 10 mg L^{-1}	[213]
Mesoporous titania networks consisting of interconnected anatase nanowires	Templating by BC, Dip coating and calcination	-10^{-5} M RhB solution - 250 W high-pressure mercury lamp	[173]
TiO$_2$ nanoparticles/Ag fibers	Electrophoretic deposition Method	- 5 mL of a 5 µM MB solution -UV (8 W low pressure mercury lamp, intensity 11 mW/cm^2)	[214]

Various strategies have been explored to improve the photocatalytic efficiency of TiO$_2$ NWs or nanotubes, such as tuning its crystallite size and structure [215, 216], doping with metal or nonmetal elements [217], sensitizing with other small bandgap semiconductor materials [218, 219], synthesizing branched structures [220], and postgrowth hydrogen annealing [221]. Doping materials with Sn or Sn^{4+} is an attractive approach [222], especially for TiO$_2$, as the small lattice mismatch between SnO$_2$ and TiO$_2$ leads to good structural compatibility and stability. Xu et al. [223], reported the one-pot hydrothermal with different ratios of SnCl$_4$ and tetrabutyl titanate, and a high acidity of the reactant solution synthesis and controlled Sn-doped TiO$_2$ nanowire arrays. The obtained Sn/TiO$_2$ NWs are single crystalline with a rutile structure. The application of ceramic photocatalytic membranes for degradation of various organic pollutants has been summarized in Table 8.

Abbreviations

Acid Orange (AO 7)	Polyacrylonitrile (PAN)
Advanced oxidation processes (AOPs)	Polyamide 6 (PA6)
Aliphatic polyamide (PA)	Polycarbonate (PC)
3-Aminopropyltrimethoxysilane (APTMS)	Polydopamine (PDA)
Ammonium persulfate (APS)	Polyetheretherketone (PEEK)
Aromatic polyamide (Aramids)	Polyether imide (PEI)
Atomic layer deposition (ALD)	Polyether sulfone (PES)
Bisphenol A (BPA)	Poly(ether urethane) (PEU)
Brilliant Green (BG)	Polyethylene (PE),
Bovine Serum Albumin (BSA)	polyethylene glycol(PEG)
Camphor-10-sulfonic acid (CSA)	Polyethylene oxide (PEO)
Carbon covered alumina (CCA)	Polyimide (PI)
Cellulose acetate (CA)	Poly(L-lactide) (PLLA)
Cellulose esters (CE)	Poly(piperazine amide) (P-PIP-A)
Cellulose triacetate (CTA)	Polypropylene (PP)
Chemical vapour deposition (CVD)	Polysulfone (PS)
4-Chlorophenol (4-CP)	Polytetrafluoroethylene (PTFE)
2-Chloroethyl phenyl sulfide (CEPS)	Polyurethane (PU)
Cimetidine (CMT)	Poly(vinylidene fluoride) (PVDF),
Ciprofloxacin (CIP)	Praseodymium (Pr)
Congo red (CR)	Reactive Black 5 (RB5)
Eosin Yellow (EY)	Reactive Orange 84 (RO 84)
Erbium (Er)	Reactive Red ED-2B (RR ED-2B)
Hydrophilic surface modifying	Reactive Red 11 (RR 11)
macromolecules (LSMM)	Reactive yellow 105 (RY-105)
Indigo Carmin (IC)	Reverse osmosis (RO)
Graphene oxide (GO)	Rhodamine B (RhB)
Malachite Green (MG)	Safranin-O (SO)

Materials from the Lavoisier Institute (MIL)	Silver cyanamide (Ag_2NCN)
Membrane distillation (MD)	Sulfamethoxazole (SMX)
Metal organic frameworks (MOFs)	Sulfonated polysulfone (S-PS)
Methacrylic acid (MAA)	Tetrabutyl titanate (TBT)
Methylene blue (MB)	Tetracycline (TC)
Methyl orange (MO)	tetraethoxysilane (TEOS)
Methyl violet 2B (MV)	Trifluoroethyl acrylate (TFA)
Microfiltration (MF)	Trifluoroethylene (TrFE)
Nanofiltration (NF)	Ultrafiltration (UF)
Oxygenated graphitic carbon nitride (OGCN)	Ultraviolet (UV)
Photocatalytic membrane reactor (PMR)	

Acknowledgment

The authors are grateful to AGH University of Science and Technology, Alzahra University, Polish Academy of Sciences, Razi University and Shenzhen University.

References

[1] O. Monfort, G. Plesch, Bismuth vanadate-based semiconductor photocatalysts: a short critical review on the efficiency and the mechanism of photodegradation of organic pollutants, Environmental Science and Pollution Research, 25 (2018) 19362-19379. https://doi.org/10.1007/s11356-018-2437-9

[2] S.-T. Yang, S. Chen, Y. Chang, A. Cao, Y. Liu, H. Wang, Removal of methylene blue from aqueous solution by graphene oxide, J. Colloid Interface Sci., 359 (2011) 24-29. https://doi.org/10.1016/j.jcis.2011.02.064

[3] J. Zhang, M. Yan, X. Yuan, M. Si, L. Jiang, Z. Wu, H. Wang, G. Zeng, Nitrogen doped carbon quantum dots mediated silver phosphate/bismuth vanadate Z-scheme photocatalyst for enhanced antibiotic degradation, Journal of Colloid and Interface Science, 529 (2018) 11-22. https://doi.org/10.1016/j.jcis.2018.05.109

[4] D.L. Zhao, T.-S. Chung, Applications of carbon quantum dots (CQDs) in membrane technologies: A review, Water research, (2018). https://doi.org/10.1016/j.watres.2018.09.040

[5] J. Luo, J. Chen, R. Guo, Y. Qiu, W. Li, X. Zhou, X. Ning, L. Zhan, Rational construction of direct Z-scheme LaMnO3/g-C3N4 hybrid for improved visible-light photocatalytic tetracycline degradation, Separation and Purification Technology, 211 (2019) 882-894. https://doi.org/10.1016/j.seppur.2018.10.062

[6] W. Zhang, Z. Zhang, S. Kwon, F. Zhang, B. Stephen, K.K. Kim, R. Jung, S. Kwon, K.-B. Chung, W. Yang, Photocatalytic improvement of Mn-adsorbed g-C3N4, Applied Catalysis B: Environmental, 206 (2017) 271-281. https://doi.org/10.1016/j.apcatb.2017.01.034

[7] R. Xie, L. Zhang, H. Xu, Y. Zhong, X. Sui, Z. Mao, Fabrication of Z-scheme photocatalyst Ag–AgBr@Bi20TiO32 and its visible-light photocatalytic activity for the degradation of isoproturon herbicide, Journal of Molecular Catalysis A: Chemical, 406 (2015) 194-203. https://doi.org/10.1016/j.molcata.2015.05.028

[8] J. Hou, C. Yang, Z. Wang, S. Jiao, H. Zhu, Hydrothermal synthesis of CdS/CdLa 2 S 4 heterostructures for efficient visible-light-driven photocatalytic hydrogen production, RSC Advances, 2 (2012) 10330-10336. https://doi.org/10.1039/c2ra21641h

[9] S. Kohtani, E. Yoshioka, H. Miyabe, Photocatalytic hydrogenation on semiconductor particles, in: Hydrogenation, IntechOpen, 2012. https://doi.org/10.5772/45732

[10] A.S. Hassanien, A.A. Akl, Effect of Se addition on optical and electrical properties of chalcogenide CdSSe thin films, Superlattices and Microstructures, 89 (2016) 153-169. https://doi.org/10.1016/j.spmi.2015.10.044

[11] R. Saravanan, V.K. Gupta, E. Mosquera, F. Gracia, Preparation and characterization of V2O5/ZnO nanocomposite system for photocatalytic application, Journal of Molecular Liquids, 198 (2014) 409-412. https://doi.org/10.1016/j.molliq.2014.07.030

[12] B.M. Pirzada, N.A. Mir, N. Qutub, O. Mehraj, S. Sabir, M. Muneer, Synthesis, characterization and optimization of photocatalytic activity of TiO2/ZrO2 nanocomposite heterostructures, Materials Science and Engineering: B, 193 (2015) 137-145. https://doi.org/10.1016/j.mseb.2014.12.005

[13] N. Kamarulzaman, M.F. Kasim, N.F. Chayed, Elucidation of the highest valence band and lowest conduction band shifts using XPS for ZnO and Zn0.99Cu0.01O band gap changes, Results in Physics, 6 (2016) 217-230. https://doi.org/10.1016/j.rinp.2016.04.001

[14] W. Zhang, C. Hu, W. Zhai, Z. Wang, Y. Sun, F. Chi, S. Ran, X. Liu, Y. Lv, Novel Ag3PO4/CeO2 pn hierarchical heterojunction with enhanced photocatalytic performance, Materials Research, 19 (2016) 673-679. https://doi.org/10.1590/1980-5373-MR-2016-0009

[15] H. Heng, Q. Gan, P. Meng, X. Liu, The visible-light-driven type III heterojunction H3PW12O40/TiO2-In2S3: A photocatalysis composite with enhanced photocatalytic activity, Journal of Alloys and Compounds, 696 (2017) 51-59. https://doi.org/10.1016/j.jallcom.2016.11.116

[16] S. Issarapanacheewin, K. Wetchakun, S. Phanichphant, W. Kangwansupamonkon, N. Wetchakun, A novel CeO2/Bi2WO6 composite with highly enhanced photocatalytic activity, Materials Letters, 156 (2015) 28-31. https://doi.org/10.1016/j.matlet.2015.04.139

[17] X. Zeng, Z. Wang, G. Wang, T.R. Gengenbach, D.T. McCarthy, A. Deletic, J. Yu, X. Zhang, Highly dispersed TiO2 nanocrystals and WO3 nanorods on reduced graphene oxide: Z-scheme photocatalysis system for accelerated photocatalytic water disinfection, Applied Catalysis B: Environmental, 218 (2017) 163-173. https://doi.org/10.1016/j.apcatb.2017.06.055

[18] J. Cao, C. Zhou, H. Lin, B. Xu, S. Chen, Surface modification of m-BiVO4 with wide band-gap semiconductor BiOCl to largely improve the visible light induced photocatalytic activity, Applied Surface Science, 284 (2013) 263-269. https://doi.org/10.1016/j.apsusc.2013.07.092

[19] T.Q. Nguyen, A.K. Thapa, V.K. Vendra, J.B. Jasinski, G.U. Sumanasekera, M.K. Sunkara, High rate capacity retention of binder-free, tin oxide nanowire arrays using thin titania and alumina coatings, Rsc Advances, 4 (2014) 3312-3317. https://doi.org/10.1039/C3RA46003G

[20] G.-H. He, C.-J. Liang, Y.-D. Ou, D.-N. Liu, Y.-P. Fang, Y.-H. Xu, Preparation of novel Sb2O3/WO3 photocatalysts and their activities under visible light irradiation, Materials Research Bulletin, 48 (2013) 2244-2249. https://doi.org/10.1016/j.materresbull.2013.02.055

[21] H. Yuan, J. Liu, J. Li, Y. Li, X. Wang, Y. Zhang, J. Jiang, S. Chen, C. Zhao, D. Qian, Designed synthesis of a novel BiVO4–Cu2O–TiO2 as an efficient visible-light-responding photocatalyst, Journal of Colloid and Interface Science, 444 (2015) 58-66. https://doi.org/10.1016/j.jcis.2014.12.034

[22] M.A. Zarepour, M. Tasviri, Facile fabrication of Ag decorated TiO2 nanorices: Highly efficient visible-light-responsive photocatalyst in degradation of contaminants, Journal of Photochemistry and Photobiology A: Chemistry, 371 (2019) 166-172. https://doi.org/10.1016/j.jphotochem.2018.11.007

[23] H.L. Hoşgün, M.T.A. Aydın, Synthesis, characterization and photocatalytic activity of boron-doped titanium dioxide nanotubes, Journal of Molecular Structure, 1180 (2019) 676-682. https://doi.org/10.1016/j.molstruc.2018.12.056

[24] M. Li, Z. Xing, J. Jiang, Z. Li, J. Kuang, J. Yin, N. Wan, Q. Zhu, W. Zhou, In-situ Ti3+/S doped high thermostable anatase TiO2 nanorods as efficient visible-light-driven photocatalysts, Materials Chemistry and Physics, 219 (2018) 303-310. https://doi.org/10.1016/j.matchemphys.2018.08.051

[25] B. Singaram, J. Jeyaram, R. Rajendran, P. Arumugam, K. Varadharajan, Visible light photocatalytic activity of tungsten and fluorine codoped TiO 2 nanoparticle for an efficient dye degradation, Ionics, (2018) 1-12. https://doi.org/10.1007/s11581-018-2628-x

[26] M. Ge, J. Li, L. Liu, Z. Zhou, Template-free synthesis and photocatalytic application of rutile TiO2 hierarchical nanostructures, Industrial & Engineering Chemistry Research, 50 (2011) 6681-6687. https://doi.org/10.1021/ie1023113

[27] S. Mozia, D. Darowna, R. Wróbel, A.W. Morawski, A study on the stability of polyethersulfone ultrafiltration membranes in a photocatalytic membrane reactor, Journal of Membrane Science, 495 (2015) 176-186. https://doi.org/10.1016/j.memsci.2015.08.024

[28] K. Fischer, P. Schulz, I. Atanasov, A. Abdul Latif, I. Thomas, M. Kühnert, A. Prager, J. Griebel, A. Schulze, Synthesis of High Crystalline TiO2 Nanoparticles on a Polymer Membrane to Degrade Pollutants from Water, Catalysts, 8 (2018) 376. https://doi.org/10.3390/catal8090376

[29] S. Singh, H. Mahalingam, P.K. Singh, Polymer-supported titanium dioxide photocatalysts for environmental remediation: A review, Applied Catalysis A: General, 462-463 (2013) 178-195. https://doi.org/10.1016/j.apcata.2013.04.039

[30] J. Nikkola, J. Sievänen, M. Raulio, J. Wei, J. Vuorinen, C.Y. Tang, Surface modification of thin film composite polyamide membrane using atomic layer deposition method, Journal of membrane science, 450 (2014) 174-180. https://doi.org/10.1016/j.memsci.2013.09.005

[31] B.S. Lalia, C. Garlisi, G. Palmisano, R. Hashaikeh, Photocatalytic activity of an electrophoretically deposited composite titanium dioxide membrane using carbon cloth as a conducting substrate, RSC Advances, 6 (2016) 64219-64227. https://doi.org/10.1039/C6RA07390E

[32] A.A. Muleja, B.B. Mamba, Development of calcined catalytic membrane for potential photodegradation of Congo red in aqueous solution, Journal of Environmental Chemical Engineering, 6 (2018) 4850-4863. https://doi.org/10.1016/j.jece.2018.07.004

[33] M.I. Shekh, D.M. Patel, K.P. Patel, R.M. Patel, Electrospun Nanofibers of Poly(NPEMA-co.-CMPMA): Used as Heavy Metal Ion Remover and Water Sanitizer, Fibers and Polymers, 17 (2016) 358-370. https://doi.org/10.1007/s12221-016-5861-9

[34] G. Abdi, A. Alizadeh, S. Zinadini, G. Moradi, Removal of dye and heavy metal ion using a novel synthetic polyethersulfone nanofiltration membrane modified by magnetic graphene oxide/metformin hybrid, Journal of membrane science, 552 (2018) 326-335. https://doi.org/10.1016/j.memsci.2018.02.018

[35] F. Jin, W. Lv, C. Zhang, Z. Li, R. Su, W. Qi, Q.-H. Yang, Z. He, High-performance ultrafiltration membranes based on polyethersulfone–graphene oxide composites, Rsc Advances, 3 (2013) 21394-21397. https://doi.org/10.1039/c3ra42908c

[36] A.L. Chibac, T. Buruiana, V. Melinte, E.C. Buruiana, Photocatalysis applications of some hybrid polymeric composites incorporating TiO2 nanoparticles and their combinations with SiO2/Fe2O3, Beilstein journal of nanotechnology, 8 (2017) 272-286. https://doi.org/10.3762/bjnano.8.30

[37] K. Monsef, M. Homayoonfal, F. Davar, Coating carboxylic and sulfate functional groups on ZrO2 nanoparticles: Antifouling enhancement of nanocomposite membranes during water treatment, Reactive and Functional Polymers, 131 (2018) 299-314. https://doi.org/10.1016/j.reactfunctpolym.2018.08.003

[38] N. Maximous, G. Nakhla, W. Wan, K. Wong, Preparation, characterization and performance of Al2O3/PES membrane for wastewater filtration, Journal of Membrane Science, 341 (2009) 67-75. https://doi.org/10.1016/j.memsci.2009.05.040

[39] R.R. Darabi, M. Jahanshahi, M. Peyravi, A support assisted by photocatalytic Fe3O4/ZnO nanocomposite for thin-film forward osmosis membrane, Chemical Engineering Research and Design, 133 (2018) 11-25. https://doi.org/10.1016/j.cherd.2018.02.029

[40] N. Ma, X. Quan, Y. Zhang, S. Chen, H. Zhao, Integration of separation and photocatalysis using an inorganic membrane modified with Si-doped TiO2 for water purification, Journal of Membrane Science, 335 (2009) 58-67. https://doi.org/10.1016/j.memsci.2009.02.040

[41] Y. Yang, H. Zhang, P. Wang, Q. Zheng, J. Li, The influence of nano-sized TiO2 fillers on the morphologies and properties of PSF UF membrane, Journal of Membrane Science, 288 (2007) 231-238. https://doi.org/10.1016/j.memsci.2006.11.019

[42] K. Fischer, R. Gläser, A. Schulze, Nanoneedle and nanotubular titanium dioxide – PES mixed matrix membrane for photocatalysis, Applied Catalysis B: Environmental, 160-161 (2014) 456-464. https://doi.org/10.1016/j.apcatb.2014.05.054

[43] K. Fischer, M. Kühnert, R. Gläser, A. Schulze, Photocatalytic degradation and toxicity evaluation of diclofenac by nanotubular titanium dioxide–PES membrane in a static and continuous setup, RSC Advances, 5 (2015) 16340-16348. https://doi.org/10.1039/C4RA16219F

[44] S.L. Phua, L. Yang, C.L. Toh, D. Guoqiang, S.K. Lau, A. Dasari, X. Lu, Simultaneous Enhancements of UV Resistance and Mechanical Properties of Polypropylene by Incorporation of Dopamine-Modified Clay, ACS Applied Materials & Interfaces, 5 (2013) 1302-1309. https://doi.org/10.1021/am3024405

[45] H. Wu, Y. Liu, L. Mao, C. Jiang, J. Ang, X. Lu, Doping polysulfone ultrafiltration membrane with TiO2-PDA nanohybrid for simultaneous self-cleaning and self-protection, Journal of Membrane Science, 532 (2017) 20-29. https://doi.org/10.1016/j.memsci.2017.03.010

[46] K. Feng, L. Hou, B. Tang, P. Wu, A self-protected self-cleaning ultrafiltration membrane by using polydopamine as a free-radical scavenger, Journal of Membrane Science, 490 (2015) 120-128. https://doi.org/10.1016/j.memsci.2015.04.056

[47] H. Zangeneh, A.A. Zinatizadeh, S. Zinadini, M. Feyzi, D.W. Bahnemann, Preparation and characterization of a novel photocatalytic self-cleaning PES nanofiltration membrane by embedding a visible-driven photocatalyst boron doped-TiO2SiO2/CoFe2O4 nanoparticles, Separation and Purification Technology, 209 (2019) 764-775. https://doi.org/10.1016/j.seppur.2018.09.030

[48] H. Zangeneh, A.A.L. Zinatizadeh, M. Habibi, M. Akia, M. Hasnain Isa, Photocatalytic oxidation of organic dyes and pollutants in wastewater using different modified titanium dioxides: A comparative review, Journal of Industrial and Engineering Chemistry, 26 (2015) 1-36. https://doi.org/10.1016/j.jiec.2014.10.043

[49] Y. Cong, J. Zhang, F. Chen, M. Anpo, D. He, Preparation, Photocatalytic Activity, and Mechanism of Nano-TiO2 Co-Doped with Nitrogen and Iron (III), The Journal of Physical Chemistry C, 111 (2007) 10618-10623. https://doi.org/10.1021/jp0727493

[50] A. Moslehyani, A.F. Ismail, M.H.D. Othman, T. Matsuura, Design and performance study of hybrid photocatalytic reactor-PVDF/MWCNT nanocomposite membrane system for treatment of petroleum refinery wastewater, Desalination, 363 (2015) 99-111. https://doi.org/10.1016/j.desal.2015.01.044

[51] Z. Xu, T. Wu, J. Shi, K. Teng, W. Wang, M. Ma, J. Li, X. Qian, C. Li, J. Fan, Photocatalytic antifouling PVDF ultrafiltration membranes based on synergy of graphene oxide and TiO2 for water treatment, Journal of Membrane Science, 520 (2016) 281-293. https://doi.org/10.1016/j.memsci.2016.07.060

[52] N.N. Patel, M.I. Shekh, K.P. Patel, R.M. Patel, Electrospun nano silver embedded polystyrene composite nanofiber as a possible water disinfectant, Indian Journal of Chemistry Section a-Inorganic Bio-Inorganic Physical Theoretical & Analytical Chemistry, 58 (2019) 288-293.   http://nopr.niscair.res.in/handle/123456789/45793

[53] Y. Sakatani, H. Ando, K. Okusako, H. Koike, J. Nunoshige, T. Takata, J.N. Kondo, M. Hara, K. Domen, Metal ion and N co-doped TiO 2 as a visible-light photocatalyst, Journal of materials research, 19 (2004) 2100-2108. https://doi.org/10.1557/JMR.2004.0269

[54] A.T. Kuvarega, N. Khumalo, D. Dlamini, B.B. Mamba, Polysulfone/N,Pd co-doped TiO2 composite membranes for photocatalytic dye degradation, Separation and Purification Technology, 191 (2018) 122-133. https://doi.org/10.1016/j.seppur.2017.07.064

[55] Z.A. Mohd Hir, A.H. Abdullah, Z. Zainal, H.N. Lim, Visible light-active hybrid film photocatalyst of polyethersulfone–reduced TiO2: photocatalytic response and radical trapping investigation, Journal of Materials Science, 53 (2018) 13264-13279. https://doi.org/10.1007/s10853-018-2570-3

[56] M.M. Mahlambi, O.T. Mahlangu, G.D. Vilakati, B.B. Mamba, Visible Light Photodegradation of Rhodamine B Dye by Two Forms of Carbon-Covered Alumina Supported TiO2/Polysulfone Membranes, Industrial & Engineering Chemistry Research, 53 (2014) 5709-5717. https://doi.org/10.1021/ie4038449

[57] Z.A.M. Hir, P. Moradihamedani, A.H. Abdullah, M.A. Mohamed, Immobilization of TiO2 into polyethersulfone matrix as hybrid film photocatalyst for effective degradation of methyl orange dye, Materials Science in Semiconductor Processing, 57 (2017) 157-165. https://doi.org/10.1016/j.mssp.2016.10.009

[58] Q. Wang, C. Yang, G. Zhang, L. Hu, P. Wang, Photocatalytic Fe-doped TiO2/PSF composite UF membranes: Characterization and performance on BPA removal under visible-light irradiation, Chemical Engineering Journal, 319 (2017) 39-47. https://doi.org/10.1016/j.cej.2017.02.145

[59] N. Salim, N. Nor, J. Jaafar, A. Ismail, T. Matsuura, M. Qtaishat, M. Othman, M. Rahman, F. Aziz, N. Yusof, Performance of PES/LSMM-OGCN Photocatalytic

Membrane for Phenol Removal: Effect of OGCN Loading, Membranes, 8 (2018) 42. https://doi.org/10.3390/membranes8030042

[60] R. Zhang, Y. Cai, X. Zhu, Q. Han, T. Zhang, Y. Liu, Y. Li, A. Wang, A novel photocatalytic membrane decorated with PDA/RGO/Ag3PO4 for catalytic dye decomposition, Colloids and Surfaces A: Physicochemical and Engineering Aspects, 563 (2019) 68-76. https://doi.org/10.1016/j.colsurfa.2018.11.069

[61] A.T. Kuvarega, B.B. Mamba, Photocatalytic Membranes for Efficient Water Treatment, Semiconductor Photocatalysis - Materials, Mechanisms and Applications, (2016) 523-539. https://doi.org/10.5772/62584

[62] A. Rahimpour, S. Madaeni, A. Taheri, Y. Mansourpanah, Coupling TiO2 nanoparticles with UV irradiation for modification of polyethersulfone ultrafiltration membranes, Journal of Membrane Science, 313 (2008) 158-169. https://doi.org/10.1016/j.memsci.2007.12.075

[63] T.-H. Bae, T.-M. Tak, Effect of TiO2 nanoparticles on fouling mitigation of ultrafiltration membranes for activated sludge filtration, Journal of Membrane Science, 249 (2005) 1-8. https://doi.org/10.1016/j.memsci.2004.09.008

[64] S.-Y. Kwak, S.H. Kim, S.S. Kim, Hybrid Organic/Inorganic Reverse Osmosis (RO) Membrane for Bactericidal Anti-Fouling. 1. Preparation and Characterization of TiO2 Nanoparticle Self-Assembled Aromatic Polyamide Thin-Film-Composite (TFC) Membrane, Environmental Science & Technology, 35 (2001) 2388-2394. https://doi.org/10.1021/es0017099

[65] M.-L. Luo, J.-Q. Zhao, W. Tang, C.-S. Pu, Hydrophilic modification of poly (ether sulfone) ultrafiltration membrane surface by self-assembly of TiO2 nanoparticles, Applied Surface Science, 249 (2005) 76-84. https://doi.org/10.1016/j.apsusc.2004.11.054

[66] J.-H. Li, Y.-Y. Xu, L.-P. Zhu, J.-H. Wang, C.-H. Du, Fabrication and characterization of a novel TiO2 nanoparticle self-assembly membrane with improved fouling resistance, Journal of Membrane Science, 326 (2009) 659-666. https://doi.org/10.1016/j.memsci.2008.10.049

[67] A. Rahimpour, M. Jahanshahi, A. Mollahosseini, B. Rajaeian, Structural and performance properties of UV-assisted TiO2 deposited nano-composite PVDF/SPES membranes, Desalination, 285 (2012) 31-38. https://doi.org/10.1016/j.desal.2011.09.026

[68] S.-J. You, G.U. Semblante, S.-C. Lu, R.A. Damodar, T.-C. Wei, Evaluation of the antifouling and photocatalytic properties of poly(vinylidene fluoride) plasma-grafted poly(acrylic acid) membrane with self-assembled TiO2, Journal of Hazardous Materials, 237-238 (2012) 10-19. https://doi.org/10.1016/j.jhazmat.2012.07.071

[69] H.-S. Choi, Y.-S. Kim, Y. Zhang, S. Tang, S.-W. Myung, B.-C. Shin, Plasma-induced graft co-polymerization of acrylic acid onto the polyurethane surface, Surface and Coatings Technology, 182 (2004) 55-64. https://doi.org/10.1016/S0257-8972(03)00880-6

[70] R.A. Damodar, S.-J. You, H.-H. Chou, Study the self cleaning, antibacterial and photocatalytic properties of TiO2 entrapped PVDF membranes, Journal of Hazardous Materials, 172 (2009) 1321-1328. https://doi.org/10.1016/j.jhazmat.2009.07.139

[71] O. Tahiri Alaoui, Q.T. Nguyen, C. Mbareck, T. Rhlalou, Elaboration and study of poly(vinylidene fluoride)–anatase TiO2 composite membranes in photocatalytic degradation of dyes, Applied Catalysis A: General, 358 (2009) 13-20. https://doi.org/10.1016/j.apcata.2009.01.032

[72] J.-H. Li, B.-F. Yan, X.-S. Shao, S.-S. Wang, H.-Y. Tian, Q.-Q. Zhang, Influence of Ag/TiO2 nanoparticle on the surface hydrophilicity and visible-light response activity of polyvinylidene fluoride membrane, Applied Surface Science, 324 (2015) 82-89. https://doi.org/10.1016/j.apsusc.2014.10.080

[73] M.I. Shekh, N.N. Patel, K.P. Patel, R.M. Patel, A. Ray, Nano silver-embedded electrospun nanofiber of poly(4-chloro-3-methylphenyl methacrylate): use as water sanitizer, Environmental Science and Pollution Research, 24 (2017) 5701-5716. https://doi.org/10.1007/s11356-016-8254-0

[74] S. Teixeira, P.M. Martins, S. Lanceros-Méndez, K. Kühn, G. Cuniberti, Reusability of photocatalytic TiO2 and ZnO nanoparticles immobilized in poly(vinylidene difluoride)-co-trifluoroethylene, Applied Surface Science, 384 (2016) 497-504. https://doi.org/10.1016/j.apsusc.2016.05.073

[75] J. Zhao, Y. Yang, C. Li, L.-a. Hou, Fabrication of GO modified PVDF membrane for dissolved organic matter removal: Removal mechanism and antifouling property, Separation and Purification Technology, 209 (2019) 482-490. https://doi.org/10.1016/j.seppur.2018.07.050

[76] G. Abdi, M. Ashokkumar, A. Alizadeh, Ultrasound-assisted oxidative-adsorptive desulfurization using highly acidic graphene oxide as a catalyst-adsorbent, Fuel, 210 (2017) 639-645.

[77] G. Abdi, A. Alizadeh, J. Amirian, S. Rezaei, G. Sharma, Polyamine-modified magnetic graphene oxide surface: Feasible adsorbent for removal of dyes, Journal of Molecular Liquids, 289 (2019) 111118.

[78] S. Muchtar, M.Y. Wahab, L.-F. Fang, S. Jeon, S. Rajabzadeh, R. Takagi, S. Mulyati, N. Arahman, M. Riza, H. Matsuyama, Polydopamine-coated poly(vinylidene fluoride)

membranes with high ultraviolet resistance and antifouling properties for a photocatalytic membrane reactor, Journal of Applied Polymer Science, 136 (2019) 47312. https://doi.org/10.1002/app.47312

[79] Z. Yi, J. Ye, N. Kikugawa, T. Kako, S. Ouyang, H. Stuart-Williams, H. Yang, J. Cao, W. Luo, Z. Li, Y. Liu, R.L. Withers, An orthophosphate semiconductor with photooxidation properties under visible-light irradiation, Nature Materials, 9 (2010) 559. https://doi.org/10.1038/nmat2780

[80] F. Wang, W. Li, S. Gu, H. Li, X. Wu, X. Liu, Samarium and Nitrogen Co-Doped Bi2WO6 Photocatalysts: Synergistic Effect of Sm3+/Sm2+ Redox Centers and N-Doped Level for Enhancing Visible-Light Photocatalytic Activity, Chemistry – A European Journal, 22 (2016) 12859-12867. https://doi.org/10.1002/chem.201602168

[81] M.-A. Lavergne, C. Chanéac, D. Portehault, S. Cassaignon, O. Durupthy, Optimized Design of Pt-Doped Bi2WO6 Nanoparticle Synthesis for Enhanced Photocatalytic Properties, European Journal of Inorganic Chemistry, 2016 (2016) 2159-2165. https://doi.org/10.1002/ejic.201501208

[82] X. Ding, K. Zhao, L. Zhang, Enhanced Photocatalytic Removal of Sodium Pentachlorophenate with Self-Doped Bi2WO6 under Visible Light by Generating More Superoxide Ions, Environmental Science & Technology, 48 (2014) 5823-5831. https://doi.org/10.1021/es405714q

[83] R. Tang, H. Su, Y. Sun, X. Zhang, L. Li, C. Liu, S. Zeng, D. Sun, Enhanced photocatalytic performance in Bi2WO6/SnS heterostructures: Facile synthesis, influencing factors and mechanism of the photocatalytic process, Journal of colloid and interface science, 466 (2016) 388-399. https://doi.org/10.1016/j.jcis.2015.12.054

[84] F. Chen, D. Li, B. Luo, M. Chen, W. Shi, Two-dimensional heterojunction photocatalysts constructed by graphite-like C3N4 and Bi2WO6 nanosheets: Enhanced photocatalytic activities for water purification, Journal of Alloys and Compounds, 694 (2017) 193-200. https://doi.org/10.1016/j.jallcom.2016.09.326

[85] H. Huang, K. Liu, K. Chen, Y. Zhang, Y. Zhang, S. Wang, Ce and F comodification on the crystal structure and enhanced photocatalytic activity of Bi2WO6 photocatalyst under visible light irradiation, The Journal of Physical Chemistry C, 118 (2014) 14379-14387. https://doi.org/10.1021/jp503025b

[86] Q.-S. Wu, Y. Cui, L.-M. Yang, G.-Y. Zhang, D.-Z. Gao, Facile in-situ photocatalysis of Ag/Bi2WO6 heterostructure with obviously enhanced performance, Separation and Purification Technology, 142 (2015) 168-175. https://doi.org/10.1016/j.seppur.2014.12.039

[87] S.M. López, M. Hidalgo, J. Navío, G. Colón, Novel Bi2WO6–TiO2 heterostructures for Rhodamine B degradation under sunlike irradiation, Journal of hazardous materials, 185 (2011) 1425-1434. https://doi.org/10.1016/j.jhazmat.2010.10.065

[88] M.-S. Gui, W.-D. Zhang, Y.-Q. Chang, Y.-X. Yu, One-step hydrothermal preparation strategy for nanostructured WO3/Bi2WO6 heterojunction with high visible light photocatalytic activity, Chemical Engineering Journal, 197 (2012) 283-288. https://doi.org/10.1016/j.cej.2012.05.032

[89] H. Li, Y. Cui, W. Hong, High photocatalytic performance of BiOI/Bi2WO6 toward toluene and Reactive Brilliant Red, Applied Surface Science, 264 (2013) 581-588. https://doi.org/10.1016/j.apsusc.2012.10.068

[90] X. Huang, H. Chen, One-pot hydrothermal synthesis of Bi2O2CO3/Bi2WO6 visible light photocatalyst with enhanced photocatalytic activity, Applied Surface Science, 284 (2013) 843-848. https://doi.org/10.1016/j.apsusc.2013.08.019

[91] Q.-S. Wu, Y. Feng, G.-Y. Zhang, Y.-Q. Sun, Y.-Y. Xu, D.-Z. Gao, α-Fe2O3 modified Bi2WO6 flower-like mesostructures with enhanced photocatalytic performance, Materials Research Bulletin, 49 (2014) 440-447. https://doi.org/10.1016/j.materresbull.2013.09.031

[92] J. Zhai, H. Yu, H. Li, L. Sun, K. Zhang, H. Yang, Visible-light photocatalytic activity of graphene oxide-wrapped Bi2WO6 hierarchical microspheres, Applied Surface Science, 344 (2015) 101-106. https://doi.org/10.1016/j.apsusc.2015.03.100

[93] X. Meng, Z. Zhang, Synthesis and characterization of plasmonic and magnetically separable Ag/AgCl-Bi2WO6@ Fe3O4@ SiO2 core-shell composites for visible light-induced water detoxification, Journal of colloid and interface science, 485 (2017) 296-307. https://doi.org/10.1016/j.jcis.2016.09.045

[94] T. Wang, S. Zhong, S. Zou, F. Jiang, L. Feng, X. Su, Novel Bi2WO6-coupled Fe3O4 Magnetic Photocatalysts: Preparation, Characterization and Photodegradation of Tetracycline Hydrochloride, Photochemistry and photobiology, 93 (2017) 1034-1042. https://doi.org/10.1111/php.12739

[95] Y. Zhu, Y. Wang, Q. Ling, Y. Zhu, Enhancement of full-spectrum photocatalytic activity over BiPO4/Bi2WO6 composites, Applied Catalysis B: Environmental, 200 (2017) 222-229. https://doi.org/10.1016/j.apcatb.2016.07.002

[96] Y. Li, L. Zhu, Evaluation of the antifouling and photocatalytic properties of novel poly(vinylidene fluoride) membranes with a reduced graphene oxide–Bi2WO6 active layer, Journal of Applied Polymer Science, 134 (2017) 45426. https://doi.org/10.1002/app.45426

Materials Research Forum LLC

https://doi.org/10.21741/9781644901359-1

[97] X. Deng, Z. Li, H. García, Visible Light Induced Organic Transformations Using Metal-Organic-Frameworks (MOFs), Chemistry – A European Journal, 23 (2017) 11189-11209. https://doi.org/10.1002/chem.201701460

[98] F. Li, D. Wang, Q.-J. Xing, G. Zhou, S.-S. Liu, Y. Li, L.-L. Zheng, P. Ye, J.-P. Zou, Design and syntheses of MOF/COF hybrid materials via postsynthetic covalent modification: An efficient strategy to boost the visible-light-driven photocatalytic performance, Applied Catalysis B: Environmental, 243 (2019) 621-628. https://doi.org/10.1016/j.apcatb.2018.10.043

[99] J. Qiu, L. Yang, M. Li, J. Yao, Metal nanoparticles decorated MIL-125-NH2 and MIL-125 for efficient photocatalysis, Materials Research Bulletin, 112 (2019) 297-306. https://doi.org/10.1016/j.materresbull.2018.12.038

[100] X. Li, Y. Pi, L. Wu, Q. Xia, J. Wu, Z. Li, J. Xiao, Facilitation of the visible light-induced Fenton-like excitation of H2O2 via heterojunction of g-C3N4/NH2-Iron terephthalate metal-organic framework for MB degradation, Applied Catalysis B: Environmental, 202 (2017) 653-663. https://doi.org/10.1016/j.apcatb.2016.09.073

[101] L. Hu, Y. Zhang, W. Lu, Y. Lu, H. Hu, Easily recyclable photocatalyst Bi2WO6/MOF/PVDF composite film for efficient degradation of aqueous refractory organic pollutants under visible-light irradiation, Journal of Materials Science, 54 (2019) 6238-6257. https://doi.org/10.1007/s10853-018-03302-w

[102] N. Li, Y. Tian, J. Zhang, Z. Sun, J. Zhao, J. Zhang, W. Zuo, Precisely-controlled modification of PVDF membranes with 3D TiO2/ZnO nanolayer: enhanced anti-fouling performance by changing hydrophilicity and photocatalysis under visible light irradiation, Journal of Membrane Science, 528 (2017) 359-368. https://doi.org/10.1016/j.memsci.2017.01.048

[103] M.I. Shekh, K.P. Patel, R.M. Patel, Electrospun ZnO Nanoparticles Doped Core–Sheath Nanofibers: Characterization and Antimicrobial Properties, Journal of Polymers and the Environment, 26 (2018) 4376-4387. https://doi.org/10.1007/s10924-018-1310-8

[104] D. Zhang, F. Dai, P. Zhang, Z. An, Y. Zhao, L. Chen, The photodegradation of methylene blue in water with PVDF/GO/ZnO composite membrane, Materials Science and Engineering: C, 96 (2019) 684-692. https://doi.org/10.1016/j.msec.2018.11.049

[105] C. Yang, J. Wang, L. Mei, X. Wang, Enhanced photocatalytic degradation of rhodamine B by Cu2O coated silicon nanowire arrays in presence of H2O2, Journal of Materials Science & Technology, 30 (2014) 1124-1129. https://doi.org/10.1016/j.jmst.2014.03.023

[106] J. Du, Y. Tian, N. Li, J. Zhang, W. Zuo, Enhanced antifouling performance of ZnS/GO/PVDF hybrid membrane by improving hydrophilicity and photocatalysis, Polymers for Advanced Technologies, 30 (2019) 351-359. https://doi.org/10.1002/pat.4472

[107] M.R.U.D. Biswas, W.-C. Oh, Synthesis of BiVO4-GO-PVDF nanocomposite: An excellent, newly designed material for high photocatalytic activity towards organic dye degradation by tuning band gap energies, Solid State Sciences, 80 (2018) 22-30. https://doi.org/10.1016/j.solidstatesciences.2018.03.021

[108] W.A. Yee, A.C. Nguyen, P.S. Lee, M. Kotaki, Y. Liu, B.T. Tan, S. Mhaisalkar, X. Lu, Stress-induced structural changes in electrospun polyvinylidene difluoride nanofibers collected using a modified rotating disk, Polymer, 49 (2008) 4196-4203. https://doi.org/10.1016/j.polymer.2008.07.032

[109] J.Z. Tan, N.M. Nursam, F. Xia, Y.B. Truong, I.L. Kyratzis, X. Wang, R.A. Caruso, Electrospun PVDF–TiO 2 with tuneable TiO 2 crystal phases: synthesis and application in photocatalytic redox reactions, Journal of Materials Chemistry A, 5 (2017) 641-648. https://doi.org/10.1039/C6TA08266A

[110] J.A. Lee, K.C. Krogman, M. Ma, R.M. Hill, P.T. Hammond, G.C. Rutledge, Highly reactive multilayer-assembled TiO2 coating on electrospun polymer nanofibers, Advanced Materials, 21 (2009) 1252-1256. https://doi.org/10.1002/adma.200802458

[111] P. Dong, Z. Huang, X. Nie, X. Cheng, Z. Jin, X. Zhang, Plasma enhanced decoration of nc-TiO2 on electrospun PVDF fibers for photocatalytic application, Materials Research Bulletin, 111 (2019) 102-112. https://doi.org/10.1016/j.materresbull.2018.11.007

[112] E.-J. Lee, A.K. An, T. He, Y.C. Woo, H.K. Shon, Electrospun nanofiber membranes incorporating fluorosilane-coated TiO2 nanocomposite for direct contact membrane distillation, Journal of Membrane Science, 520 (2016) 145-154. https://doi.org/10.1016/j.memsci.2016.07.019

[113] S. Ramasundaram, A. Son, M.G. Seid, S. Shim, S.H. Lee, Y.C. Chung, C. Lee, J. Lee, S.W. Hong, Photocatalytic applications of paper-like poly(vinylidene fluoride)–titanium dioxide hybrids fabricated using a combination of electrospinning and electrospraying, Journal of Hazardous Materials, 285 (2015) 267-276. https://doi.org/10.1016/j.jhazmat.2014.12.004

[114] T. He, H. Ma, Z. Zhou, W. Xu, F. Ren, Z. Shi, J. Wang, Preparation of ZnS–Fluoropolymer nanocomposites and its photocatalytic degradation of methylene blue,

Materials Research Forum LLC
https://doi.org/10.21741/9781644901359-1

Polymer Degradation and Stability, 94 (2009) 2251-2256.
https://doi.org/10.1016/j.polymdegradstab.2009.08.012

[115] S. Ramasundaram, M.G. Seid, J.W. Choe, E.-J. Kim, Y.C. Chung, K. Cho, C. Lee, S.W. Hong, Highly reusable TiO2 nanoparticle photocatalyst by direct immobilization on steel mesh via PVDF coating, electrospraying, and thermal fixation, Chemical Engineering Journal, 306 (2016) 344-351. https://doi.org/10.1016/j.cej.2016.07.077

[116] T. He, A. Bahi, W. Zhou, F. Ko, Electrospun Nanofibrous Ag–TiO2/Poly (vinylidene fluoride)(PVDF) Membranes with Enhanced Photocatalytic Activity, Journal of Nanoscience and Nanotechnology, 16 (2016) 7388-7394.
https://doi.org/10.1016/j.polymdegradstab.2009.08.012

[117] C. Wang, Y. Wu, J. Lu, J. Zhao, J. Cui, X. Wu, Y. Yan, P. Huo, Bioinspired synthesis of photocatalytic nanocomposite membranes based on synergy of Au-TiO2 and polydopamine for degradation of tetracycline under visible light, ACS applied materials & interfaces, 9 (2017) 23687-23697. https://doi.org/10.1021/acsami.7b04902

[118] L. Aoudjit, P.M. Martins, F. Madjene, D.Y. Petrovykh, S. Lanceros-Mendez, Photocatalytic reusable membranes for the effective degradation of tartrazine with a solar photoreactor, Journal of Hazardous Materials, 344 (2018) 408-416.
https://doi.org/10.1016/j.jhazmat.2017.10.053

[119] X. Meng, P. Yao, Y. Xu, H. Meng, X. Zhang, Fabrication of organic–inorganic hybrid membranes composed of poly (vinylidene fluoride) and silver cyanamide and their high photocatalytic activity under visible light irradiation, RSC Advances, 6 (2016) 61920-61926. https://doi.org/10.1039/C6RA10434G

[120] J. Zhao, C. Liao, J. Liu, X. Shen, H. Tong, Development of mesoporous titanium dioxide hybrid poly (vinylidene fluoride) ultrafiltration membranes with photocatalytic properties, Journal of Applied Polymer Science, 133 (2016).
https://doi.org/10.1002/app.43427

[121] N.A. Almeida, P.M. Martins, S. Teixeira, J.A.L. da Silva, V. Sencadas, K. Kühn, G. Cuniberti, S. Lanceros-Mendez, P.A. Marques, TiO 2/graphene oxide immobilized in P (VDF-TrFE) electrospun membranes with enhanced visible-light-induced photocatalytic performance, Journal of materials science, 51 (2016) 6974-6986.
https://doi.org/10.1007/s10853-016-9986-4

[122] M. Wang, G. Yang, P. Jin, H. Tang, H. Wang, Y. Chen, Highly hydrophilic poly (vinylidene fluoride)/meso-titania hybrid mesoporous membrane for photocatalytic membrane reactor in water, Scientific reports, 6 (2016) 19148.
https://doi.org/10.1038/srep19148

[123] N.A.M. Nor, J. Jaafar, A.F. Ismail, M.A. Mohamed, M.A. Rahman, M.H.D. Othman, W.J. Lau, N. Yusof, Preparation and performance of PVDF-based nanocomposite membrane consisting of TiO2 nanofibers for organic pollutant decomposition in wastewater under UV irradiation, Desalination, 391 (2016) 89-97. https://doi.org/10.1016/j.desal.2016.01.015

[124] P. Martins, V. Gomez, A. Lopes, C. Tavares, G. Botelho, S. Irusta, S. Lanceros-Mendez, Improving photocatalytic performance and recyclability by development of Er-doped and Er/Pr-codoped TiO2/poly (vinylidene difluoride)–trifluoroethylene composite membranes, The Journal of Physical Chemistry C, 118 (2014) 27944-27953. https://doi.org/10.1021/jp509294v

[125] S. Guo, H. Yoshioka, Y. Kato, H. Kakehi, M. Miura, N. Isu, A. Manseri, H. Sawada, B. Ameduri, Photocatalytic activity of vinylidene fluoride-containing copolymers/anatase titanium oxide/silica nanocomposites, European Polymer Journal, 58 (2014) 79-89. https://doi.org/10.1016/j.eurpolymj.2014.04.022

[126] Y. Zhang, G. Zhang, S. Liu, C. Zhang, X. Xu, "Naked" TiO2 capsulated in nanovoid microcapsule of poly(vinylidene fluoride) supporter with enhanced photocatalytic activity, Chemical Engineering Journal, 204-206 (2012) 217-224. https://doi.org/10.1016/j.cej.2012.07.124

[127] P. Raja, M. Bensimon, U. Klehm, P. Albers, D. Laub, L. Kiwi-Minsker, A. Renken, J. Kiwi, Highly dispersed PTFE/Co3O4 flexible films as photocatalyst showing fast kinetic performance for the discoloration of azo-dyes under solar irradiation, Journal of Photochemistry and Photobiology A: Chemistry, 187 (2007) 332-338. https://doi.org/10.1016/j.jphotochem.2006.10.033

[128] W. Kang, J. Ju, H. He, F. Li, L. Tao, Y. Dong, B. Cheng, Photocatalytic Degradation Performance of TiO2/PTFE Membrane Catalyst to Methylene Blue, Chemistry Letters, 45 (2016) 1440-1443. https://doi.org/10.1246/cl.160732

[129] D.E. Tsydenov, A.V. Vorontsov, Influence of Nafion loading on hydrogen production in a membrane photocatalytic system, Journal of Photochemistry and Photobiology A: Chemistry, 297 (2015) 8-13. https://doi.org/10.1016/j.jphotochem.2014.09.014

[130] P. Raja, M. Bensimon, U. Klehm, P. Albers, D. Laub, L. Kiwi-Minsker, A. Renken, J. Kiwi, Highly dispersed PTFE/Co3O4 flexible films as photocatalyst showing fast kinetic performance for the discoloration of azo-dyes under solar irradiation, Journal of Photochemistry and Photobiology A: Chemistry, 187 (2007) 332-338. https://doi.org/10.1016/j.jphotochem.2006.10.033

[131] Z. Ding, Y. Dong, B. Li, Preparation of a modified PTFE fibrous photo-Fenton catalyst and its optimization towards the degradation of organic dye, International Journal of Photoenergy, 2012 (2012). https://doi.org/10.1155/2012/121239

[132] M. Lombardi, P. Palmero, M. Sangermano, A. Varesano, Electrospun polyamide-6 membranes containing titanium dioxide as photocatalyst, Polymer International, 60 (2011) 234-239. https://doi.org/10.1002/pi.2932

[133] Z. Liu, Y.-E. Miao, M. Liu, Q. Ding, W.W. Tjiu, X. Cui, T. Liu, Flexible polyaniline-coated TiO2/SiO2 nanofiber membranes with enhanced visible-light photocatalytic degradation performance, Journal of Colloid and Interface Science, 424 (2014) 49-55. https://doi.org/10.1016/j.jcis.2014.03.009

[134] S.B. Teli, S. Molina, A. Sotto, E.G.a. Calvo, J.d. Abajob, Fouling resistant polysulfone–PANI/TiO2 ultrafiltration nanocomposite membranes, Industrial & Engineering Chemistry Research, 52 (2013) 9470-9479. https://doi.org/10.1021/ie401037n

[135] D. Pathania, G. Sharma, A. Kumar, N.C. Kothiyal, Fabrication of nanocomposite polyaniline zirconium(IV) silicophosphate for photocatalytic and antimicrobial activity, Journal of Alloys and Compounds, 588 (2014) 668-675. https://doi.org/10.1016/j.jallcom.2013.11.133

[136] H. Zhang, R. Zong, J. Zhao, Y. Zhu, Dramatic visible photocatalytic degradation performances due to synergetic effect of TiO2 with PANI, Environmental Science and Technology, 42 (2008) 3803-3807. https://doi.org/10.1021/es703037x

[137] C. Leng, J. Wei, Z. Liu, R. Xiong, C. Pan, J. Shi, Facile synthesis of PANI-modified CoFe 2 O 4–TiO 2 hierarchical flower-like nanoarchitectures with high photocatalytic activity, Journal of nanoparticle research, 15 (2013) 1643. https://doi.org/10.1007/s11051-013-1643-0

[138] H. Ahmadizadegan, S. Esmaielzadeh, Investigating the effect of ultrasonic irradiation on preparation and properties of conductive nanocomposites, Solid State Sciences, 85 (2018) 9-20. https://doi.org/10.1016/j.solidstatesciences.2018.08.012

[139] M.R. Karim, H.W. Lee, I.W. Cheong, S.M. Park, W. Oh, J.H. Yeum, Conducting polyaniline-titanium dioxide nanocomposites prepared by inverted emulsion polymerization, Polymer Composites, 31 (2010) 83-88. https://doi.org/10.1002/pc.20769

[140] M. Shang, W. Wang, S. Sun, J. Ren, L. Zhou, L. Zhang, Efficient visible light-induced photocatalytic degradation of contaminant by spindle-like PANI/BiVO4, The Journal of Physical Chemistry C, 113 (2009) 20228-20233. https://doi.org/10.1021/jp9067729

[141] L. Geng, Y. Zhao, X. Huang, S. Wang, S. Zhang, S. Wu, Characterization and gas sensitivity study of polyaniline/SnO2 hybrid material prepared by hydrothermal route, Sensors and Actuators B: Chemical, 120 (2007) 568-572. https://doi.org/10.1016/j.snb.2006.03.009

[142] M. Rabia, H. Mohamed, M. Shaban, S. Taha, Preparation of polyaniline/PbS core-shell nano/microcomposite and its application for photocatalytic H 2 electrogeneration from H 2 O, Scientific reports, 8 (2018) 1107. https://doi.org/10.1038/s41598-018-19326-w

[143] Y. Lin, D. Li, J. Hu, G. Xiao, J. Wang, W. Li, X. Fu, Highly efficient photocatalytic degradation of organic pollutants by PANI-modified TiO 2 composite, Journal of Physical Chemistry C, 116 (2012) 5764-5772. https://doi.org/10.1021/jp211222w

[144] X. Li, C. Shi, J. Wang, J. Wang, M. Li, H. Qiu, H. Sun, K. Ogino, Polyaniline-doped TiO 2/PLLA fibers with enhanced visible-light photocatalytic degradation performance, Fibers and Polymers, 18 (2017) 50-56. https://doi.org/10.1007/s12221-017-6895-3

[145] J. Li, Q. Ma, X. Dong, D. Li, X. Xi, W. Yu, J. Wang, G. Liu, Novel electrospun bilayered composite fibrous membrane endowed with tunable and simultaneous quadrifunctionality of electricity–magnetism at one layer and upconversion luminescence–photocatalysis at the other layer, RSC Advances, 6 (2016) 96084-96092. https://doi.org/10.1039/C6RA20591G

[146] S. Neubert, D. Pliszka, V. Thavasi, E. Wintermantel, S. Ramakrishna, Conductive electrospun PANi-PEO/TiO2 fibrous membrane for photo catalysis, Materials Science and Engineering: B, 176 (2011) 640-646. https://doi.org/10.1016/j.mseb.2011.02.007

[147] S.H. Kim, S.-Y. Kwak, B.-H. Sohn, T.H. Park, Design of TiO2 nanoparticle self-assembled aromatic polyamide thin-film-composite (TFC) membrane as an approach to solve biofouling problem, Journal of Membrane Science, 211 (2003) 157-165. https://doi.org/10.1016/S0376-7388(02)00418-0

[148] H. Zhang, L. Yang, Immobilization of nanoparticle titanium dioxide membrane on polyamide fabric by low temperature hydrothermal method, Thin Solid Films, 520 (2012) 5922-5927. https://doi.org/10.1016/j.tsf.2012.04.057

[149] N. Daels, M. Radoicic, M. Radetic, S.W.H. Van Hulle, K. De Clerck, Functionalisation of electrospun polymer nanofibre membranes with TiO2 nanoparticles in view of dissolved organic matter photodegradation, Separation and Purification Technology, 133 (2014) 282-290. https://doi.org/10.1016/j.seppur.2014.06.040

[150] J. Geltmeyer, H. Teixido, M. Meire, T. Van Acker, K. Deventer, F. Vanhaecke, S. Van Hulle, K. De Buysser, K. De Clerck, TiO2 functionalized nanofibrous membranes for removal of organic (micro)pollutants from water, Separation and Purification Technology, 179 (2017) 533-541. https://doi.org/10.1016/j.seppur.2017.02.037

[151] E. Cossich, R. Bergamasco, M.T. Pessoa de Amorim, P.M. Martins, J. Marques, C.J. Tavares, S. Lanceros-Méndez, V. Sencadas, Development of electrospun photocatalytic TiO2-polyamide-12 nanocomposites, Materials Chemistry and Physics, 164 (2015) 91-97. https://doi.org/10.1016/j.matchemphys.2015.08.029

[152] E.A. de Campos, S.D. de Campos, A.A. Roos, B.V. de Souza, J.M. Schneider, M.B. Uliana, R.C. de Oliveira, Titanium Dioxide Dispersed on Cellulose Acetate and its Application in Methylene Blue Photodegradation, Polymers and Polymer Composites, 21 (2013) 423-430. https://doi.org/10.1177/096739111302100703

[153] X. Yang, J. Ma, J. Ling, N. Li, D. Wang, F. Yue, S. Xu, Cellulose acetate-based SiO2/TiO2 hybrid microsphere composite aerogel films for water-in-oil emulsion separation, Applied Surface Science, 435 (2018) 609-616. https://doi.org/10.1016/j.apsusc.2017.11.123

[154] L. Zhang, W. Wang, J. Yang, Z. Chen, W. Zhang, L. Zhou, S. Liu, Sonochemical synthesis of nanocrystallite Bi2O3 as a visible-light-driven photocatalyst, Applied Catalysis A: General, 308 (2006) 105-110. https://doi.org/10.1016/j.apcata.2006.04.016

[155] G. Liu, S. Li, Y. Lu, J. Zhang, Z. Feng, C. Li, Controllable synthesis of α-Bi2O3 and γ-Bi2O3 with high photocatalytic activity by α-Bi2O3→ γ-Bi2O3→ α-Bi2O3 transformation in a facile precipitation method, Journal of Alloys and Compounds, 689 (2016) 787-799. https://doi.org/10.1016/j.jallcom.2016.08.047

[156] H.-Y. Jiang, K. Cheng, J. Lin, Crystalline metallic Au nanoparticle-loaded α-Bi 2 O 3 microrods for improved photocatalysis, Physical Chemistry Chemical Physics, 14 (2012) 12114-12121. https://doi.org/10.1039/c2cp42165h

[157] S. Jiang, L. Wang, W. Hao, W. Li, H. Xin, W. Wang, T. Wang, Visible-light photocatalytic activity of S-doped α-Bi2O3, The Journal of Physical Chemistry C, 119 (2015) 14094-14101. https://doi.org/10.1021/jp5117036

[158] H.-Y. Jiang, G. Liu, T. Wang, P. Li, J. Lin, J. Ye, In situ construction of α-Bi 2 O 3/gC 3 N 4/β-Bi 2 O 3 composites and their highly efficient photocatalytic performances, RSC Advances, 5 (2015) 92963-92969. https://doi.org/10.1039/C5RA18420G

[159] S. Juntrapirom, D. Tantraviwat, S. Suntalelat, O. Thongsook, S. Phanichphant, B. Inceesungvorn, Visible light photocatalytic performance and mechanism of highly

efficient SnS/BiOI heterojunction, Journal of colloid and interface science, 504 (2017) 711-720. https://doi.org/10.1016/j.jcis.2017.06.019

[160] C. Liu, Y. Yang, J. Li, S. Chen, W. Li, X. Tang, An in situ transformation approach for fabrication of BiVO4/WO3 heterojunction photoanode with high photoelectrochemical activity, Chemical Engineering Journal, 326 (2017) 603-611. https://doi.org/10.1016/j.cej.2017.05.179

[161] L.M. Pastrana-Martínez, S. Morales-Torres, J.L. Figueiredo, J.L. Faria, A.M. Silva, Graphene oxide based ultrafiltration membranes for photocatalytic degradation of organic pollutants in salty water, Water research, 77 (2015) 179-190. https://doi.org/10.1016/j.watres.2015.03.014

[162] G. He, C. Xing, X. Xiao, R. Hu, X. Zuo, J. Nan, Facile synthesis of flower-like Bi12O17Cl2/β-Bi2O3 composites with enhanced visible light photocatalytic performance for the degradation of 4-tert-butylphenol, Applied Catalysis B: Environmental, 170 (2015) 1-9. https://doi.org/10.1016/j.apcatb.2015.01.015

[163] M. Zhao, L. Dong, Q. Zhang, H. Dong, C. Li, H. Tang, Novel plate-stratiform nanostructured Bi 12 O 17 Cl 2 with visible-light photocatalytic performance, Powder Diffraction, 31 (2016) 2-7. https://doi.org/10.1017/S0885715615000901

[164] L.-C. Tien, Y.-L. Lin, S.-Y. Chen, Synthesis and characterization of Bi12O17Cl2 nanowires obtained by chlorination of α-Bi2O3 nanowires, Materials Letters, 113 (2013) 30-33. https://doi.org/10.1016/j.matlet.2013.09.064

[165] Z. Yu, X. Min, F. Li, D. Yin, Y. Peng, G. Zeng, A mussel-inspired method to fabricate a novel reduced graphene oxide/Bi12O17Cl2 composites membrane for catalytic degradation and oil/water separation, Polymers for Advanced Technologies, 30 (2019) 101-109. https://doi.org/10.1002/pat.4448

[166] H. Tang, C.M. Hessel, J. Wang, N. Yang, R. Yu, H. Zhao, D. Wang, Two-dimensional carbon leading to new photoconversion processes, Chemical Society Reviews, 43 (2014) 4281-4299. https://doi.org/10.1039/C3CS60437C

[167] A. Alizadeh, G. Abdi, M.M. Khodaei, M. Ashokkumar, J. Amirian, Graphene oxide/Fe 3 O 4/SO 3 H nanohybrid: a new adsorbent for adsorption and reduction of Cr (vi) from aqueous solutions, RSC Advances, 7 (2017) 14876-14887. https://doi.org/10.1039/C7RA01536D

[168] Z. Zhou, X. Peng, L. Zhong, L. Wu, X. Cao, R.C. Sun, Electrospun cellulose acetate supported Ag@AgCl composites with facet-dependent photocatalytic properties on degradation of organic dyes under visible-light irradiation, Carbohydrate Polymers, 136 (2016) 322-328. https://doi.org/10.1016/j.carbpol.2015.09.009

[169] R. Konwarh, N. Karak, M. Misra, Electrospun cellulose acetate nanofibers: The present status and gamut of biotechnological applications, Biotechnology Advances, 31 (2013) 421-437. https://doi.org/10.1016/j.biotechadv.2013.01.002

[170] Y. Li, J. Tian, C. Yang, B. Hsiao, Nanocomposite Film Containing Fibrous Cellulose Scaffold and Ag/TiO2 Nanoparticles and Its Antibacterial Activity, Polymers, 10 (2018) 1052. https://doi.org/10.3390/polym10101052

[171] H.S. Barud, R.M.N. Assunção, M.A.U. Martines, J. Dexpert-Ghys, R.F.C. Marques, Y. Messaddeq, S.J.L. Ribeiro, Bacterial cellulose–silica organic–inorganic hybrids, Journal of Sol-Gel Science and Technology, 46 (2008) 363-367. https://doi.org/10.1007/s10971-007-1669-9

[172] H.S. Barud, C. Barrios, T. Regiani, R.F.C. Marques, M. Verelst, J. Dexpert-Ghys, Y. Messaddeq, S.J.L. Ribeiro, Self-supported silver nanoparticles containing bacterial cellulose membranes, Materials Science and Engineering: C, 28 (2008) 515-518. https://doi.org/10.1016/j.msec.2007.05.001

[173] D. Zhang, L. Qi, Synthesis of mesoporous titania networks consisting of anatase nanowires by templating of bacterial cellulose membranes, Chemical Communications, (2005) 2735-2737. https://doi.org/10.1039/b501933h

[174] N.M. Bedford, A.J. Steckl, Photocatalytic Self Cleaning Textile Fibers by Coaxial Electrospinning, ACS Applied Materials & Interfaces, 2 (2010) 2448-2455. https://doi.org/10.1021/am1005089

[175] X. Zhang, W. Chen, Z. Lin, J. Shen, Photocatalytic degradation of a methyl orange wastewater solution using titanium dioxide loaded on bacterial cellulose, Synthesis and Reactivity in Inorganic, Metal-Organic, and Nano-Metal Chemistry, 41 (2011) 1141-1147. https://doi.org/10.1080/15533174.2011.591359

[176] E.A. de Campos, S.D. de Campos, A.A. Roos, B.V.C. de Souza, J.M. Schneider, M.B. Uliana, R.C. de Oliveira, Titanium Dioxide Dispersed on Cellulose Acetate and its Application in Methylene Blue Photodegradation, Polymers and Polymer Composites, 21 (2013) 423-430. https://doi.org/10.1177/096739111302100703

[177] W. Zheng, S. Chen, S. Zhao, Y. Zheng, H. Wang, Zinc sulfide nanoparticles template by bacterial cellulose and their optical properties, Journal of Applied Polymer Science, 131 (2014). https://doi.org/10.1002/app.40874

[178] A. Wittmar, H. Thierfeld, S. Köcher, M. Ulbricht, Routes towards catalytically active TiO 2 doped porous cellulose, RSC Advances, 5 (2015) 35866-35873. https://doi.org/10.1039/C5RA03707G

[179] S.-D. Wang, Q. Ma, H. Liu, K. Wang, L.-Z. Ling, K.-Q. Zhang, Robust electrospinning cellulose acetate@ TiO 2 ultrafine fibers for dyeing water treatment by photocatalytic reactions, RSC Advances, 5 (2015) 40521-40530. https://doi.org/10.1039/C5RA03797B

[180] H. Bai, X. Zan, J. Juay, D.D. Sun, Hierarchical heteroarchitectures functionalized membrane for high efficient water purification, Journal of Membrane Science, 475 (2015) 245-251. https://doi.org/10.1016/j.memsci.2014.10.036

[181] I. Chauhan, P. Mohanty, In situ decoration of TiO2 nanoparticles on the surface of cellulose fibers and study of their photocatalytic and antibacterial activities, Cellulose, 22 (2015) 507-519. https://doi.org/10.1007/s10570-014-0480-3

[182] M.A. Mohamed, W.N.W. Salleh, J. Jaafar, A.F. Ismail, M. Abd Mutalib, S.M. Jamil, Incorporation of N-doped TiO2 nanorods in regenerated cellulose thin films fabricated from recycled newspaper as a green portable photocatalyst, Carbohydrate Polymers, 133 (2015) 429-437. https://doi.org/10.1016/j.carbpol.2015.07.057

[183] A.S. Monteiro, R.R. Domeneguetti, M. Wong Chi Man, H.S. Barud, E. Teixeira-Neto, S.J.L. Ribeiro, Bacterial cellulose–SiO2@TiO2 organic–inorganic hybrid membranes with self-cleaning properties, Journal of Sol-Gel Science and Technology, 89 (2019) 2-11. https://doi.org/10.1007/s10971-018-4744-5

[184] A. Rajeswari, E. Jackcina Stobel Christy, A. Pius, New insight of hybrid membrane to degrade Congo red and Reactive yellow under sunlight, Journal of Photochemistry and Photobiology B: Biology, 179 (2018) 7-17. https://doi.org/10.1016/j.jphotobiol.2017.12.024

[185] B. Li, J. Chu, Y. Li, M. Meng, Y. Cui, Q. Li, Y. Feng, Preparation and Performance of Visible-Light-Driven Bi2O3/ZnS Heterojunction Functionalized Porous CA Membranes for Effective Degradation of Rhodamine B, physica status solidi (a), 215 (2018) 1701061. https://doi.org/10.1002/pssa.201701061

[186] Q. Chen, Z. Yu, F. Li, Y. Yang, Y. Pan, Y. Peng, X. Yang, G. Zeng, A novel photocatalytic membrane decorated with RGO-Ag-TiO2 for dye degradation and oil–water emulsion separation, Journal of Chemical Technology & Biotechnology, 93 (2018) 761-775. https://doi.org/10.1002/jctb.5426

[187] A. Rajeswari, A. Pius, Preparation, Characterization and Application of Nano ZnO - Blended Polymeric Membrane, Materials Today: Proceedings, 5 (2018) 16814-16820. https://doi.org/10.1016/j.matpr.2018.05.149

[188] W. Li, T. Li, G. Li, L. An, F. Li, Z. Zhang, Electrospun H4SiW12O40/cellulose acetate composite nanofibrous membrane for photocatalytic degradation of tetracycline

and methyl orange with different mechanism, Carbohydrate Polymers, 168 (2017) 153-162. https://doi.org/10.1016/j.carbpol.2017.03.079

[189] F. Li, Z. Yu, H. Shi, Q. Yang, Q. Chen, Y. Pan, G. Zeng, L. Yan, A Mussel-inspired method to fabricate reduced graphene oxide/g-C3N4 composites membranes for catalytic decomposition and oil-in-water emulsion separation, Chemical Engineering Journal, 322 (2017) 33-45. https://doi.org/10.1016/j.cej.2017.03.145

[190] A. Rajeswari, S. Vismaiya, A. Pius, Preparation, characterization of nano ZnO-blended cellulose acetate-polyurethane membrane for photocatalytic degradation of dyes from water, Chemical Engineering Journal, 313 (2017) 928-937. https://doi.org/10.1016/j.cej.2016.10.124

[191] L. Song, B. Zhu, V. Jegatheesan, S.R. Gray, M.C. Duke, S. Muthukumaran, Effect of Hybrid Photocatalysis and Ceramic Membrane Filtration Process for Humic Acid Degradation, in: M. Pannirselvam, L. Shu, G. Griffin, L. Philip, A. Natarajan, S. Hussain (Eds.) Water Scarcity and Ways to Reduce the Impact: Management Strategies and Technologies for Zero Liquid Discharge and Future Smart Cities, Springer International Publishing, Cham, 2019, pp. 95-113. https://doi.org/10.1007/978-3-319-75199-3_6

[192] M.A. Anderson, M.J. Gieselmann, Q. Xu, Titania and alumina ceramic membranes, Journal of Membrane Science, 39 (1988) 243-258. https://doi.org/10.1016/S0376-7388(00)80932-1

[193] R. Goei, T.-T. Lim, Asymmetric TiO2 hybrid photocatalytic ceramic membrane with porosity gradient: Effect of structure directing agent on the resulting membranes architecture and performances, Ceramics International, 40 (2014) 6747-6757. https://doi.org/10.1016/j.ceramint.2013.11.137

[194] T. Yang, H. Xiong, F. Liu, Q. Yang, B. Xu, C. Zhan, Effect of UV/TiO2 pretreatment on fouling alleviation and mechanisms of fouling development in a cross-flow filtration process using a ceramic UF membrane, Chemical Engineering Journal, 358 (2019) 1583-1593. https://doi.org/10.1016/j.cej.2018.10.149

[195] Q. Li, R. Jia, J. Shao, Y. He, Photocatalytic degradation of amoxicillin via TiO2 nanoparticle coupling with a novel submerged porous ceramic membrane reactor, Journal of Cleaner Production, 209 (2019) 755-761. https://doi.org/10.1016/j.jclepro.2018.10.183

[196] B. Van der Bruggen, C. Vandecasteele, T. Van Gestel, W. Doyen, R. Leysen, A review of pressure-driven membrane processes in wastewater treatment and drinking water production, Environmental progress, 22 (2003) 46-56. https://doi.org/10.1002/ep.670220116

[197] H. Imai, Y. Takei, K. Shimizu, M. Matsuda, H. Hirashima, Direct preparation of anatase TiO2 nanotubes in porous alumina membranes, Journal of Materials Chemistry, 9 (1999) 2971-2972. https://doi.org/10.1039/a906005g

[198] S. Liu, K. Li, Preparation TiO2/Al2O3 composite hollow fibre membranes, Journal of Membrane Science, 218 (2003) 269-277. https://doi.org/10.1016/S0376-7388(03)00184-4

[199] N. Ma, Y. Zhang, X. Quan, X. Fan, H. Zhao, Performing a microfiltration integrated with photocatalysis using an Ag-TiO2/HAP/Al2O3 composite membrane for water treatment: Evaluating effectiveness for humic acid removal and anti-fouling properties, Water research, 44 (2010) 6104-6114. https://doi.org/10.1016/j.watres.2010.06.068

[200] X.P. Cao, D. Li, W.H. Jing, W.H. Xing, Y.Q. Fan, Synthesis of visible-light responsive C, N and Ce co-doped TiO 2 mesoporous membranes via weak alkaline sol–gel process, Journal of Materials Chemistry, 22 (2012) 15309-15315. https://doi.org/10.1039/c2jm31576a

[201] L. Liu, Z. Liu, H. Bai, D.D. Sun, Concurrent filtration and solar photocatalytic disinfection/degradation using high-performance Ag/TiO2 nanofiber membrane, Water research, 46 (2012) 1101-1112. https://doi.org/10.1016/j.watres.2011.12.009

[202] X. Zhang, A.J. Du, P. Lee, D.D. Sun, J.O. Leckie, TiO2 nanowire membrane for concurrent filtration and photocatalytic oxidation of humic acid in water, Journal of Membrane Science, 313 (2008) 44-51. https://doi.org/10.1016/j.memsci.2007.12.045

[203] X. Zhang, D.K. Wang, D.R.S. Lopez, J.C. Diniz da Costa, Fabrication of nanostructured TiO2 hollow fiber photocatalytic membrane and application for wastewater treatment, Chemical Engineering Journal, 236 (2014) 314-322. https://doi.org/10.1016/j.cej.2013.09.059

[204] S. Liu, K. Li, Preparation TiO2/Al2O3 composite hollow fibre membranes, Journal of Membrane Science, 218 (2003) 269-277. https://doi.org/10.1016/S0376-7388(03)00184-4

[205] J.E. Koresh, A. Soffer, Mechanism of permeation through molecular-sieve carbon membrane. Part 1.—The effect of adsorption and the dependence on pressure, Journal of the Chemical Society, Faraday Transactions 1: Physical Chemistry in Condensed Phases, 82 (1986) 2057-2063. https://doi.org/10.1039/f19868202057

[206] S. Liu, K. Li, R. Hughes, Preparation of porous aluminium oxide (Al2O3) hollow fibre membranes by a combined phase-inversion and sintering method, Ceramics International, 29 (2003) 875-881. https://doi.org/10.1016/S0272-8842(03)00030-0

[207] X. Tan, S. Liu, K. Li, Preparation and characterization of inorganic hollow fiber membranes, Journal of Membrane Science, 188 (2001) 87-95. https://doi.org/10.1016/S0376-7388(01)00369-6

[208] X. Chen, S.S. Mao, Titanium dioxide nanomaterials: synthesis, properties, modifications, and applications, Chemical reviews, 107 (2007) 2891-2959. https://doi.org/10.1021/cr0500535

[209] S.P. Albu, A. Ghicov, J.M. Macak, R. Hahn, P. Schmuki, Self-Organized, Free-Standing TiO2 Nanotube Membrane for Flow-through Photocatalytic Applications, Nano Letters, 7 (2007) 1286-1289. https://doi.org/10.1021/nl070264k

[210] M. Guo, B. Ding, X. Li, X. Wang, J. Yu, M. Wang, Amphiphobic nanofibrous silica mats with flexible and high-heat-resistant properties, The Journal of Physical Chemistry C, 114 (2009) 916-921. https://doi.org/10.1021/jp909672r

[211] D. Vu, Z. Li, H. Zhang, W. Wang, Z. Wang, X. Xu, B. Dong, C. Wang, Adsorption of Cu (II) from aqueous solution by anatase mesoporous TiO2 nanofibers prepared via electrospinning, Journal of colloid and interface science, 367 (2012) 429-435. https://doi.org/10.1016/j.jcis.2011.09.088

[212] H. Choi, E. Stathatos, D.D. Dionysiou, Sol–gel preparation of mesoporous photocatalytic TiO2 films and TiO2/Al2O3 composite membranes for environmental applications, Applied Catalysis B: Environmental, 63 (2006) 60-67. https://doi.org/10.1016/j.apcatb.2005.09.012

[213] X. Zhang, C. Shao, Z. Zhang, J. Li, P. Zhang, M. Zhang, J. Mu, Z. Guo, P. Liang, Y. Liu, In situ generation of well-dispersed ZnO quantum dots on electrospun silica nanotubes with high photocatalytic activity, ACS applied materials & interfaces, 4 (2012) 785-790. https://doi.org/10.1021/am201420b

[214] Z. Hosseini, N. Taghavinia, N. Sharifi, M. Chavoshi, M. Rahman, Fabrication of high conductivity TiO2/Ag fibrous electrode by the electrophoretic deposition method, The Journal of Physical Chemistry C, 112 (2008) 18686-18689. https://doi.org/10.1021/jp8046054

[215] P. Roy, S. Berger, P. Schmuki, TiO2 nanotubes: synthesis and applications, Angewandte Chemie International Edition, 50 (2011) 2904-2939.

[216] K. Shankar, J.I. Basham, N.K. Allam, O.K. Varghese, G.K. Mor, X. Feng, M. Paulose, J.A. Seabold, K.-S. Choi, C.A. Grimes, Recent advances in the use of TiO2 nanotube and nanowire arrays for oxidative photoelectrochemistry, The Journal of Physical Chemistry C, 113 (2009) 6327-6359.

[217] S. Hoang, S. Guo, N.T. Hahn, A.J. Bard, C.B. Mullins, Visible light driven photoelectrochemical water oxidation on nitrogen-modified TiO2 nanowires, Nano letters, 12 (2011) 26-32. https://doi.org/10.1021/nl2028188

[218] H. Tada, M. Fujishima, H. Kobayashi, Photodeposition of metal sulfide quantum dots on titanium (IV) dioxide and the applications to solar energy conversion, Chemical Society Reviews, 40 (2011) 4232-4243. https://doi.org/10.1039/c0cs00211a

[219] Y.J. Hwang, C. Hahn, B. Liu, P. Yang, Photoelectrochemical properties of TiO2 nanowire arrays: a study of the dependence on length and atomic layer deposition coating, Acs Nano, 6 (2012) 5060-5069. https://doi.org/10.1021/nn300679d

[220] I.S. Cho, Z. Chen, A.J. Forman, D.R. Kim, P.M. Rao, T.F. Jaramillo, X. Zheng, Branched TiO2 nanorods for photoelectrochemical hydrogen production, Nano Letters, 11 (2011) 4978-4984. https://doi.org/10.1021/nl2029392

[221] G. Wang, H. Wang, Y. Ling, Y. Tang, X. Yang, R.C. Fitzmorris, C. Wang, J.Z. Zhang, Y. Li, Hydrogen-treated TiO2 nanowire arrays for photoelectrochemical water splitting, Nano letters, 11 (2011) 3026-3033. https://doi.org/10.1021/nl201766h

[222] Y. Ling, G. Wang, D.A. Wheeler, J.Z. Zhang, Y. Li, Sn-doped hematite nanostructures for photoelectrochemical water splitting, Nano letters, 11 (2011) 2119-2125. https://doi.org/10.1021/nl200708y

[223] M. Xu, P. Da, H. Wu, D. Zhao, G. Zheng, Controlled Sn-doping in TiO2 nanowire photoanodes with enhanced photoelectrochemical conversion, Nano letters, 12 (2012) 1503-1508. https://doi.org/10.1021/nl2042968

Photocatalysis
Materials Research Foundations **100** (2021) 57-76

Materials Research Forum LLC
https://doi.org/10.21741/9781644901359-2

Chapter 2

Photocatalytic Heavy Metal Detoxification from Water Systems

P. Senthil Kumar[*] and G. Janet Joshiba

Department of Chemical Engineering, Sri Sivasubramaniya Nadar College of Engineering, Chennai 603110, India

* senthilchem8582@gmail.com; senthilkumarp@ssn.edu.in

Abstract

Heavy metals are one of the greatest elevating threat to mankind and other living organisms and it is released into the environment due to increasing dumpsites, transports, and industrial sectors. The industrial wastewater containing heavy metal ions easily enters into the food chain through the air, water, and soil; it results in bioaccumulation and biomagnifications of metal ions in human beings. It causes severe chronic health disorders affecting the nervous system, circulatory system, digestive system and other sensitive organs of the human body. Many conventional techniques such as adsorption, coagulation, flocculation, electrochemical treatment, and biological treatment are used for the reduction of heavy metal ions in the aqueous system. The photocatalysis method is one of the emerging effective ways for eliminating the toxic metal ions from the aqueous solution. This chapter elaborates the principle, mechanism and various methods utilized in the photocatalytic reduction of heavy metal ions from the wastewater.

Keywords

Photocatalytic, Heavy Metal, Environment, Mechanism, Wastewater

Contents

Materials Research Forum LLC
https://doi.org/10.21741/9781644901359-2

1. Introduction

Our universe is basically composed of three fundamental factors such as air, water, and soil in which the entire ecosystem depend upon for survival and growth. The term environment is characterized as the collection of various substantial conditions encompassing the biotic and abiotic factors present in the ecosystem [1]. Water is one of the fundamental requirement for the growth and development of all living beings. Life in the earth cannot be imagined without the presence of water and it always plays a crucial role in the sustenance of the living organisms. The ingress of various hazardous contaminants from the environment into the water bodies damages the quality of the drinking water by making it unfit for drinking and causes water deficiency. The water related problems have become a major concern in many developing countries and it causes various irreversible negative impacts on the environment. Some of the major contaminants released from the industries which affect the wellness of the environment are heavy metals, dyes, pesticides, oils and hydrocarbons, and these contaminants as a great threat to aquatic and terrestrial ecosystem [2]. Heavy metals are an indigenous component of the surface of the earth; furthermore, it is also released during the natural activities such as volcanic eruption and disintegration of rocks. They are a group of metals and metalloids that possess higher solidity multiple times than that of water [3]. The accelerated stride in the

Photocatalysis   Materials Research Forum LLC
Materials Research Foundations **100** (2021) 57-76   https://doi.org/10.21741/9781644901359-2

proliferation of industrialization, urbanization, and land utilization sectors are the major reason contributing to the evolution of toxic metal pollution. The heavy metal pollution has become one of the greatest threats globally because of their hazardous consequences on the living beings and ecosystem [4]. Heavy metals generally refer to a group of metals and metalloids which possess atomic weight higher than 4 g/cm^3 which is probably five times or denser than the water. The heavy metals are lethal metallic components of higher atomic weight and density; it is also highly pernicious even in trace concentrations [1]. The heavy metals possess specific gravity more than 5.0 $g/cm^3$and their atomic weight lies between the ranges of 63.5 – 200.6. The heavy metals are branched into three major types such as essential, valuable and pernicious metal ions. The noxious and pernicious metal ions are highly capable of causing serious damages to the wellness of the aquatic and terrestrial living organisms. Due to the increased utilization of heavy metal in various industrial and automobile sectors, their discharge into the environment got elevated. In addition, the complex non-degradable nature of the heavy metal retains it in the ecosystem leading to bioaccumulation of toxic metals in the food items [5]. Various anthropogenic activities such as industrial, domestic and commercial sources emit an enormous amount of toxic metal ions into the environment, furthermore, these contaminants endure strongly in the surrounding air, water and soil gradually invades the human body through the food cycle and gets bio-magnified. Some of the heavy metals which are predominantly used in various industrial applications are Copper, Arsenic, Silver, Mercury, Lead, Iron, Cadmium, Platinum, Zinc, Boron, Chromium, Antimony, Selenium and Cobalt [1,6]. In light of their prompt danger to human wellbeing and nature, WHO has declared four toxic metals such as Lead, Arsenic, Cadmium, and mercury in their list of top 10 hazardous chemical substances which are the greatest threat to mankind [3].The heavy metals can be segregated into two types such as essential and toxic heavy metals. Some of the trace elements such as Zn, Fe, and Cu are required for the enhanced metabolism and growth of the human body; in addition, these essential heavy metals play a vital role in the enzymatic activities taking part in the human body. Certain other heavy metals such Arsenic, Lead, Cadmium, etc., fall under the category of hazardous metal ions which are not required for the function of the human body, in addition, these toxic metals are extremely destructive posing serious threats to the ecosystem [4]. The density and toxicity of a compound are directly proportional to each other, the compounds with higher density possess acute toxicity even at trace concentrations. In the current era, the health complications caused because of these heavy metals are elevated, on the other hand, the utilization of the heavy metals have also been increased in various applications such as agriculture, industries, municipal and household sectors [7]. In order to reduce the catastrophic aftereffect of the heavy metals, Government has enunciated several rules and regulations for the industries and domestic sectors for the proper disposal of heavy metals. Government authorities have made stringent laws for the treatment of industrial effluent before discharging it into the environment, furthermore has also derived strict permissible limits for the heavy metals to be present in drinking water, industrial effluents and all other drainage systems [2]. The application of phosphate fertilizers in agricultural land emits an enormous concentration of cadmium heavy metal which is one of the most hazardous compounds in damaging the lives [6]. The conventional treatment techniques such as Adsorption, Coagulation, Flocculation, Chemical precipitation, Biological treatment, Ion exchange, Flotation, Advanced oxidation process and Photo catalysis have been used for the treatment of heavy metal [2, 8]. Among, all the techniques the Photocatalysis is seemed to be an efficient technique in the elimination of heavy metal ions, in addition, this technique seemed to possess the capability to overcome all the limitations occurring in conventional methods such as higher cost, lesser performance and secondary waste production. The Photo catalysis technique basically utilizes the solar energy as a main source and it is used in various appliances such as dye degradation, production of hydrogen, reduction of carbon dioxide and generation of heavy metal ions of high valence [9]. This chapter explains the impact of heavy metal ions on the environment and it also demonstrates the effective detoxification of heavy metals present in the aqueous system using the photocatalysis technique.

1.1   Sources of heavy metal pollution

Heavy metal is an integral component of earth's crust and it is naturally present in rocks, sediments, water, soil, microorganism, etc. The heavy metal ores remain in the sediments and rocks in various chemical forms such as sulfides, oxides, or both sulfide and oxide ores [1]. According to the reports stated by UNEP 2004, the persistent nature of the heavy metals leads to the bioaccumulation and biomagnifications of toxic metal ions inside the human body. The natural and anthropogenic sources are the two main reasons contributing to the discharge of toxic metals into the environment and the various sources of heavy metal pollution is depicted in the fig 1. More commonly, heavy metals are present in natural sources such as rocks, soils, and sediments. Furthermore, the human activities such as mining, smelting, and weathering release an enormous amount of toxic heavy metal ion into the atmosphere and the drain released from these processes consist of toxic chemicals, heavy metals and acids [6].

The effluent released from some industries such as textile, leather, electroplating, distilleries, and chemical is composed of elevated levels of toxic metals like Mn, Cu, Pb, Fe, Cr and Ni which can cause dreadful health disorders to mankind. The sources of heavy metals are classified into two types such as point sources and non-point sources. The combustion, agricultural dissipation, soil abrasion, fertilizers usage, and merchandise

Materials Research Forum LLC

https://doi.org/10.21741/9781644901359-2

repository are some of the non-point sources of the heavy metals. On the other hand, the industries indulging in transport, leather, mining, textiles, chemicals, electroplating, thermal power, smelting, and e-waste are some of the major sources of a point source of heavy metal pollution. The heavy metal particles gushed from its sources gets compiled in the soil surfaces, further, the deposited heavy metals get acquired in the root through soil and consumption of this heavy metal accumulated plant by plants and humans result in bioaccumulation and biomagnifications [4]. The feasibility of the heavy metal is based on the certain physical, chemical and biological parameters such as adsorption, detachment, temperature, solubility, interaction and type of phase in which it is clubbed to [7].

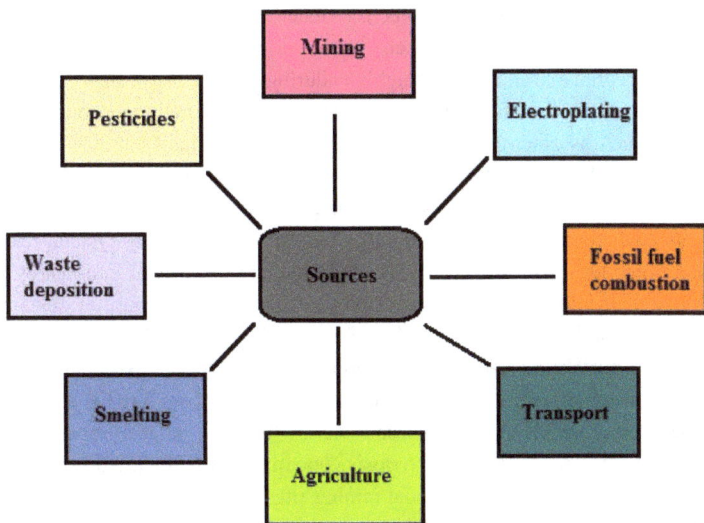

Figure 1: Various sources of heavy metal pollution.

1.2 Bio-importance of heavy metals

Heavy metals occupy an important position in industries for the production of various goods used by humans in their everyday life [1]. The heavy metals act as micronutrients and it also plays a vital role in stabilizing the redox process, regulating the osmotic pressure and enhancement of enzymatic reactions [6]. They occupy an important place in enhancing various physiological and biochemical reactions [7]. The heavy metals such as Lead,

Photocatalysis   Materials Research Forum LLC
Materials Research Foundations **100** (2021) 57-76   https://doi.org/10.21741/9781644901359-2

Mercury, and Aluminium are used in the manufacturing process of products such as lamps, utensils and medical equipment [1]. Arsenic at a limited concentration serves as one of the good trace elements with good nutrition. Magnesium plays an important role in the circulatory system as an electrolytic substance and it is highly required by pregnant ladies and feeding mothers. Zinc supports in maintaining the copper concentration in the human body, in addition, it is highly required in increasing the reproductive activity in the males and enhancing the enzymatic activity [1].

1.3   Deleterious Environmental and health effects

European Commission has declared that the chemical substances which are highly noxious, obdurate and acquired in the living cells. Generally, several organic contaminants breakdown into the two major end products such as water and Carbon dioxide but the heavy metals are non-degradable and possess highly enduring properties [6]. Basically, the heavy metals are highly toxic and life-damaging components present in the environment [1]. The heavy metals such as zinc, copper, and iron remain as an indispensable ingredient in enhancing the metabolic activity of the human body [4]. The pernicious nature of heavy metals affects the living beings vigorously when utilized above the required levels, further; it causes dreadful diseases which can affect the intestines, circulatory system, respiratory system and other sensitive organs [1]. The emission of heavy metal into nearby aquatic sources through the industrial effluent results in damaging the maritime creatures by inducing oxygen deficiency and eutrophication in the water resources. In addition, the dissipated heavy metals are reformed into hydrated ions which are exceedingly deadly to the organisms surviving in the water [2]. The uptake of metal ions into the cells spoils the DNA by denaturing the DNA and nuclear proteins, further it induces structural changes and collapse the DNA leading to apoptosis, uncontrolled cell growth and cell cycle modification [7]. The region near the coal mining sites is the predominant sites which are highly affected by the heavy metals. The heavy metals emitted from the mines seeps into the underground and reach the water table resulting in the water pollution [1]. The ionic mode of toxic metals such as Silver, Cadmium, Arsenic and Lead easily interacts with the human body easily and also it is very difficult to separate it from the biomolecules. Further, this heavy metals enter the food chain of a living organism and gets bio magnified [2]. The consumption of heavy metal accumulated green leafy vegetables results in health disorders such as mental retardation, stomach cancer, nutritional deficiency, immunological disorder and depletion of bones [4]. The consumption of lethal heavy metals through food, air and water damages the cell development by production of free radicals which induce oxidative stress to the cells [2]. The heavy metals on adsorption into the cell structure of the living beings affect the internal components such as mitochondria, endoplasmic reticulum, nucleus, cell membrane, etc. It also damages the enzymatic activity which is indulged in dealing with cell repair, metabolism, and detox activity [7].

2.   Various heavy metal removal technologies

In order to eliminate these toxic metals from our ecosystem, several treatment methodologies have been adopted by industries and domestic sectors. Basically, the effluent treatment process is divided into three categories such as physical, chemical and biological treatment. Several conventional technologies such as Adsorption, Precipitation, Electrodialysis, Filtration, Extraction, Ion exchange, Membrane reactors, Electrowinning, Floatation and Advanced oxidation process have been followed in treating the heavy metal contaminated industrial effluents. Several legislations and laws have been enunciated to restrict the emission of heavy metals into the society, but the effluent treatment is not effectively implemented in several industries due to some limitations such as higher utilization of water, huge energy expenditure and persistence of the pollutants [2, 4]. In recent times, the photocatalysis is known to be a developing technique for the sequestration of heavy metal ions. It utilizes a very abundant renewable energy source (i.e.) solar energy and it is used to activate the catalyst for the production of electron-hole pairs which helps in the sequestration of toxic metal ions [10]. The secondary waste emission is a major drawback found in many conventional technologies used in the wastewater treatment sector. Further, to combat this drawback and also to provide an environmentally hygienic technology for the elimination of toxic pollutants the photocatalysis technique is highly preferred among the researchers. Highly non-degradable contaminants such as petroleum and chlorinated hydrocarbons are tedious to degrade in conventional treatment technologies, whereas in the photocatalysis it is efficiently degraded into simpler compounds such as CO_2 and H_2O using the TiO_2/UV catalyst.

2.1   Advanced oxidation process

Advanced oxidation process (AOP) is an emerging eco-friendly technique used in various industrial and research sectors widely. This AOP can be effectively used for the treatment of all kinds of contaminants such as pesticides, insecticides, dyes, heavy metals, aromatic hydrocarbons, chlorinated hydrocarbons, air contaminants and other complicated organic compounds [12]. In order to reduce the hazardous effects of toxic metals, the Advanced Oxidation Process (AOP) was introduced into the effluent treatment to eliminate the heavy metal ions from the environment. In this process, the hydroxyl radicals are originated through which the harmful pollutants are sequestered or reduce into lower reactive compounds [10]. The hydroxyl ions produced formed from the excitation of the electron is the most essential compound required the photocatalysis operation. The hydroxyl ions

(OH^-) possess a shortened life span, further; they actively participate in effective degradation of organic and chemical compounds into simpler compounds such as CO_2 and H_2O. The photocatalysis is one such advanced oxidation process used widely by majorly utilizing sunlight as the main source for pollutant degradation [12].

2.2   Photocatalysis

The term photocatalysis is derived from the amalgamation of two words such as "Photo" and "catalysis" where photo means light and catalysis meaning chemical reaction. Catalysis is the phenomenon where a compound helps in speeding up the rate of a chemical reaction without getting modified or expended in the reaction. Photocatalysis is a process which utilizes light to actuate a substance which modifies the reaction rate of a chemical reaction without being included itself. Also, the photocatalyst is the substance which can adjust the rate of synthetic response utilizing light illumination. In general, the Chlorophyll pigment present in plants is an ordinary photocatalyst. The distinction between chlorophyll photocatalyst to man-made nano TiO_2photocatalyst, typically chlorophyll catches daylight to transform water and carbon dioxide into oxygen and glucose, on the other hand, photocatalyst utilizes strong oxidizing chemicals and electron-hole pairs to breakdown the organic compounds into carbon dioxide and water within the sight of photocatalyst, light, and water [11]. Photocatalysis is one of the outstanding technology which works by utilizing solar energy. It is seemed to be an assuring technique in the reduction of heavy metal ions, dye degradation, CO_2 debasement, and energy exchange operations [9].

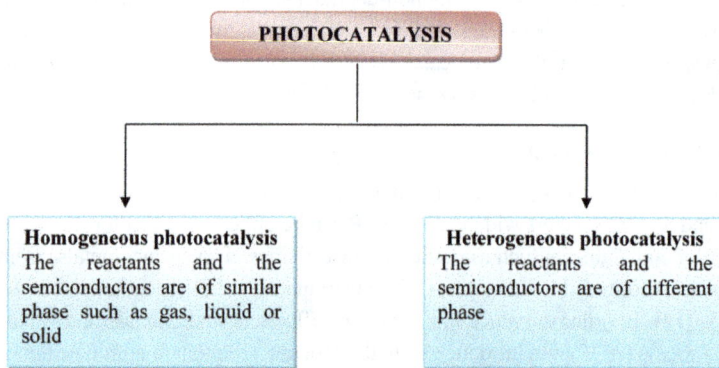

Figure 2: Types of photocatalysis.

Materials Research Forum LLC
https://doi.org/10.21741/9781644901359-2

Several semiconductor compounds such as carbon nitride, metal sulfides, metal oxides are used conventionally as active photocatalytic materials [9]. There are various photocatalysts such as GaP, ZnO, ZnS, TiO_2, WO_3, CdS, etc. which are used as photodegraders and reduction of harmful pollutants in the wastewater treatment system. [10]. Over a period of time, many new catalysts such as ligands, organic metal, and inorganic metal frameworks have been formulated for the efficient photocatalysis operation under the UV light radiation [9]. Generally, the photocatalyst is semiconducting materials which generate the electron-hole pairs for breaking down the complicated organic compound into CO_2 and H_2O in the presence of UV or Visible irradiation. Based on the physical state of the compounds, the photocatalysis process is majorly divided into two main types as explained in the figure 2.

By virtue of its semiconductor structure, the TiO_2 is preferred as one of the predominantly used photocatalysts in the wastewater treatment sector. In addition, this TiO_2 catalyst comprises an empty conduction band and electron-filled valence band with a band gap energy of about 2-3.5 eV. The hydroxyl radical synthesized during the photocatalytic reaction is a strong oxidizing agent which is effectively utilized in the remediation of harmful environmental pollutants such as phenols, pesticides, heavy metals, dyes, etc., [13]. The photocatalysis is mainly used for purposes such as air purification, industrial effluent treatment, self- purification, energy conservation and antifouling applications [12]. The application of photocatalysis method is explained in the fig 3.

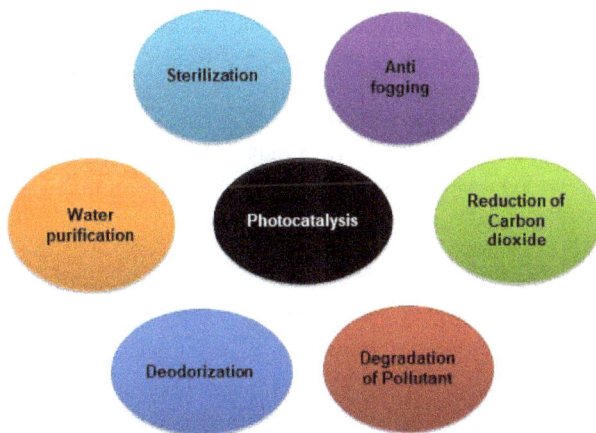

Figure 3: Various applications of photocatalysis.

Photocatalysis   Materials Research Forum LLC
Materials Research Foundations **100** (2021) 57-76   https://doi.org/10.21741/9781644901359-2

2.3   Principle & mechanism behind photocatalysis

The fundamentals of photocatalysis are based on the combination of two major subjects such as photochemistry and catalysis. It requires both the light source and a semiconductor source to originate the chemical operation. Generally, the photons are derived from two light sources such as UV light and visible light sources based upon the type of catalyst utilized in the photocatalysis operation. The semiconductor materials are represented by two bands such as valence band and conduction band. The photon from the light source stimulates the Valence band electron resulting in the transfer of the excited electron from the valence band to the conduction band creating a positive charge in the valence band called hole [10]. The photocatalysis reaction follows a mechanism which involves four steps such as:

- The photons from the light energy are used to provoke the electron-hole pairs
- Segregation of the stimulated charges
- The transition of electrons and holes to the photocatalyst surface to generate heat energy
- The uptake of charges on the photocatalyst surface for redox reactions

Under the influence of the photons emitted from the visible light, the electron present in the valence bans gets stimulated to the conduction band. The stimulation of electron results in the formation of electron-hole pairs which is formed on the surface of the photocatalyst. The formed electron-hole pairs can further work in two capabilities. That is: The formed carrier charges may rejoin and produce energy in the form of heat or they may encounter other electron donor or acceptor and form hydroxyl free radicals for photooxidation reaction [12]. The photocatalysis reaction can end in two ways such as photo-oxidation or photoreduction reaction based on the presence of the electron and hole. The photo oxidation reaction takes place in the presence of electron scavengers, whereas the photo reduction reaction takes place in the presence of the hole scavengers [13].

3.   Photocatalysis for the heavy metal recovery

Heavy metal sequestration using photocatalysis process is generally associated with the reduction of heavy metal into simpler form at a lower oxidation state. The main special case is arsenic, which exists in the anionic structure and needs oxidation to be changed over to a high oxidation state. From the beginning, Several AOPs such as TiO_2/UV, O_3/UV, H_2O_2/UV, $O_3/H_2O_2/UV$ are used in the heavy metal sequestration. In recent times, among the various AOPs used generally the Heterogeneous photocatalytic oxidation and reduction technique is preferred at a higher rate [10]. When the photons from the light source are incident on the semiconductor material present in the photocatalytic reactor, the electron-hole pairs are produced on the conduction band and valence band of the semiconductor. The transfer between the electron-hole pairs from the conduction band to valence band produced oxidizing species which aids in the reduction of heavy metals. Several semiconductors such as ZnO, CdS, ZnS, CeO_2, TiO_2, etc. have been used conventionally for the effective reduction of heavy metals. Out of the various semiconductors used in photocatalyst used in recent times for pollutant removal, the TiO_2 was found to be the most extensively used semiconductor because of some of its properties such as feasibility, durability, non-poisonous nature, and water immiscibility. The photocatalyst is also seemed to be efficient in degrading the endocrine disruptors in the wastewater [14, 15]. The mechanism behind the photocatalysis process is depicted in the figure 4. Some of the characteristics which make the photocatalysis efficient than other techniques are as follows:

- The photocatalysis process does not involve extravagant chemicals for the operation

- This set up works under atmospheric temperature and pressure conditions

- Photocatalysis undergo reduction reaction same as that of the AOPs

- It is an economical method

- It is an enduring, nontoxic and eco-friendly technique

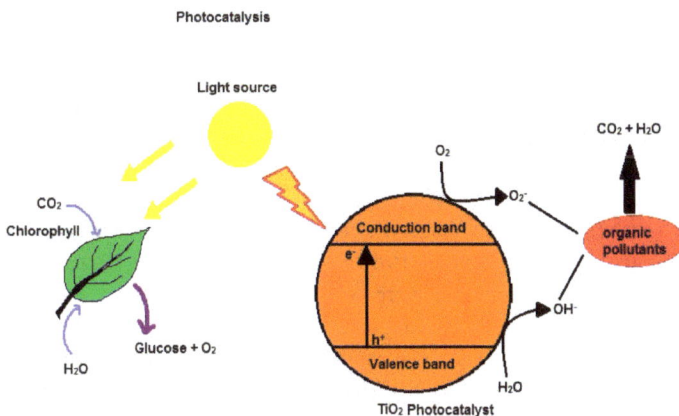

Figure 4: Mechanism behind the photocatalysis process.

Photocatalysis   Materials Research Forum LLC
Materials Research Foundations **100** (2021) 57-76   https://doi.org/10.21741/9781644901359-2

Heterogeneous photocatalysts are claimed to be the easy, efficient and economically feasible method for reduction of chromium [16]. Habila et al. have investigated the degradation studies of heavy metal ions composed organic mixture using the core double shell $Fe_3O_4@SiO_2@TiO_2$ magnetic nanoparticle coated semiconductor photocatalyst. In this work, the TiO_2 catalyst is coated with the magnetic iron oxide nanoparticle which is eco-friendly and easy to synthesize. As the combination of Fe_3O_4 nanoparticles and TiO_2 catalyst lessens the efficacy of the photocatalysis reaction, the SiO_2 is sandwiched between the catalyst and the nanoparticles. The SiO_2 layer enhanced the efficiency of the photocatalyst effectively; also it helped in increasing the mechanical and thermal stability of the semiconductor, thus the core double shell material utilizing photocatalysts degraded the heavy metal ions with high efficiency [17]. Many conventional methods used in the effluent treatment primitively do not completely degrade the harmful contaminants, whereas the photocatalysis effectively removed several organic and inorganic contaminants found in the environment. The ZnO semiconductor material is considered to be other effective photocatalytic material for the elimination of heavy metal ions. In addition, the ZnO is a low cost and higher surface area possessing material for the enhanced degradation of heavy metal ions [18]. The heterogeneous photocatalysis process is applied for various purposes such as dehydrogenation, oxidation, transfer of hydrogen ion, air pollutants elimination, heavy metal deposition, water purification, and sterilization. Furthermore, this method can also be performed for both the liquid phase and gaseous phase [19]. Various heavy metal ions eliminated using the photocatalysis methods are explained below in a detailed manner.

3.1 Chromium (Cr)

Chromium in normally present in various food items such as organic products, vegetables, grains, and meats, Furthermore, It goes into the human body mostly through the nourishment chain. Chromium is one of the vital heavy metal used in various industrial and research sectors as a textile, tannery, electroplating, metallurgy and wood preservatives. Basically, the chromium occurs in two various forms such as Cr(III) and Cr(VI), whereas the Cr (VI) is found to be the more noxious form of chromium damaging the living organisms and ecosystem. In drinking water, the dimension of chromium is commonly low, however, industrial wastewater can contain the unsafe chromium (IV) ions. Due to its solubility, motility, and toxicity, the chromium heavy metal ions were considered to be one of the biggest danger to the ecosystem globally. According to the World Health Organization (WHO), the permissible limit of Cr(VI) was determined to be 0.05 mg L^{-1}. When comparing with the Cr(VI) ions the Cr(III) was found to be the less toxic and less motile form of chromium. Several photocatalyst semiconductor materials such as TiO_2, ZnO, WO_3, ZnS, CdS and $SnIn_4S_8$ are used in the effective reduction of chromium ions

Materials Research Forum LLC
https://doi.org/10.21741/9781644901359-2

present in the industrial effluents [10, 20]. Primitively, the Cr(VI) ions are reduced to Cr(III) ions using some of the chemical compounds such as ferrous sulphate, sodium metabisulfite, sodium thiosulfate, sulphur dioxide [18]. The Neodymium doped TiO_2 semiconductor which is fabricated using the sol-gel method is utilized in the reduction reaction of Cr (VI) ions under the incidence of UV light in the work done by Rengaraj et al. The results concluded that the doping of Neodymium on the TiO_2 surface progressively augment the reduction of Cr(VI) ions into the Cr(III) ions [21]. The bismuth carbonate doped with iron oxide (II) is fabricated and utilized effectively for the reduction of most toxic Cr (VI) metal ions to less toxic Cr(II) ions [22]. Hou et al have investigated the elimination of chromium ions from the aqueous solution using photocatalysis and capacitive deionization. In this study the MIL-53 (Fe) semiconductor under the illumination of visible light is used for the reduction of Cr(VI) to Cr(III) ions. The results concluded that the higher concentration of Cr(VI) ions of about 72.2% is eliminated effectively using the photocatalysis technique and it revealed that it is the best effective method for the treatment of chromium contaminated wastewater [9].

3.2   Arsenic (As)

Arsenic is one of the harmful rare earth elements released from manmade sources such as mining, wood preservatives, and biocides, however, it also chiefly originates from some natural sources such as volcano and disintegration of minerals in land and water sources. Arsenic is one of the heavy metal which is highly abundant in the earth's crust, further; it exists in two forms such as trivalent arsenite and pentavalent arsenate. The consumption of Arsenic elements in human beings leads to severe disorders such as arsenicosis, palmoplantar keratosis, cancer, and pigmentation variation. The WHO has established the permissible limits of Arsenic as 10 µg L^{-1} [b]. The As(III) is highly toxic form because it easily attaches to the proteins causing cell disruptions[10, 20]. The highly toxic Arsenite As(III) is oxidized to less toxic As(V) form of ion utilizing the photocatalytic oxidation method in the work performed by Dutta et al. In this work, the OH radical which is produced by the benzoic acid is used as the main oxidant for the conversion of As(III) and As(V) ions, the photocatalysis of Arsenic using OH radical is compared with and without TiO_2 semiconductor under the UV light. The results concluded that the Arsenic is effectively reduced into less toxic forms using the OH radicals, further, using this method the permissible limits of Arsenic which is formulated by WHO can be maintained effectively in the industries [23]. Majumder and Chaudhari have examined the potential of solar photocatalytic oxidation of Arsenic from ground water using the Tartarate and Citrate photocatalysts. The results revealed that Arsenic oxidation using this catalyst has effectively reduced the Arsenic concentration from 250-260 µg/L to below 50 µg/L and the As removal efficiency of about 80% is observed in this study [24]. In the work carried

Materials Research Forum LLC
https://doi.org/10.21741/9781644901359-2

out by Reyna et al the elimination of Arsenic species using the ZnO catalysts under the UV irradiation is investigated. The ZnO catalyst prepared by the sol-gel method is compared with the conventional TiO_2degussa P25 suspension. The results revealed that higher removal efficiency is determined in the ZnO catalyst when compared with TiO_2 catalysts [25].

3.3   Cadmium (Cd)

Cadmium is one of the remarkable toxic heavy metal ion released from the domestic and industrial sectors and it is also considered as a growing threat to human beings and other living organisms. Around, 0.1 mg/kg of cadmium ions are present in the layer of the earth's crust, and nearly 15 mg of Cd/kg is deposited in the earth's environment as rocks and sediments. These metal ions are mostly emitted into the environment through industrial sectors such as textile industries, battery industries, and metallurgical industry. In the United States, for instance, the day-by-day cadmium admission is about 0.4 mg/kg/day, not exactly 50% of the US EPA's oral reference portion. These lower permissible limits are set to decrease the toxic level of cadmium waste generated from plating industry, all the more as of late, to the presentation of general confinements on cadmium utilization in specific nations [7]. Visa and Duta have investigated about the elimination of heavy metals such as Cd^{2+} and Cu^{2+} using the TiO_2 catalyst which is obtained by the hydrothermal processing of TiO_2 Degussa powder and Fly ash powder. The results concluded that this novel substrate effectively degrades the heavy metal ions present in the solution mixture [26]. Le et al have investigated the elimination of cadmium ions using ZnO particles which was fabricated using the solid precipitation technique. The results revealed that only less than 15% of cadmium ions are removed (19). The TiO_2 catalyst is seemed to be an effective photocatalyst for the elimination of cadmium ions and the removal efficiency was about 99.8 was observed in the study conducted by Fatehizadeh et al. The reactor utilized in this study is made of three parts such as ultraviolet source, reaction cell and mixing unit, Further, the removal efficiency of cadmium is affected by parameters such as pH, dose and dispersion of air [27]. The Activated sewage sludge-derived activated carbon coupled with TiO_2nanomaterial is also used as photocatalytic material for cadmium removal in the research work performed by Rashed et al. The results declared that the cadmium metal ions are removed with an efficiency of about 90% and the activated sludge activated carbon TiO_2 enhanced the cadmium removal in the bi-pollutant solution [28]. Davis and Green have examined the photocatalytic oxidation potential of the Cadmium-EDTA complex using the TiO_2 semiconductor. The results of this study declared that the effective removal of Cadmium using TiO_2 was observed on the optimum concentration of about 2×10^{-5} to 10^{-3} M at a pH range of about 3 to 8, Further, the cadmium ions are released as Cd^{2+} after the mineralization process [29].

3.4   Copper (Cu^{2+})

Copper is obtained from the mineral ores of Chalcopyrite, CuFeS$_2$, Further, it is also obtained as a major contaminant from the mining, hydrometallurgical, pyrometallurgical industry as effluents and fumes. The copper poisoning causes health disorders such as paralysis, nausea, depression, pneumonia, diarrhea, etc. [1]. Copper is utilized as an important cofactor of various enzymes such as peroxidase, catalase, monoamine oxidase, ferroxidases, etc. which deals with the oxidative stress reactions. In addition, it plays a vital role in the metabolism of carbohydrate, haemoglobin production and collagen crosslinking reactions [7]. Chaudary et al. investigated the elimination of copper metal ions using the TiO$_2$ or H$_2$O$_2$ catalyst. The results declared that the copper ions are recovered efficiently at a pH range of about 1.5 – 4.5, in addition, the removal efficiency can be enhanced by coupling the TiO$_2$ nanomaterial with the activated carbon concentrated cell system [30]. The higher removal efficiency of about 85% of copper metal ions is observed in the photocatalytic degradation of copper ions using the ZnO nanoparticles semiconductors [19]. In the work carried out by Habila et al., the copper metal ions are effectively recovered using Fe$_3$O$_4$@SiO$_2$@ TiO$_2$ with a higher adsorption capacity of about 125 mg g^{-1}. Further, the higher removal efficiency was observed at pH 5 [17]. In the research work carried out by Kanakaraju et al., the copper metal ion is recovered using the aggregation of the TiO$_2$/ZnO semiconductor with the calcium alginate beads. The results showed that the copper metal recovery with 50:50 ratio of calcium alginate beads and TiO$_2$/ZnO photocatalysis showed higher removal efficiency even in three consecutive cycles and it also served as an environmentally friendly method for metal recovery [31]. The chelated copper ion complexes such as copper EDTA and copper cyanide complexes are reduced using the photo anodes which are made up of TiO$_2$ nanotubes [32].

3.5   Lead (Pb^{2+})

Lead is a bluish grey metal which is found naturally in the layer of the earth's crust. The industrial activities and anthropogenic sources are the main sources of lead pollution. Some of the anthropogenic activities such as combustion of fossil fuels, agriculture, and domestic sources contribute to the emission of lead into the environment. The mining and transportation play a vital role in increasing the level of the lead metal ion. Some of the industries such as battery industry, metal plating industry, paint industry, textile industries, glass industries, chemical industries, and instrument manufacturing industries utilize lead as an important raw material [7]. Sreekanthan et al. investigated the photocatalysis reduction of lead ions using Cu-TiO$_2$ nanotubes. In this work, the TiO$_2$ nanotubes are coated with the Cu^{2+} ions through the wet impregnation method. The results concluded that the combination of Cu^{2+}ions with the TiO$_2$ catalyst effectively increased the photocatalytic reduction of Pb(II) ions, the higher removal efficiency was observed in an optimum ion concentration of about 20 and 60 ppm at alkaline condition under the incidence of UV light [33]. The Titania based binary oxide materials are utilized in the photocatalytic reduction of the lead ions in the research work performed by Mishra et al. The Titania sample is combined with the zirconia and silica at equal Ration to enhance the mechanical strength and removal efficiency of the heavy metal ions effectively. Sodium formate is used as the hole scavenger in this method, furthermore, it resulted in heavy metal removal of about 89% in a time period of 60 min [34]. The photocatalytic reduction of Pb(II) using TiO_2 and Pt-TiO_2 powders is investigated by Marunni et al. The investigation results inferred that the metal removal efficacy was observed higher using Pt-TiO_2 when compared using the TiO_2 material [35].

3.6   Mercury

Mercury is one of the notable toxic elements which is rated as the top third element in the ASTRD's Priority list of Hazardous Substances. It causes serious health disorders to humans and living beings. The mercury heavy metal ions is a substantial metal which comes under the transition element series of the periodic table and it occurs in three different forms such as organic, inorganic and elemental form. The toxicity of the mercury ions depends on the chemical form in which it exists. In the optimum room temperature, the natural mercury exists as a fluid which has a high vapor pressure. In addition, Mercury exists as a cation with oxidation conditions of +1 (mercurous) or +2 (mercuric). Mercury is a far-reaching ecological toxicant and contamination which instigates extreme adjustments in the body tissues and causes a wide scope of antagonistic wellbeing impacts. Various chemical forms of mercury which is vigorously affecting the humans and other living beings are mercury vapor (Hg^0), inorganic mercurous (Hg^+), mercuric (Hg^{2+}), and the natural mercury mixes. The industrial activities such as hardware, battery manufacturing, dentistry, beverage industry, nuclear reactors, and the pharmaceutical industry contribute greatly to the mercury contamination [7].

Conclusion & Future perspectives

The reduction of heavy metals present in wastewater using the photocatalysis technique is broadly explained in this chapter. This technique seems to be outstanding method in reducing the negative impacts exerted by the toxic metal ions on the environment. The photocatalysis method has overcome several limitation and drawbacks that are found in the advanced oxidation process of heavy metal reduction. The AOPs are capable only of converting the toxic heavy metal ions into oxidative species, whereas the photocatalysis can perform oxidation and also reduction reaction to convert the toxic metal ions. In the photocatalysis process the complexed heavy metal ions into less toxic heavy metal species which do not affect the wellness of the environment. Some of the factors such as pH, initial metal ion concentration, type of the semiconductor, the intensity of light, the mass of photocatalyst have a strong impact on the photocatalysis reduction of toxic metal ions. Photocatalyst advancement would assist us in utilizing the bottomless sun based radiation. Due to its effectiveness in wastewater treatment, much advancement has been made in the photocatalysis such as submersion type and distribution type for heavy metal treatment in industries. This chapter declares that the photocatalysis process is an effective method for treating heavy metal ions in the aqueous solution.

References

[1] J. O. Duruibe, M. O. C. Ogwuegbu, J. N. Egwurugwu, Heavy metal pollution and human biotoxic effects, International Journal of Physical Sciences, 2 (2007) 112-118.

[2] C. F. Carolin, P.S. Kuma, A. Saravanan, G. J. Joshiba, Mu. Naushad, Efficient Techniques for the Removal of Toxic Heavy Metals from Aquatic Environment: A Review, Journal of Environmental Chemical Engineering 5 (2017) 2782-2799. https://doi.org/10.1016/j.jece.2017.05.029

[3] A. Roychowdary, R. Dutta, D. Sarkar, Heavy Metal Pollution and Remediation, Green chemistry, in: B. Torok, T. Dransfield, Green chemistry, Elsevier, USA, 2018, pp 359-373. https://doi.org/10.1016/B978-0-12-809270-5.00015-7

[4] P. K. Rai, S. S. Lee, M. Zhang, Y. F. Tsangd, K. H. Kim, Heavy metals in food crops: Health risks, fate, mechanisms, and management, Environment International, 125 (2019) 365- 385. https://doi.org/10.1016/j.envint.2019.01.067

[5] S. Rangabhashiyam, P. Balasubramanian, Characteristics, performances, equilibrium and kinetic modeling aspects of heavy metal removal using algae, Bioresource technology reports, 5 (2019) 261-279. https://doi.org/10.1016/j.biteb.2018.07.009

[6] A. S. Mohammed, A. Kapri, R. Goel, Heavy Metal Pollution: Source, Impact, and Remedies, In: M. Khan, A. Zaidi, R. Goel, J. Musarrat, Biomanagement of Metal-Contaminated Soils, Environmental pollution, Springer, Dordrecht, 2011, pp 1-28. https://doi.org/10.1007/978-94-007-1914-9_1

[7] P. B. Tchounwou, C. G. Yedjou, A. K. Patlolla, D. J. Sutton, Heavy Metals Toxicity and the Environment, HHS Author Manuscripts, 101 (2012)133–164. https://doi.org/10.1007/978-3-7643-8340-4_6

[8] L. Fang,L. Li,Z. Qu, H. Xu,J. Xu,N. Yan, A novel method for the sequential removal and separation of multiple heavy metals from wastewater, Journal of Hazardous Materials, 342 (2018) 617–624. https://doi.org/10.1016/j.jhazmat.2017.08.072

[9] S. Hou, X. Xu, M. Wang, T. Lu, C. Q. Sun, L. Pan, Synergistic conversion and removal of total Cr from aqueous solution by photocatalysis and capacitive deionization, Chemical engineering journal, 337 (2018) 398- 404. https://doi.org/10.1016/j.cej.2017.12.120

[10] P. Chowdhury, A. Elkamel, A. K. Ray, Photocatalytic Processes for the Removal of ToxicMetal Ions, in: S. Sharma, Heavy Metals In Water : Presence, Removal and Safety, Royal Society of Chemistry, 2014, pp 25-43. https://doi.org/10.1039/9781782620174-00025

[11] R. Ameta, M. S. Solanki, S. Benjamin, S. C. Ameta, Photocatalysis, Advanced Oxidation Processes for Waste Water Treatment Emerging Green Chemical Technology, 2018, pp 135-175. https://doi.org/10.1016/B978-0-12-810499-6.00006-1

[12] http://www.greenearthnanoscience.com/what-is-photocatalyst.php

[13] E. T. Wahyuni, N. H. Aprilita, H. Hatimah, A. M. Wulandari, M. Mudasir, Removal of Toxic Metal Ions in Water by Photocatalytic Method, American Chemical Science Journal 5 (2015) 194-201.

[14] M.A. Barakat, New trends in removing heavy metals from industrial wastewater, Arabian Journal of Chemistry, 4 (2011) 361–377. https://doi.org/10.1016/j.arabjc.2010.07.019

[15] V. Belgiorn, L. Rizzo, D. Fatta, C. D. Rocca, G. Lofrano, A. Nikolaou, V. Naddeo, Review on endocrine disrupting-emerging compounds in urban wastewater: occurrence and removal by photocatalysis and ultrasonic irradiation for wastewater reuse, Desalination, 215 (2007) 166–176. https://doi.org/10.1016/j.desal.2006.10.035

[16] R. K. Gautam, S. K. Sharma, S. Mahiya, M. C. Chattopadhaya, Contamination of Heavy metals in aquatic media: Transport, Toxicity and Technologies for remediation, In: S. K. Sharma, Heavy Metals In Water: Presence, Removal and Safety, Royal society of chemistry, 2015. https://doi.org/10.1039/9781782620174-00001

[17] M. A. Habila, Z. A. ALOthman, A. M. El-Toni, J. P. Labis, M. Soylak, Synthesis and application of $Fe_3O_4@SiO_2@TiO_2$ for Photocatalytic Decomposition of Organic Matrix Simultaneously with Magnetic Solid Phase Extraction of Heavy Metals prior to ICP-MS analysis, Talanta, 154 (2016) 539-547. https://doi.org/10.1016/j.talanta.2016.03.081

Materials Research Forum LLC
https://doi.org/10.21741/9781644901359-2

[18] F. A. Harraz, O. E. Abdel-Salam, A. A. Mostafa, R. M. Mohamed, M. Hanafy, Rapid synthesis of titania–silica nanoparticles photocatalyst by a modified sol–gel method for cyanide degradation and heavy metals removal, Journal of Alloys and Compounds, 551 (2013) 1–7. https://doi.org/10.1016/j.jallcom.2012.10.004

[19] A. T. Le, S. Y. Pung, S. Sreekantan, A. Matsuda, D. P. Huynh, Mechanisms of removal of heavy metal ions by ZnO particles, Heliyon, 5 (2019). https://doi.org/10.1016/j.heliyon.2019.e01440

[20] M. I. Litter, Mechanisms of removal of heavy metals and arsenic from water by TiO2-heterogeneous photocatalysis, Pure Applied Chemistry, 8 7(2015) 557–567. https://doi.org/10.1515/pac-2014-0710

[21] S. Rengaraj, S. Venkataraj, J. W. Yeon, Y. Kim, X. Z. Li, G. K. H. Pang, Preparation, characterization and application of Nd–TiO$_2$ photocatalyst for the reduction of Cr(VI) under UV light illumination, Applied Catalysis B: Environmental, 77 (2007) 157-165. https://doi.org/10.1016/j.apcatb.2007.07.016

[22] M. N. Subramaniam, P. S. Goh, W. J. Lau, A. F. Ismail, The Roles of Nanomaterials in Conventional and Emerging Technologies for Heavy Metal Removal: A State-of-the-Art Review, Nanomaterials, 9 (2009). https://doi.org/10.3390/nano9040625

[23] P. K. Dutta, S. K. Pehkonen, V. K. Sharma, A. K. Ray, Photocatalytic Oxidation of Arsenic(III): Evidence of Hydroxyl Radicals, Environmental Science & Technology, 39 (2005) 1827-1834. https://doi.org/10.1021/es0489238

[24] A. Majumder, M. Chaudari, Solar photocatalytic oxidation and removal of Arsenic from ground water, Indian Journal of Engineering & Material sciences, 12 (2005) 122-128.

[25] N. R. Reyna, L. H. Reyes, J. L. Guzman-Mar, Y. Cai, K. O. Shea, A. H. Ramirez, Photocatalytical removal of inorganic and organic arsenic species from aqueous solution using zinc oxide semiconductor, Photochemical Photobiological Sciences, 12 (2013) 653-659. https://doi.org/10.1039/C2PP25231G

[26] M. Visa and A. Duta, TiO$_2$/fly ash novel substrate for simultaneous removal of heavymetals and surfactants, Chemical Engineering Journal, 223 (2013) 860–868. https://doi.org/10.1016/j.cej.2013.03.062

[27] S. Rahimi, M. Ahmadian, R. Barati, N. Yousefi, S. P. Moussavi, K. Rahimi, S. Reshadat, S. R. Ghasemi, N. R. Gilan, A. Fatehizadeh, Photocatalytic removal of cadmium (II) and lead (II) from simulated wastewater at continuous and batch system,

Int Jounal of Environmental Health Engineering, 3 (2014).
https://doi.org/10.4103/2277-9183.139756

[28] M. N. Rashed, M. A. Eltaher, A. N. A. Abdou, Adsorption and photocatalysis for
methyl orange and Cd removal from wastewater using TiO_2/sewage sludge-based
activated carbon nanocomposites, Royal society open science, 4(2017).
https://doi.org/10.1098/rsos.170834

[29] A. P. Davis, D. L. Green, Photocatalytic Oxidation of Cadmium-EDTA with
Titanium Dioxide, Environmental Science & Technology, 33 (1999) 609-617.
https://doi.org/10.1021/es9710619

[30] A. J.Chaudhary, M. Hassan, S. M. Grimes, Simultaneous recovery of metals and
degradation of organic species: Copper and 2,4,5-trichlorophenoxyacetic acid (2,4,5-
T), Journal of Hazardous Materials, 165 (2009) 825 – 831.
https://doi.org/10.1016/j.jhazmat.2008.10.066

[31] D. Kanakaraju, S. Ravichandar, Y. C. Lim, Combined effects of adsorption and
photocatalysis by hybrid TiO_2/ZnO-calcium alginate beads for the removal of copper,
Journal of Environmental sciences, 55 (2017) 214-223.
https://doi.org/10.1016/j.jes.2016.05.043

[32] Y. Zhu, W. Fan, T. Zhou, X. Li, Removal of chelated heavy metals from aqueous
solution: A review of current methods and mechanisms, Science of the Total
Environment, 678 (2019) 253–266. https://doi.org/10.1016/j.scitotenv.2019.04.416

[33] S. Sreekantan, C. W. Lai, S. M. Zaki, The Influence of Lead Concentration on
Photocatalytic Reduction of Pb(II) Ions Assisted by Cu-TiO_2 Nanotubes, International
Journal of Photoenergy, 2014. https://doi.org/10.1155/2014/839106

[34] T.Mishra, J.Hait, N. Aman, R.K.Jana, S.Chakravarty, Effect of UV and visible light
on photocatalytic reduction of lead and cadmium over titania based binary oxide
materials, Journal of Colloid and Interface Science, 316 (2007) 80-84.
https://doi.org/10.1016/j.jcis.2007.08.037

[35] L. Murruni, G. Leyya, M. I. Litter, Photocatalytic removal of Pb(II) over TiO_2 and
Pt–TiO_2 powders, Catalysis Today, 129 (2007) 127-135.
https://doi.org/10.1016/j.cattod.2007.06.058

Chapter 3

Carbon Nanotubes based Nanocomposites as Photocatalysts in Water Treatment

Tong Ling Tan[1] and Chin Wei Lai[2,*]

[1]Institute of Advanced Technology, Materials Processing and Technology Laboratory (MPTL), Universiti Putra Malaysia, 43400, Selangor, Malaysia

[2]Nanotechnology & Catalysis Research Centre (NANOCAT), Institute for Advanced Studies (IAS), University of Malaya, 50603 Kuala Lumpur, Malaysia

*cwlai@um.edu.my

Abstract

The shortage of worldwide clean water and the increasing water demand are now ubiquitous problems around the world. Thus, efficient water treatment is an important research topic, of which phocatalysis is known as simplest and efficient technique utilized in the photocatalytic degradation of all major water pollutants, including heavy metal ion, organic and inorganic pollutants. In this context, the use of one- dimensional carbon nanotubes-based nanocomposites in water treatment have been widely demonstrated to be capable of removing persistent organic compounds due to their unique physical and electronic properties, large surface area, tunable morphology, biocompatible and chemical-environmental-thermal stability. This chapter begins with the discussion of the importance and properties of carbon nanotubes, and then briefs about the types and methods of preparation of carbon nanotubes-based nanocomposites in detail. The next section emphasizes the fundamentals of photocatalysis phenomenon and its proposed mechanism for the photocatalytic degradation of pollutants. The last section highlights the recent development in the carbon-based nanocomposites as photocatalyst in water treatment systems, supported by comprehensive literature account. Finally, the remaining challenges and perspectives for using carbon nanotubes-based nanocomposites are discussed.

Keywords

Carbon Nanotubes, Nanocomposites, Photocatalyst, Water Treatment

Materials Research Forum LLC
https://doi.org/10.21741/9781644901359-3

Contents

1.   Introduction

Water covers more than 70% of the Earth's surface and is the major constituent for all living organisms' well-being and sustenance. Today, an inadequate access to clean water and increasing levels of water pollution are among the most concerning global issues around the world [1]. The water pollution in our environment was aggravated by the rapid population growth and intense urbanization which degrading the water quality of lakes, rivers and coastal shorelines [2-4]. The unsafe water is jeopardizing our health and causes about 2 million deaths a year, either through dehydration or acute toxicity [5, 6]. As reported by United Nations, it is estimated that by 2025, two-thirds of the world population would be under water stress conditions and more than 2.8 billion people in 48 countries will be facing absolute water scarcity, and by 2050, 4 billion people in 54 countries will be suffering from water-insufficient conditions [7] as shown in Figure 1.

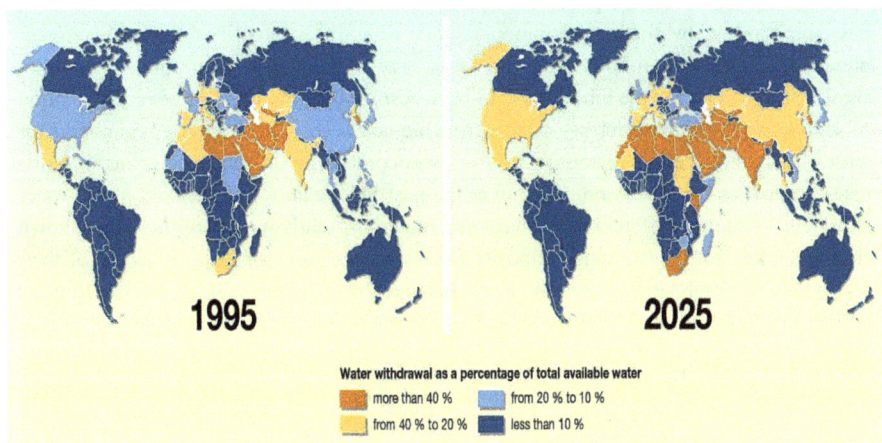

Figure 1: Freshwater stress map displayed percentage ranges of water withdrawal with respect to the total available water. Reprinted (adapted) from http://www.grida.no/resources/5625 with permission of the GRID-Arendal.

Meanwhile, the water shortage is exacerbated by water pollution. There are various water pollutants, namely heavy metals, organic and inorganic contaminants to emerging endocrine disrupting pollutants that can provoke carcinogenic and even mutagenic effects in aquatic organisms and humans [8-11]. Based on the United Nation- Water statistic report, 70% of the industrial wastes are dumped untreated into water in the developing countries while 2 million tons of human waste are disposed of in water courses every day resulting in a limited clean water supply [12]. Therefore, intensive efforts are needed in engineered water treatments systems to overcome the clean water pollution problems.

There are various physico-chemical methods including membrane filtration, coagulation, flocculation, ozonation, adsorption and photocatalytic degradation for removing water pollutants from wastewater [13-16]. Among these processes, the photocatalytic degradation method is recognized as a high potential alternative technology for alleviating the refractory organic pollutants and hazardous waste in water. This is due to advantages such as simple, environmental-friendly, inexpensive technology and ease of application at ambient temperature and pressure [17-19]. Generally, the principle of photocatalysis is to convert the photon energy by using light absorption into the chemical energy in the forms of products. It involves the acceleration of photoreaction with the presence of semiconducting oxide photocatalysts (e.g: TiO_2, ZnO, WO_3, Fe_2O_3, CdS, ZnS, $SrTiO_3$, etc.)

[11, 20, 21]. However, the photocatalytic efficiency of the semiconductor photocatalyst has been limited by the two common factors which are the fastest recombination of photogenerated electron-hole pairs and large energy band gap. Therefore, in order to significantly overcome the limited visible-light absorption and rapid charge recombination of semiconductor photocatalysts, an effective method is by combining the semiconductor with the carbonaceous materials into a nanocomposite. Carbonaceous-metal oxide nanocomposites have been proposed over the past few years for application in the water treatment. Among different carbonaceous materials, carbon nanotubes have shown significant key role in the nanocomposite for the photocatalysis process, because of their several merits: high surface area, high mechanical strength, chemical inertness and good adsorbent properties, which are beneficial to improve water treatment [1, 2, 14].

This chapter provides insight into the properties of carbon nanotubes as well as the fundamental principle of the photocatalysis mechanism, followed by the summary of the various synthesis techniques for carbon-nanotube-based nanocomposites. The characteristics and advantages of existing synthesis methods are discussed. Detailed discussions are provided on the summary of recent advances of water treatment using carbon nanotubes-based nanocomposites as photocatalyst. The remaining challenges are also mentioned.

2.   Properties of carbon nanotubes

Carbon nanotubes (CNT) is a carbon family member with one dimensional (1D) structure that rolled up into tubular with diameter of a few nanometers and up to few millimeters in length. It has highly organized helical arrangement of hexagonal arrays with open or closed ends capped with a hemisphere of the fullerene structure. Since the first discovery by Iijima in 1991, CNT have gained much attention due to their high surface to volume ratio, high aspect ratio, tunable physical, chemical, electrical, and structural properties [22, 23]. CNTs possess good structural integrity and chemical inertness support with relatively high oxidation stability which could endow CNTs more excellent performances as support for photocatalytic active materials [24, 25]. Eventually, the remarkable properties of CNT make them potentially valuable to be applied in a wide range of applications (sensors, electronics, water treatment etc.).

Photocatalysis   Materials Research Forum LLC
Materials Research Foundations **100** (2021) 77-112   https://doi.org/10.21741/9781644901359-3

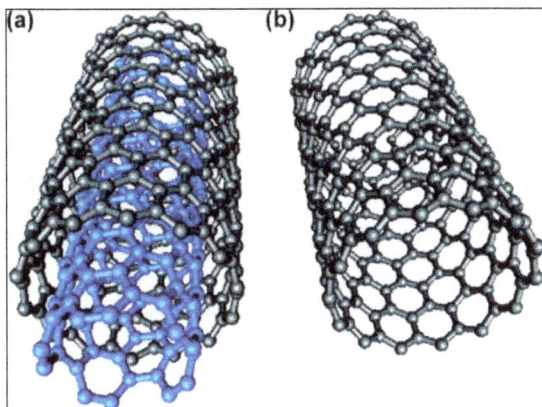

Figure 2: Structure representations of (a) MWCNTs and (b) SWCNTs. Reprinted (adapted) with permission from [26]. Copyright (2009) American Chemical Society.

The oxygen functional groups such as hydroxyl and carboxyl are introduced onto the CNTs surface during the synthesis and purification process. These functional groups can influence the maximum adsorption capacity of CNTs and make them more hydrophilic and suitable for the adsorption. Up to now, CNTs can be easily scaled-up to batch-scale production by chemical vapor deposition (CVD) method due to simplicity and economy.

Moreover, CNTs may acts as effective electron sinks with a high electrical conductivity and high electron storage capacity which can store up to 1 electron per 32 carbon atoms during photoexcitation [27-30].

Basically, there are two main types of CNTs: Single walled carbon nanotubes (SWCNTs) and Multi walled carbon nanotubes (MWCNTs) that differ in the arrangement of their graphene cylinders (Figure 2). SWCNTs consist of a single layer of graphene cylinders; while MWCNTs consist of multiple layers of graphite rolled into form a tube shape. There are three different crystallographic configurations of SWCNTs such as zigzag, armchair, and chiral structure, depending on how the graphene sheet is rolled up to a cylinder way as shown in Figure 3. A pair of indices (n, m) is used to describe the chiral vector in the honeycomb crystal lattice for the SWCNTs structure along the two directions. The SWCNTs are named as zigzag nanotubes when m=0 which means the C-C bonds are parallel to the tube axis; while the SWCNTs are named as armchair nanotubes when n=m (C-C bonds of each hexagonal are perpendicular to the tube axis). For the chiral nanotubes, C-C bonds are adjoining at an angle to the tube axis.

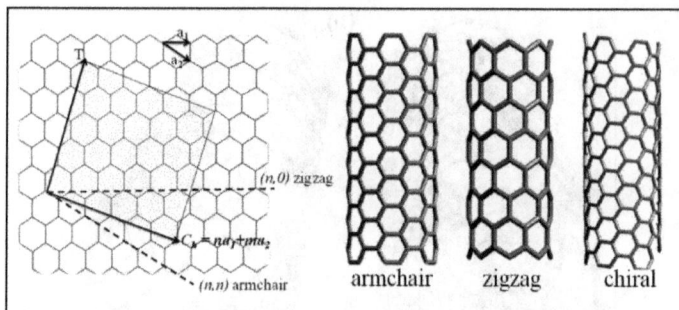

Figure 3: Three different types of single walled carbon nanotubes. Reprinted (adapted) with permission from [32]. Copyright (2011) IntechOpen.

MWCNTs structures can be categorized into two structural models based on their arrangements of graphite layers: Russian doll model and Parchment model. For Russian doll model, the layers of graphene sheets are arranged within a concentric structure and the inner nanotube has a smaller diameter than the outer nanotube. On the other hand, for Parchment model, a graphene sheet is rolled up around itself manifold times exactly like rolled up scroll of paper. Typically, SWCNTs can have diameters of 0.8 to 2 nm while MWCNTs having diameters of 5 to 20 nm. As reported, the total specific surface area of SWCNTs are in the range of 400-900 m^2/g while for MWCNTs are between 200-400 m^2/g, which provides CNTs with high surface active site to volume ratio as compared to conventional granular carbon [29, 31]. Nonetheless, MWCNTs exhibit advantages over SWCNTs, such as low cost per unit since the synthesis process of SWCNTs is extremely complicated and the yields are small as compared to MWCNTs. The properties of SWCNTs and MWCNTs are summarized in Table 1.

Table 1: Characteristics and textural properties of SWCNTs and MWCNTs [33].

Properties	SWCNTs	MWCNTs
Layer of graphite	Single layer	Multiple layers
Diameters	0.8-2 nm	5-20 nm
Specific gravity, g/cm^3	0.8	1.8
Elastic Modulus, TPa	~1	~0.3-1
Strength, GPa	50-500	10-60
Thermal conductivity, $W\ m^{-1}\ K^{-1}$	3000	3000
Total specific surface area, m^2/g	400-900	200-400
Cost	High	Low

Environmental applications based on CNTs offer more realistic possibilities and have many advantages over other carbon materials in terms of costs, structural and electrical properties. Indeed, the adsorption capability and high adsorption efficiency of CNTs makes them excellent prospects for widespread use in a variety of industries all around the world. Based on the intrinsic properties (high thermal conductivity, good chemical and environmental stability) of CNTs shown in Figure 4, CNTs can act as promising blocks for carbon-based composite, giving great potential in water treatment applications. Nano-scaled composite materials of MWCNTs are in great interest due to its high chemical inert nature, non-swelling effect and rigidity [28, 29, 34].

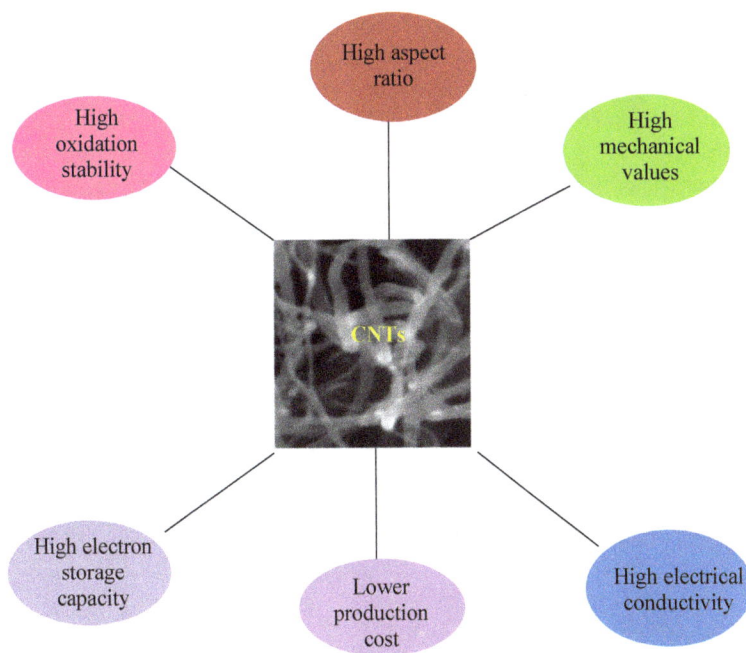

Figure 4: Properties of CNTs.

Photocatalysis   Materials Research Forum LLC
Materials Research Foundations **100** (2021) 77-112   https://doi.org/10.21741/9781644901359-3

3.   Synthesis Methods of CNTs-based Nanocomposites

Over the past few years, various synthesis methods to form CNTs-based nanocomposites have been explored. The incorporation of the CNTs can control the physiochemical properties of metal oxide nanoparticle photocatalyst to meet the specific requirement for a given application purpose. Additionally, it may also exhibit some new properties caused by the interaction between the two types of components. To date, there are several synthesis methods that can be utilized to produce CNTs-based nanocomposites photocatalyst. These techniques include liquid phase deposition method, simple mixing, ultrasonic irradiation, chemical vapor deposition, electrodeposition, hydrothermal, solvothermal and sol-gel method. However, in these methods, functionalization of CNTs is a crucial step that should be carried out before the synthesis and preparation of CNT-based nanocomposites material. This step is essential to remove the impurities, like metal catalyst and amorphous carbon as the adsorption ability of CNTs can be influenced by coating with CNTs. Usually, the functionalization of CNTs are executed by oxidation treatment with the oxidizing agents, namely nitric acid, sulfuric acid, a mixture of sulfuric acid and nitric acid, potassium permanganate, hydrogen peroxide and so forth. Different oxidizing agents used would resulting in creating different functional groups on the CNT surface such as carboxylic ($-COOH$), alcoholic ($-OH$), aldehydic ($-COH$), ketonic ($-C=O$), and esteric ($-C=O-O$) oxygenated functional groups as shown in Figure 5 [35].

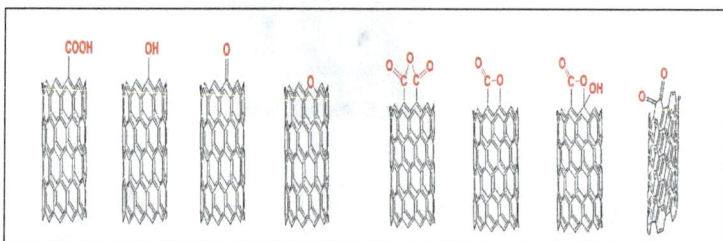

Figure 5: Three different types of single walled carbon nanotubes. Reprinted (adapted) with permission from [35] Copyright (2011) IntechOpen.

The chemical affinity of CNTs to metal oxide can be enhanced after the chemical functionalization of the CNTs sidewalls and tips. An example is the fabrication of CNTs/ metal oxide nanocomposites with good interface adhesion after oxidizing treatment of

CNTs as illustrated in Figure 6. Metal oxides particles are well-attached on the surface of oxidized CNTs.

To date, researchers are looking for the facile and greener methods especially with shorten reaction time, minimal steps involved, less energy and less toxic by-products for the synthesis of nanocomposites materials. Therefore, different synthesis techniques applied will result in different structure and morphology of the CNTs-based nanocomposites. In the following section, the most commonly techniques such as simple mixing, chemical vapor deposition, electrodeposition, hydrothermal and sol-gel methods to yield the desired CNTs-based nanocomposites are briefly described.

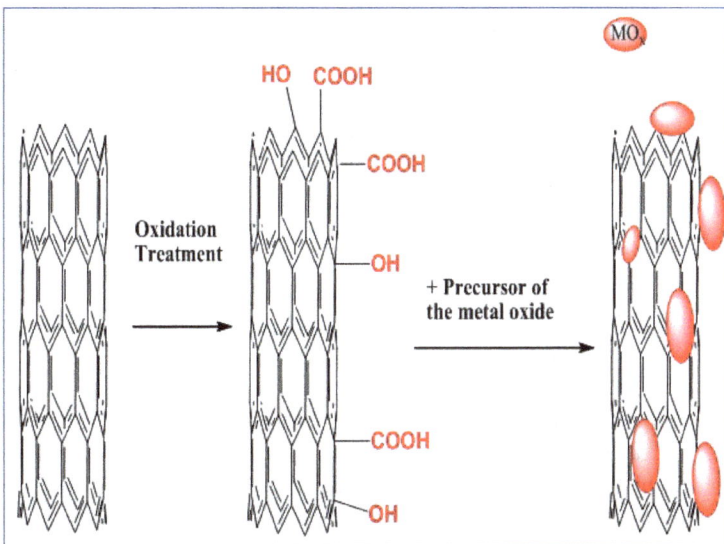

Figure 6: Schematic representation of the preparation of treated CNTs/metal oxide nanocomposites. Reprinted (adapted) with permission from [35] Copyright (2011) IntechOpen.

3.1   Simple mixing

Simple mixing has been reported as one of the most fundamental routes for the synthesis of CNTs based Metal oxide (MO) nanocomposites. Under simple mixing method, a certain amount of MO powder and CNTs were dispersed and mixed in water under slow stirring rate at room temperature. This method usually performed in solution and water is used to be solvent. According to Ashkarran et al., they have prepared CNT-TiO$_2$ using three different simple mixing methods: (i) simple mixing of CNTs with TiO_2 nanoparticles, (ii) simple mixing of CNTs with TiO_2 nanoparticles followed by heat treatment and (iii) simple mixing of CNTs with TiO_2 nanoparticles followed by UV illumination. From the TEM image, they have found that TiO_2 nanoparticles are well decorated on the CNTs surface in all three simple mixing methods [36].

Thus, this implies that it is possible to decorate CNTs with TiO_2 nanoparticles by simple mixing. The attachment of MO onto the surface of CNTs is through van der Waals interactions. Xu et al. synthesized CNTs/MO nanocomposites using simple mechanical mixing method for gas phase degradation of benzene. In respect to this method, it is simple and less time-consuming; however, it has been criticized for the low level of interaction where there is no intimate contact or chemical bonding between CNTs and MO. The type of bonding exists between CNTs and MO in this method is physisorption and MO may detach from CNTs surface through mechanical process such as sonication as reported by Yu et al. Significantly, the interaction between CNTs and MO is one of the key factors for controlling electron transfer and photocatalytic activity in CNTs/metal oxide nanocomposite structure [37-39].

3.2   Chemical vapor deposition (CVD)

Chemical vapor deposition (CVD) techniques are among the most common methods to synthesis CNTs based MO nanocomposites that utilize the growth of a solid material from gaseous phase through chemical reaction at the surface of a substrate at controlled flow rates. The chemical reactions that occurred on the heated surface of the substrate resulting in the deposition of coated thin on the surface. Due to the high purity of reagent used in this method, it can provide MO with a good control of the size, shape, purity, composition and easy scalability. In contrast to high pressure or high temperature synthesis, medium temperature at around 600-800 °C and a slightly reduced atmospheric pressure is required by this technique.

According to Tavakkoli et al., maghemite (γ-Fe_2O_3) nanoparticles are decorated on carbon nanotubes (CNTs) in a one-step synthesis using a low cost floating catalyst vapor deposition method for hydrogen evolution reaction. This study shows the growth of carbon encapsulated iron nanoparticles are highly dependent on the amount of iron source, ferrocene and an average size of 5.4 nm for the carbon encapsulated metallic iron nanoparticles are obtained from the HRTEM image [40]. Uniform coating of MO on CNTs can be achieved by this technique but when it scaled down to few nanometers, uniformity and defect-free coatings are difficult to be achieved due to the fast deposition. Moreover, this technique is not simple and specialized equipment is needed [41, 42].

Photocatalysis   Materials Research Forum LLC
Materials Research Foundations **100** (2021) 77-112   https://doi.org/10.21741/9781644901359-3

3.3   Electrodeposition

Electrodeposition is one of the most powerful techniques for the deposition of MO nanoparticles via reduction of metal complexes by electrons. The nanocomposites can be produced through the migration of charged particles dispersed in a liquid medium under an electric field, followed by the particles are coagulated at an electrode. In the early stages of electrodeposition, the limiting step corresponds to electron transfer from work electrode for metallic ions in solution. The size and the coverage of MO nanoparticles on the sidewalls of CNTs can be controlled by electrodeposition parameters, including concentration of the metal salt, nucleation potential and deposition time. The main advantage of this technique is that the electrodeposited nanoparticles show higher purity as well as a good adhesion to the surface of CNTs. In detail, a strong adhesion could be provided via a simple van der Waals interaction between the CNTs and the MO nanoparticles.

The work done by Zhao et al., showed that, the deposition of TiO_2 is carried out by using $TiCl_3$ as a precursor and electrolyte through galvanostatic oxidation with 1 mA/cm^2 at pH 2.5 with HCl/Na_2CO_3. As a result, it has been reported that this technique occurs to the same extent on both the sidewalls of the tubes and the tips [43]. Another work done by Nguyen and Bui, high density gold nanoparticles are electrodeposited into a three dimensional framework attached CNTs in 0.2 M hydrogen tetrachlororoaurate ($HAuCl_4.3H_2O$) solution at an applied potential of -1.2 V for 200 s. The SEM image indicated that the gold nanoparticles are densely distributed on the whole surface of the CNTs [44]. However, electrodeposition technique is difficult to produce bulk quantities of samples and this becoming the major drawback of this method.

3.4   Hydrothermal

Hydrothermal techniques have been employed in the synthesis of CNTs based MO nanocomposites in recent years. This method is highly dependent on the chemical reactions and solubility changes of precursors in a Teflon-sealed autoclave under controlled temperature and/or pressure for growing the nanocrystals [45].

Kalubarme et al., synthesized CNT/cerium oxide (CeO_2) composite with one-pot hydrothermal method in the presence of potassium hydroxide (KOH) and capping agent polyvinylpyrrolidone. The precursor solutions (functionalized CNTs, Cerium nitrate, KOH and polyvinylpyrrolidone) were mixed and put in a Teflon stainless steel autoclave at 170 °C for 24 h reaction. They indicated that two steps of reaction are involved for the growth of CeO_2; (1) Ce^{3+} is oxidized into Ce^{4+} when reacting with hydroxyl ion followed by (2) spherical particles of CeO_2 will grow as hexagonal nanocrystals by the effect of capping agent under the hydrothermal condition. The capping agent will give control upon the crystal size of the nanoparticles by hindering the crystal growth. The HRTEM image revealed that the CeO_2 nanoparticles exhibit hexagonal structure on the CNT support [46].

Another work done by Dai et al., stated that in a typical hydrothermal synthesis of MWCNTs/TiO_2 nanocomposites, the pristine or acid-treated CNTs were added to the aqueous solution of the titanium precursor and treated in an autoclave at temperature between 100-200 °C to produce crystalline films of TiO_2. As a result, a dense coating of spherical or slightly elongated nanoparticles are obtained. It is worth mentioning that the titanium precursors using for hydrothermal processes are mainly titanium tetrachloride ($TiCl_4$) and titanium oxysulfate ($TiOSO_4$). Typically, the final materials obtained are mostly amorphous TiO_2 with seeds of anatase regardless the chosen conditions and chemistry. Thus, a further heat treatment that occurs at 300-500 °C in air is required to fully crystallize the TiO_2 coating on CNTs and avoid the burnout of the CNTs [47]. In contrast with other methods, the advantages of the composites prepared by hydrothermal methods are found to give better results, including less agglomeration among particles, narrow particle size distribution, and controllable phase homogeneity and particle morphology. However, this method requires longer preparation times (several hours or a day) and acquires hydrothermal temperature and pressure [48, 49].

3.5   Sol-gel

Sol-gel is a versatile technique that involves the transition of a liquid colloidal sol, followed by hydrolysis and polymerization of the precursors which are usually metal alkoxides into a solid gel phase. The sol particles may interact by Van der Waals forces or hydrogen bonds while the interactions in gel system are of covalent nature and the gel process is irreversible.

The rate of hydrolysis and poly-condensation will significantly affect the structures and properties of MO. Generally, sol-gel method is the most common technique to synthesize MO on the CNTs surface because this method is cheap, easy to control the chemical composition, does not require complicated equipment and occur at ambient temperature that can produce materials with high purity and homogeneity. However, the major drawback of this synthesis is that the final product normally consists of an amorphous phase rather than the defined crystals and hence, crystallization and post annealing steps are required. Moreover, this method usually leads to heterogeneous, non-uniform coating and random aggregation of MO onto the CNTs surface. Significantly, the selection of precursor is crucial for sol-gel method due to the reason that homogeneous nucleation or heterogeneous on CNTs are influenced by the reactivity of the precursor with the solvent. The growth rate of nanoparticles is also affected where the fast condensation rates resulting in a large particle-size distribution. Kose et al., synthesized zinc oxide (ZnO)/ MWCNTs nanocomposites for lithium ion battery anodes by sol-gel technique using zinc acetate

Photocatalysis   Materials Research Forum LLC
Materials Research Foundations **100** (2021) 77-112   https://doi.org/10.21741/9781644901359-3

dehydrate and functionalized MWCNTs as the starting materials. The authors studied the effect of two chelating agents, triethanolamine and glycerin used in the sol-gel synthesis on the structure of ZnO/ MWCNTs nanocomposites. Field emission gun-scanning electron microscopy (FEG-SEM) shows that a very fine layer of ZnO was homogeneously coated on the individual MWCNTs the ZnO/ MWCNTs nanocomposites prepared with chelating agents [50].

Another example by Ellis et al., indium oxide (In_2O_3) is grown on the surface of oxidized SWCNTs using a sol-gel synthesis for ethanol sensing at room temperature. The authors reported that the degree of calcination will provide a close contact between the In_2O_3 and the nanotube surface where the In_2O_3 can be crystallized and condensed on SWCNTs after calcination. HRTEM image indicated that the crystalline In_2O_3 are attached above a bundle of SWCNTs [51]. The most common titanium precursor used for preparing TiO_2/ CNTs is titanium isopropoxide since it readily dissolves in alcohol and is not overly sensitive as revealed from the work done by Rodriguez et al. Acids or bases are added for altering the reaction rate. TiO_2 are formed by hydrolysis and poly-condensation process of titanium alkoxides.

Soroodan and Fatemi synthesized TiO_2/ CNTs composites using sol-gel method for the degradation of acetaldehyde. Benzyl alcohol as the linking agent was added to the TiO_2/ CNTs composites resulting in homogeneously coating of TiO_2 sols over the CNTs [52-54]. The comparison between the unique features of simple mixing, chemical vapor deposition, electrodeposition, hydrothermal and sol-gel methods is summarized in Table 2. Synthesis routes covers the range of techniques are tabulated in Table 3.

Table 2: Comparison of the most widely used synthesis methods for preparing CNTs/MO composite.

Synthesis method	Advantage	Disadvantage
Simple mixing	•Simple •Less time-consuming	•No intimate contact or chemical bonding between CNTs and MO nanoparticles
CVD	•Short processing time	• Requires higher deposition temperature • Complicated steps
Electrodeposition	•Provide MO with a good control of size, the shape and purity •Uniform coating of MO on CNTs	•Needs specialized equipment
	•Electrodeposited MO nanoparticles show higher purity as well as good adhesion to the surface of CNTs	•Difficult to produce bulk quantities of CNTs/MO sample
Hydrothermal	•Reduce agglomeration among MO nanoparticles •Control phase homogeneity and particle morphology	•Longer preparation time • Acquires hydrothermal temperature and pressure
Sol-gel	•Simple •Cheap •Occur at ambient temperature •Fine and well-dispersion of MO nanoparticles on CNTs surface	•Difficult controlling of hydrolysis rate

Materials Research Forum LLC

https://doi.org/10.21741/9781644901359-3

Table 3: Representative summary of CNT/MO composite synthesis routes

Synthesis routes	CNTs	MO	Synthesis conditions	Remarks	Applications	Ref.
Simple mixing	CNTs	TiO_2	•CNTs are purified with HCl •CNT/TiO_2 composite were prepared with the simple mixing, UV illumination and heat treatment	•No observable formation of chemical bonds and the CNTs are wrapped around TiO_2 anatase grains	Degradation of Rhodamine B	[36]
Simple mixing	MWCNTs	Commercial P25	• Purification of MWCNTs using HNO_3 •Dispersed in anhydrous ethanol solution	•Less individual interfacial between MWCNTs and TiO_2 •Simple mixing cannot combine P25 and MWCNTs effectively	Degradation of benzene	[38]
Simple mixing	MWCNTs	Titanium sulfate	•MWCNTs was treated using HNO_3/H_2SO_4 mixture •Acid-treated MWCNTs were dispersed in water and introduced into titanium sulfate solution	•Unable to create effective interfacial contact between TiO_2 and CNTs	Degradation of acetone	[39]

Materials Research Forum LLC
https://doi.org/10.21741/9781644901359-3

CVD	CNTs	Maghemit e (γ-Fe_2O_3)	•Synthesis temperature of 1100 °C, atmospheric pressure and in laminar flow conditions •Toluene and ethylene as the carbon sources for the CNT •Ferrocene was decomposed to Fe nanoparticles at high temperature	•A mixture of carbon encapsulated iron nanoparticles are decorated on the CNTs.	Electrocataly sts for the oxygen evolution reaction	[40]
CVD	CNTs	Titanium tetrachlori de	•CVD growth of CNTs was carried out using the supported Ni as the catalyst •CVD deposition of a TiO_2 layer on the CNTs	•The conformal coverage of TiO_2 nanoparticles on the CNTs walls	Photocatalyti c antifungal activity	[41]
CVD	CNTs	Commerc ial anatase TiO_2	•CVD growth of CNTs was carried out using the supported Fe as the catalyst and a mixture of C_2H_4 and H_2 •CNTs directly grown on TiO_2	•CNTs were found to be uniformly grown in TiO_2 particles without altering the crystalline structure	Photocatalyti c Hydrogen Production	[42]

Materials Research Forum LLC

https://doi.org/10.21741/9781644901359-3

Electrodeposition	CNTs	TiO_2	•The electrodeposition of TiO_2 was performed in a three-electrode cell with resin as a working electrode, Pt wire and Ag/AgCl were used as counter and reference electrodes •KCl solution as the electrolyte	•CNTs were coated with a uniform layer of highly crystalline anatase TiO_2 nanoparticles	Photo degradation of organic pollutant	[43]
Electrodeposition	CNTs	Gold nanoparticles	Applied potential of -1.2 V for 200 s in 0.2 M hydrogen tetrachloroaurate solution	The TEM image showed that the gold nanoparticles are homogeneously dispersed onto CNTs surface after the electrodeposition process	Hydrazine detection	[44]
Hydrothermal	CNTs	Cerium oxide	•CNTs were treated with 30% HNO_3 with stirring for 6h at 120 °C • Hydrothermal treatment at 170 °C for 24 h in an 80 ml Teflon stainless steel.	•The growth of CeO_2 taken place when come in contact with hydroxyl ion. •The hexagonal CeO_2 nanocrystals were observed to grow on the CNTs surface.	Supercapacitor	[46]

Hydrothermal	SWCNTs	Titanium sulfate	•SWCNTs were treated in the boiled nitrate solution •Hydrothermal treatment at 140 °C for 24 h	•Well-dispersed SWCNTs and TiO_2 nanoparticles having intimate contact, which inhibited the growth of TiO_2 grain	Photocatalytic hydrogen evolution	[47]
Hydrothermal	MWCNTs	Titanium isopropoxide	•Purification and functionalization of MWCNTs using HCl and HNO_3 •Hydrothermal treatment at 140 °C for 24 h	• Favors a decrease in agglomeration among particles and controlled particle morphology	Degradation of methylene blue solution	[48]
Sol-gel	MWCNTs	ZnO	• MWCNTs were functionalized using an acid mixture of HNO_3 :H_2SO_4 (1:3 v/ v) • Triethanolamine and glycerin acts as chelating agents	• The addition of chelating agents may result in enhancing thin film porosity and increasing solution viscosity. •The FEG-SEM image indicated that a mesoporous structure with a homogenous distribution of pores are formed	Lithium ion battery anodes	[50]

Sol-gel	SWCNTs	In$_2$O$_3$	• SWCNTs are oxidized using HNO$_3$. Oxidized SWCNTs, indium hydroxide (indium (III) chloride and ammonia) were sonicated, stirred and dialyzed in water for overnight.	• A core-shell morphology of In$_2$O$_3$/SWCNTs composite was observed where the CNTs are completely encapsulated by the indium oxide layer.	Ethanol Sensing	[51]
Sol-gel	MWCNTs	Titanium isopropoxide Titanium butoxide	•MWCNTs was treated using HNO$_3$/H$_2$SO$_4$ mixture •Titanium precursor, ethanol, H$_2$O, and HCl was mixed and stirred at ambient temperature	•Dispersion of TiO$_2$ nanoparticles on the surface of CNTs	Degradation of methylene blue solution	[52]
Sol-gel	MWCNTs	Titanium isopropoxide	•CNT/TiO$_2$ composite were prepared with the aiding of benzyl alcohol as a linking agent	•Homogeneous coating of TiO$_2$ sols over the CNTs	Degradation of acetaldehyde	[54]

4.   Fundamental principles of photocatalysis

Generally, photocatalysis has been considered as one of the most sustainable methodologies which has been the subject of a huge amount of studies for water purification treatment. According to International Union of Pure and Applied Chemistry (IUPAC) compendium of chemical terminology, a catalytic reaction involving light absorption by a catalyst can be defined as photocatalysis [55]. It can be carried out in various media, including gas phase, pure organic liquid phases or aqueous solutions. Among these AOPs, heterogeneous photocatalysis employing semiconductor catalysts has demonstrated its efficiency in degrading a wide range of ambiguous refractory organics into readily biodegradable compounds and substantially, mineralization to innocuous carbon dioxide and water [56-58]. Taking into account the advance photocatalysis processes involved photodegradation under UV irradiation, the materials used as photocatalysts must satisfy several functional requirements with respect to band gap energy, morphology, surface area, stability and reusability [59]. Besides, light harvesting ability of a photocatalyst is an important criterion to produce maximum photo-induced charge carriers. The basic parameter that governs the light-harvesting ability of the photocatalyst is its electronic structure, which determines its band-gap energy [60, 61]. Figure 7 illustrates the E_g values and the positions of band edges of various semiconductors.

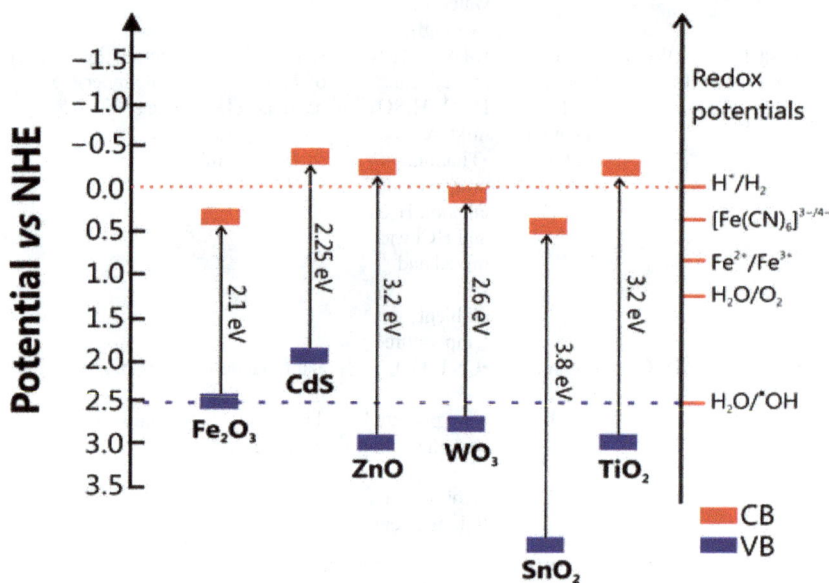

Figure 7: Schematic representation of E_g values (in eV) and position of band edges of various semiconductors Reprinted (adapted) with permission from [63] Copyright (2015) Springer Nature.

Among the advanced semiconducting materials (i.e. WO_3, Fe_2O_3, CeO_2, ZnO, and ZnS) being reported for electronics, magnetic, optical and photocatalytic applications; TiO_2 is the most widely used photocatalyst for the removal of organic pollutants from drinking

Photocatalysis   Materials Research Forum LLC
Materials Research Foundations **100** (2021) 77-112   https://doi.org/10.21741/9781644901359-3

water treatment, industrial and aqueous systems in environment clean-up. The extensive use of TiO_2 in photocatalytic building materials is attributed to the following characteristics: (a) relatively inexpensive, non-toxic and chemically stable; (b) high photocatalytic activity compared with other metal oxide photocatalysts and (c) the photogenerated holes are highly oxidizing [62].

From the literature, binary metal sulfides semiconductors such as CdS, CdSe or ZnS are regarded as toxic and insufficiently stable for catalysis in aqueous media due to photoanodic corrosion. Photo-corrosion occurs when the anion from the catalyst itself is oxidized by photogenerated holes instead of water. Besides, the band gap (3.2 eV) and band-edge positions of ZnO are similar to those of TiO_2 and it was reported that ZnO has a higher electron mobility than TiO_2. However, ZnO is unstable in water due to incongruous dissolution and will form $Zn(OH)_2$ on the ZnO particle surfaces and thus lead to catalyst inactivation. Moreover, ZnO suffer from photo corrosion induced by self-oxidation and they can react with the photogenerated holes giving the following reactions as shown in Equation 1:

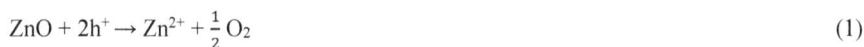

$$ZnO + 2h^+ \rightarrow Zn^{2+} + \frac{1}{2} O_2 \tag{1}$$

The Zn^{2+} released by photo-corrosion may produce a new poisoning effect on the environment, bringing about further pollution. As a result, the utilization of ZnO in environmental purification is greatly limited by these disadvantages. Thus, TiO_2 proves to be more a suitable benchmark photocatalyst for photodegradation of organic pollutants due to its high photoactivity, resistance to photo corrosion and possess a good mechanical resistance in acidic and oxidative environments. The photocatalytic properties of TiO_2 nanoparticles are derived from the illumination with photons having energy greater than the band gap energy, the photons will excite the electrons from the valence band (VB) into the conduction band (CB), resulting in the formation of photogenerated charge carriers (hole and electron). The photogenerated holes in the valence band will react with adsorbed water molecules, forming hydroxyl radicals ($\bullet OH$). $\bullet OH$ produced has the second highest oxidation potential which can non-selectively oxidize almost all electron rich organic molecules and eventually converting them into CO_2 and water. Therefore, the organic pollutants which are adsorbed on the surface of the catalyst will be oxidized by $\bullet OH$. In the meantime, electrons in the conduction band will typically participate in reduction processes, which typically react with molecular oxygen to produce superoxide radical anion ($O_2^{\bullet-}$). The presence of molecular oxygen plays a substantial role in the photoinduced processes on the irradiated TiO_2 surface, as effective charge carriers, separation is enabled. In addition, $O_2^{\cdot-}$ may not only take place in the further oxidation process but also prevents the electron-hole recombination, hence electron neutrality within the TiO_2 is maintained. Therefore, both $\cdot OH$ and $O_2^{\cdot-}$ play such an important roles in the photocatalytic reaction mechanism. The Equation (2-5) represents the mechanism of the photocatalytic degradation by TiO_2 [62, 64, 65].

$$TiO_2 + h\nu \rightarrow e_{CB}^- + h_{VB}^+ \tag{2}$$

$$H_2O + h^+{}_{vb} \rightarrow \cdot OH + H^+ \tag{3}$$

$$O_2 + e^-{}_{cb} \rightarrow O_2^{\cdot-} \tag{4}$$

$$\cdot OH + Organic\ Pollutant \rightarrow CO_2 + H_2O \tag{5}$$

It is essential for effective photocatalytic degradation of organic pollutants that the oxidation of pollutant and the reduction process of oxygen occur simultaneously in order to prevent the accumulation of electron in the conduction band and hence, the rate of recombination of e_{CB}^- and h_{VB}^+ is reduced. A schematic diagram illustrating the mechanism of photocatalysis on TiO_2 nanoparticles is shown in Figure 8.

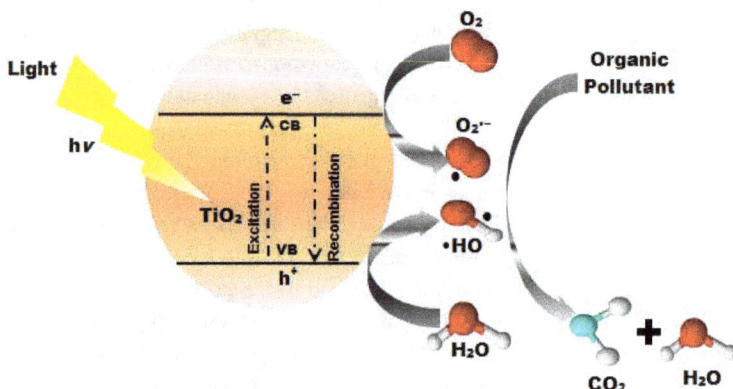

Figure 8: A schematic diagram illustrating the mechanism of photocatalysis on TiO₂ nanoparticles.

However, an obvious hindrance to the widespread use of TiO_2 nanoparticles as a photocatalyst is its poor visible light response, the rapid recombination rate of charge carriers, low adsorption capacity to hydrophobic contaminants, high aggregation tendency and difficulty of separation and recovery, which decreases the quantum efficiency of the overall reaction. CNTs is used as a one-dimensional photocatalyst for contributing to the routes of enhancing TiO_2 photocatalytic activities, e.g., TiO_2 is an n-type semiconductor, but with the presence of CNTs, photoinduced electrons may migrate freely in the direction to the CNTs surfaces where CNTs can act as extremely effective electron sinks, which might have a lower Fermi level like a metal for suppressing the electron-hole pairs recombination. [28, 29, 66, 67].

The general principles photocatalytic mechanism of CNT/MO nanocomposites are briefly introduced in the following section. Accordingly, the photocatalytic degradation process initiated via the absorption of photon by the MO semiconductor. Under such conditions, the electrons from valence band would be excited to the conduction band of MO creating the electron-hole pairs on the catalyst surface. The photogenerated electrons were then transferred to CNTs, (act as the electron acceptor). The trapped electrons would then react with the dissolved oxygen to form the reactive oxygen species, followed by suppressing the rate of electron-hole recombination. In the meantime, the photogenerated electrons on the MO surface tended to reduce the dissolved oxygen to form superoxide anion, which may subsequently react with the adsorbed water to form the hydroxyl radicals. The hydroxyl radicals (acted as a primary oxidant in the photocatalytic system) then reacted with organic pollutant molecules and mineralized it into CO_2, H_2O and mineral acids [68, 69]

The mechanism reactions are shown as below:

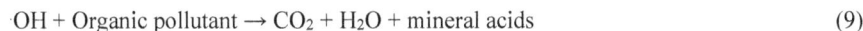

$$MO/MWCNTs \xrightarrow{h\nu} MO\ (h^+) - CNTs\ (e^-) \tag{6}$$

$$MWCNTs\ (e^-) + O_2 \rightarrow MWCNTs + O_2^{\cdot-} \tag{7}$$

$$MO\ (h^+) + H_2O/OH^- \rightarrow MO + \cdot OH \tag{8}$$

$$\cdot OH + Organic\ pollutant \rightarrow CO_2 + H_2O + mineral\ acids \tag{9}$$

5.   Applications of CNTs-based nanocomposites in water treatment

CNTs-based nanocomposites have been being one popular type of photocatalyst for water treatment, since it combines the excellent performances of MO semiconductor with CNTs. To date, removal of heavy metal ion, organic, inorganic and emerging pollutants in water have been increasingly studied using CNTs-based nanocomposites. Therefore, the water treatment applications using several MO photocatalysts supported on CNTs used in degradation of organic pollutants are extensively discussed in the following sections. Table 4 summarizes the targeted organic pollutants, light sources, catalysts and the influences of various operational parameters on the photodegradation of organic pollutants based on the recently published work. It can be seen that most organic pollutants can be photocatalytically degraded by MO within minutes to several hours under UV or near UV-visible irradiation, while longer time was needed for visible irradiation. The combination of CNTs with MO will extend the wavelength from UV to visible light range. Wongaree et al., demonstrated that a highly removal efficiency of methylene blue solution with CNT/TiO_2 nanocomposites (70 %) within 90 min under visible light irradiation as compared with pristine TiO_2 (22 %). Based on this, the lower band gap energy of CNTs/TiO_2 as compared with pristine TiO_2 resulting in a better light absorption of the composites in visible region [52]. In the similar work presented by Zhang et al., showed that the as prepared (CNTs)/Bismuth (Bi_4VO_8Cl) composite photocatalysts exhibited 30 % higher photocatalytic degradation of methyl orange (MO), hexavalent chromium Cr(IV), and bisphenol A (BPA) in water under visible light irradiation than that of pure Bi_4VO_8Cl. The enhanced photocatalytic activity by the composites are due to the synergetic effects by increased adsorption efficiency of pollutant, increased light absorption and reduced recombination rate of photogenerated charge carriers [70]. Zouzelka et al., prepared MWCNTs/TiO_2 composites for the degradation of 4-chlorophenol as compared to a P25 reference layer. The authors found that the first order rate reaction constant increased by about 100 % for the MWCNTs/TiO_2 composites, which suggested, MWCNTs acts as an electron sink and favors for photogenerated charge separation [71]. Photocatalysis using MWCNTs/TiO_2 composites are completely removed tetracycline under UVC irradiation as reported by Ahmadi et al., The authors also carried out the removal of real pharmaceutical wastewater using MWCNTs/TiO_2 composites which obtained total organic carbon (TOC) removal of 82.3 % after 4 h. The finding also demonstrated that the MWCNTs/TiO_2 composites exhibited superior catalytic capabilities over TiO_2 nanoparticles for degradation of tetracycline [72]. Another work reported by Zhu et al., ZnO/CNTs hierarchical mircosphere composites was prepared by chemical deposition route for the photocatalytic removal of methylene blue under UV and visible light irradiation. The maximum photocatalytic activity of the composites (92.3 %) is higher than that of pure ZnO (70.4 %) which suggested that CNTs play an important role for the photocatalytic removal of methylene blue [73]. An interesting work done by Zaman et al., revealed that the sonocatalytic degradation efficiency of methylene blue by SnO_2/MWCNT nanocomposite. During the sonocatalytic process, MWCNTs acts as an electron sink for accepting sonogenerated electron from the SnO_2 to promote interfacial electron transfer process. A complete removal of methylene blue was achieved in 30 min in the presence of SnO_2/MWCNT nanocomposite as compared with the bared SnO_2. A simple hydrothermal synthesis of WO_3-CNTs composite for sulfamethoxazole degradation was constructed by Zhu et al. [74]. The as-synthesized WO_3-CNT (73.3%) possessed enhanced photocatalytic performance for the sulfamethoxazole degradation than bare WO_3 (25.4 %) under visible light irradiation. Payan et al. described the fabrication of titanate nanotube (TNT)/MWCNT nanocomposites via hydrothermal method for the photodegradation of 4-chlorophenol under solar and UV illumination. For both UV and solar irradiations, TNT/MWCNT nanocomposites exhibited higher photocatalytic efficiency compared to the bare TNTs, which provided that SWCNTs has high surface areas that can create more active sites on the surface of CNTs, leading to more effectual degradation of 4-chlorophenol [75]. In other recent work, Oveisi et al., explored the Materials Institute Lavoisier-125 (Ti) (MIL-125 (Ti))/carbon nanotube (CNT) nanocomposites synthesized by hydrothermal method and tested for photocatalytic reactive black 5 degradation. The photocatalytic result showed that MIL/CNT composite (60 %) had higher photocatalytic dye degradation as compared to only MIL 125 (Ti) alone (<. 40 %) due to the generation of powerful oxidants which included free radicals in water [76]. Kamil et al., reported MWCNTs/TiO_2 nanocomposite was prepared using sol-gel method for the Bismarck brown R dye degradation. The maximum photocatalytic efficiency was achieved by MWCNTs/TiO_2 nanocomposite (82.7 %) compared to TiO_2 (47.3 %) respectively.

Table 4: Summary of organic pollutants removal by photocatalytic reaction under various operational parameters

Photocatalyst	Type of organic pollutants	Initial Conc. of Pollutant	Catalyst dosage	Light source	Irradiation time	Removal efficiency (%)	Ref.
CNT/TiO$_2$	Methylene blue	1 x 10^{-5}M	1.0 g/L	Fluorescence lamp (15W)	1.5 h	70	[52]
CNTs)/Bi$_4$VO$_8$Cl	Methyl orange (MO) Hexavalent chromium Cr(VI) + Bisphenol A (BPA)	10 mg/L 8.6 mg/L	1.0 g/L 1.5 g/L	300 W Xenon lamp	10 h 4 h	74.8 100	[70]
MWCNT/TiO$_2$	4-chlorophenol	0.1 mmol.L^{-1}	N/A	UV light (365 nm)	5 h	70	[71]
MWCNT/TiO$_2$	Tetracycline	10 mg/L	0.2 g/L	6W UVC lamps	100 min	100	[72]
CNT/ZnO	Methylene blue	30 mg/L	0.17 g/L	500W Hg lamp 500 W Xenon lamp	1 h	92.3 76	[73]
SnO$_2$/MWCNT	Methylene blue	0.02 mM	30 mg	n.d.	30 min	100	[77]
WO$_3$-CNT	Sulfamethoxazole	10 mg/L	0.50 g/L	300 W Xenon lamp	3	73.3	[74]
TNT/SWCNT	4-chlorophenol	50 mg/L	2 g/L	UVC lamp	8 h	92	[75]
MIL-125 (Ti /CNTs	Reactive black 5	20 mg/L	0. 1 g/L	UV lamp	3 h	60	[76]
MWCNT/TiO$_2$	Bismarck brown R dye	5 × 10^{-5} M	0. 5 g/L	UV lamp	1.5 h	82.7	[78]
CuO/CNT	p-chloroaniline	10 mg/L	0.375 g/L	Metal halide lamp	3 h	97	[79]
C$_3$N$_4$/CNTs	Rhodamine B	10 mg/L	0. 6 g/L	500 W Xenon lamp	2 h	98.6	[80]
TiO$_2$/CNT	Methylene blue	10 mg/L	0. 4 g/L	UV lamp	1 h	> 90%	[81]

Conclusion and future directions

In conclusion, it is notable that the CNT-based MO nanocomposites act as a potential photocatalyst for the degradation of various organic pollutants. The photocatalytic effect of CNT/MO nanocomposites occurs not only because of the adsorption of CNTs, but also because of the electron transfer between CNTs and MO, removing the organic pollutant from the solution. Despite the advances, great opportunities still exist in the exploitation of carbon-based hybrid assemblies. In this chapter, we have summarized the recent advances in the synthesis and application of the CNT-based MO nanocomposites. Despite various exploitations and extensive studies of this MWCNTs/MO nanocomposites have been undertaken, a number of great challenges still remain and require attention in the future works. In order to bring this technology to practical applications, further studies and developments are required. Many studies applying photocatalysis for water treatment have been conducted by taken into account the concentration of the targeted organic pollutant without taking their attention to the possible intermediate products produced after the degradation. In fact, some intermediate products formed during the photodegradation experiment might be more dangerous and toxic compared to the original pollutant. The underlying mechanism of the photocatalytic enhancement by CNTs/MO nanocomposites is not fully understood and need to be examined in details to gain the clear understanding and to prove guidelines for future water treatment design based on photocatalysis. Generally, this enhancement is ascribed to the extended absorption and improved charge transfer in the hybrids. Although some studies have suggested a different mechanism in the literature, the possible mechanism of photocatalytic reaction may depend on the interfacial status of CNTs/MO nanocomposites, type of bonding, type of CNTs, functionalization, surface area and so forth. Thus, all these aspects in the fundamental processing-structural-developments will need to be better understood and controlled. Broader environmental applications on this CNTs/MO nanocomposite shall be explored. For instance, the photocatalytic degradation of persistent organic pollutants that hardly found in the literature shall be investigated, which important to the environmental protection and human health. Although it is still in the infancy, a widespread and great progress in this area can be expected in the future. The previous work mainly focused on a batch scale of photocatalytic degradation for pollutant removal, but the lack of application in a larger scale. Thus, there is a need for the design of a prototype for the convenience of photocatalytic degradation of organic pollutants so that it can be applied in real practice.

References

[1] X. Liu, M. Wang, S. Zhang, B. Pan, Application potential of carbon nanotubes in water treatment: A review, 2013. https://doi.org/10.1016/S1001-0742(12)60161-2

[2] L. Ma, X. Dong, M. Chen, L. Zhu, C. Wang, F. Yang, Y. Dong, Fabrication and Water Treatment Application of Carbon Nanotubes (CNTs)-Based Composite Membranes: A Review, Membranes 7 (2017) 16. https://doi.org/10.3390/membranes7010016

[3] M.S. Mauter, I. Zucker, F. Perreault, J.R. Werber, J.-H. Kim, M. Elimelech, The role of nanotechnology in tackling global water challenges, Nature Sustainability 1 (2018) 166-175. https://doi.org/10.1038/s41893-018-0046-8

Materials Research Forum LLC
https://doi.org/10.21741/9781644901359-3

[4] A. Kumar, G. Sharma, M. Naushad, A.a.H. Al-Muhtaseb, A. García-Peñas, G.T. Mola, C. Si, F.J. Stadler, Bio-inspired and biomaterials-based hybrid photocatalysts for environmental detoxification: A review, Chemical Engineering Journal 382 (2020) 122937. https://doi.org/10.1016/j.cej.2019.122937

[5] R.P. Schwarzenbach, T. Egli, T.B. Hofstetter, U.v. Gunten, B. Wehrli, Global Water Pollution and Human Health, Annual Review of Environment and Resources 35 (2010) 109-136. https://doi.org/10.1146/annurev-environ-100809-125342

[6] V.K. Sharma, R. Zboril, R.S. Varma, Ferrates: Greener Oxidants with Multimodal Action in Water Treatment Technologies, Accounts of Chemical Research 48 (2015) 182-191. https://doi.org/10.1021/ar5004219

[7] S. Kar, R.C. Bindal, P.K. Tewari, Carbon nanotube membranes for desalination and water purification: Challenges and opportunities, Nano Today 7 (2012) 385-389. https://doi.org/10.1016/j.nantod.2012.09.002

[8] O.M. Rodriguez-Narvaez, J.M. Peralta-Hernandez, A. Goonetilleke, E.R. Bandala, Treatment technologies for emerging contaminants in water: A review, Chemical Engineering Journal 323 (2017) 361-380. https://doi.org/10.1016/j.cej.2017.04.106

[9] X. Mao, W. Tian, Y. Ren, D. Chen, S.E. Curtis, M.T. Buss, G.C. Rutledge, T.A. Hatton, Energetically efficient electrochemically tunable affinity separation using multicomponent polymeric nanostructures for water treatment, Energy & Environmental Science 11 (2018) 2954-2963. https://doi.org/10.1039/C8EE02000K

[10] I.S. Yunus, Harwin, A. Kurniawan, D. Adityawarman, A. Indarto, Nanotechnologies in water and air pollution treatment, Environmental Technology Reviews 1 (2012) 136-148. https://doi.org/10.1080/21622515.2012.733966

[11] S. Sharma, A. Bhattacharya, Drinking water contamination and treatment techniques, Applied Water Science 7 (2017) 1043-1067. https://doi.org/10.1007/s13201-016-0455-7

[12] B.R. Johnston, The political ecology of water: an introduction, Capitalism Nature Socialism 14 (2003) 73-90. https://doi.org/10.1080/10455750308565535

[13] C. Santhosh, V. Velmurugan, G. Jacob, S.K. Jeong, A.N. Grace, A. Bhatnagar, Role of nanomaterials in water treatment applications: A review, Chemical Engineering Journal 306 (2016) 1116-1137. https://doi.org/10.1016/j.cej.2016.08.053

[14] A.S. Adeleye, J.R. Conway, K. Garner, Y. Huang, Y. Su, A.A. Keller, Engineered nanomaterials for water treatment and remediation: Costs, benefits, and applicability,

Chemical Engineering Journal 286 (2016) 640-662.
https://doi.org/10.1016/j.cej.2015.10.105

[15] G. Sharma, A. Kumar, S. Sharma, A.H. Al-Muhtaseb, M. Naushad, A.A. Ghfar, T. Ahamad, F.J. Stadler, Fabrication and characterization of novel Fe^0@Guar gum-crosslinked-soya lecithin nanocomposite hydrogel for photocatalytic degradation of methyl violet dye, Separation and Purification Technology 211 (2019) 895-908. https://doi.org/10.1016/j.seppur.2018.10.028

[16] P. Dhiman, J. Chand, A. Kumar, R.K. Kotnala, K.M. Batoo, M. Singh, Synthesis and characterization of novel Fe@ZnO nanosystem, Journal of Alloys and Compounds 578 (2013) 235-241. https://doi.org/10.1016/j.jallcom.2013.05.015

[17] F. Petronella, A. Truppi, C. Ingrosso, T. Placido, M. Striccoli, M.L. Curri, A. Agostiano, R. Comparelli, Nanocomposite materials for photocatalytic degradation of pollutants, Catalysis Today 281 (2017) 85-100. https://doi.org/10.1016/j.cattod.2016.05.048

[18] S. Dong, J. Feng, M. Fan, Y. Pi, L. Hu, X. Han, M. Liu, J. Sun, J. Sun, Recent developments in heterogeneous photocatalytic water treatment using visible light-responsive photocatalysts: a review, Rsc Advances 5 (2015) 14610-14630. https://doi.org/10.1039/C4RA13734E

[19] A. Kumar, S.K. Sharma, G. Sharma, C. Guo, D.-V.N. Vo, J. Iqbal, M. Naushad, F.J. Stadler, Silicate glass matrix@Cu_2O/$Cu_2V_2O_7$ p-n heterojunction for enhanced visible light photo-degradation of sulfamethoxazole: High charge separation and interfacial transfer, Journal of Hazardous Materials 402 (2021) 123790. https://doi.org/10.1016/j.jhazmat.2020.123790

[20] M.R.D. Khaki, M.S. Shafeeyan, A.A.A. Raman, W.M.A.W. Daud, Application of doped photocatalysts for organic pollutant degradation - A review, Journal of Environmental Management 198 (2017) 78-94. https://doi.org/10.1016/j.jenvman.2017.04.099

[21] A. Kumar, A. Rana, G. Sharma, M. Naushad, A.a.H. Al-Muhtaseb, C. Guo, A. Iglesias-Juez, F.J. Stadler, High-Performance Photocatalytic Hydrogen Production and Degradation of Levofloxacin by Wide Spectrum-Responsive Ag/Fe_3O_4 Bridged $SrTiO_3$/g-C_3N_4 Plasmonic Nanojunctions: Joint Effect of Ag and Fe_3O_4, ACS Applied Materials & Interfaces 10 (2018) 40474-40490. https://doi.org/10.1021/acsami.8b12753

[22] Ihsanullah, Carbon nanotube membranes for water purification: Developments, challenges, and prospects for the future, Separation and Purification Technology 209 (2019) 307-337. https://doi.org/10.1016/j.seppur.2018.07.043

[23] B.S. Al-anzi, O.C. Siang, Recent developments of carbon based nanomaterials and membranes for oily wastewater treatment, RSC Advances 7 (2017) 20981-20994. https://doi.org/10.1039/C7RA02501G

[24] M.F. De Volder, S.H. Tawfick, R.H. Baughman, A.J. Hart, Carbon nanotubes: present and future commercial applications, science 339 (2013) 535-539. https://doi.org/10.1126/science.1222453

[25] A. Eatemadi, H. Daraee, H. Karimkhanloo, M. Kouhi, N. Zarghami, A. Akbarzadeh, M. Abasi, Y. Hanifehpour, S.W. Joo, Carbon nanotubes: properties, synthesis, purification, and medical applications, Nanoscale Res Lett 9 (2014) 393. https://doi.org/10.1186/1556-276X-9-393

[26] Y.-L. Zhao, J.F. Stoddart, Noncovalent Functionalization of Single-Walled Carbon Nanotubes, Accounts of Chemical Research 42 (2009) 1161-1171. https://doi.org/10.1021/ar900056z

[27] X. An, J.C. Yu, Graphene-based photocatalytic composites, RSC Advances 1 (2011) 1426-1434. https://doi.org/10.1039/c1ra00382h

[28] C.P. Bergmann, F.M. Machado, Carbon Nanomaterials as Adsorbents for Environmental and Biological Applications, Springer2015. https://doi.org/10.1007/978-3-319-18875-1

[29] R. Leary, A. Westwood, Carbonaceous nanomaterials for the enhancement of TiO_2 photocatalysis, Carbon 49 (2011) 741-772. https://doi.org/10.1016/j.carbon.2010.10.010

[30] F. Perreault, A.F. de Faria, M. Elimelech, Environmental applications of graphene-based nanomaterials, Chemical Society Reviews 44 (2015) 5861-5896. https://doi.org/10.1039/C5CS00021A

[31] L. Dai, D.W. Chang, J.-B. Baek, W. Lu, Carbon Nanomaterials for Advanced Energy Conversion and Storage, Small 8 (2012) 1130-1166. https://doi.org/10.1002/smll.201101594

[32] V. Choudhary, A. Gupta, Polymer/carbon nanotube nanocomposites, Carbon nanotubes-polymer nanocomposites, IntechOpen2011. https://doi.org/10.5772/18423

[33] Y.T. Ong, A.L. Ahmad, S.H.S. Zein, S.H. Tan, A review on carbon nanotubes in an environmental protection and green engineering perspective, Brazilian Journal of

Chemical Engineering 27 (2010) 227-242. https://doi.org/10.1590/S0104-66322010000200002

[34] S.B.A. Hamid, T.L. Tan, C.W. Lai, E.M. Samsudin, Multiwalled carbon nanotube/TiO$_2$ nanocomposite as a highly active photocatalyst for photodegradation of Reactive Black 5 dye, Chinese Journal of Catalysis 35 (2014) 2014-2019. https://doi.org/10.1016/S1872-2067(14)60210-2

[35] V. Gupta, T.A. Saleh, Syntheses of carbon nanotube-metal oxides composites; adsorption and photo-degradation, Carbon Nanotubes-From Research to Applications, IntechOpen2011. https://doi.org/10.5772/18009

[36] A.A. Ashkarran, M. Fakhari, H. Hamidinezhad, H. Haddadi, M.R. Nourani, TiO$_2$ nanoparticles immobilized on carbon nanotubes for enhanced visible-light photo-induced activity, Journal of Materials Research and Technology 4 (2015) 126-132. https://doi.org/10.1016/j.jmrt.2014.10.005

[37] M. Barberio, P. Barone, A. Imbrogno, S.A. Ruffolo, M. La Russa, N. Arcuri, F. Xu, Study of band gap of carbon nanotube-titanium dioxide heterostructures, Journal of Chemistry and Chemical Engineering 8 (2014) 36.

[38] Y.-J. Xu, Y. Zhuang, X. Fu, New Insight for Enhanced Photocatalytic Activity of TiO2 by Doping Carbon Nanotubes: A Case Study on Degradation of Benzene and Methyl Orange, The Journal of Physical Chemistry C 114 (2010) 2669-2676. https://doi.org/10.1021/jp909855p

[39] J. Yu, T. Ma, S. Liu, Enhanced photocatalytic activity of mesoporous TiO$_2$ aggregates by embedding carbon nanotubes as electron-transfer channel, Physical Chemistry Chemical Physics 13 (2011) 3491-3501. https://doi.org/10.1039/C0CP01139H

[40] M. Tavakkoli, T. Kallio, O. Reynaud, A.G. Nasibulin, J. Sainio, H. Jiang, E.I. Kauppinen, K. Laasonen, Maghemite nanoparticles decorated on carbon nanotubes as efficient electrocatalysts for the oxygen evolution reaction, Journal of Materials Chemistry A 4 (2016) 5216-5222. https://doi.org/10.1039/C6TA01472K

[41] S. Darbari, Y. Abdi, F. Haghighi, S. Mohajerzadeh, N. Haghighi, Investigating the antifungal activity of TiO$_2$ nanoparticles deposited on branched carbon nanotube arrays, Journal of Physics D: Applied Physics 44 (2011) 245401. https://doi.org/10.1088/0022-3727/44/24/245401

[42] P. Chen, L. Wang, P. Wang, A. Kostka, M. Wark, M. Muhler, R. Beranek, CNT-$TiO_{2-\delta}$ Composites for Improved Co-Catalyst Dispersion and Stabilized Photocatalytic Hydrogen Production, Catalysts 5 (2015) 270. https://doi.org/10.3390/catal5010270

[43] Y. Zhao, Y. Hu, Y. Li, H. Zhang, S. Zhang, L. Qu, G. Shi, L. Dai, Super-long aligned TiO2/carbon nanotube arrays, Nanotechnology 21 (2010) 505702. https://doi.org/10.1088/0957-4484/21/50/505702

[44] D.M. Nguyen, Q.B. Bui, Three-dimensional mesoporous hierarchical carbon nanotubes/nickel foam-supported gold nanoparticles as a free-standing sensor for sensitive hydrazine detection, Journal of Electroanalytical Chemistry 832 (2019) 444-452. https://doi.org/10.1016/j.jelechem.2018.11.053

[45] A. Kumar, G. Sharma, A. Kumari, C. Guo, M. Naushad, D.-V.N. Vo, J. Iqbal, F.J. Stadler, Construction of dual Z-scheme g-C_3N_4/$Bi_4Ti_3O_{12}$/$Bi_4O_5I_2$ heterojunction for visible and solar powered coupled photocatalytic antibiotic degradation and hydrogen production: Boosting via I^-/I_3^- and Bi^{3+}/Bi^{5+} redox mediators, Applied Catalysis B: Environmental 284 (2021) 119808. https://doi.org/10.1016/j.apcatb.2020.119808

[46] R.S. Kalubarme, Y.-H. Kim, C.-J. Park, One step hydrothermal synthesis of a carbon nanotube/cerium oxide nanocomposite and its electrochemical properties, Nanotechnology 24 (2013) 365401. https://doi.org/10.1088/0957-4484/24/36/365401

[47] K. Dai, X. Zhang, K. Fan, P. Zeng, T. Peng, Multiwalled carbon nanotube-TiO_2 nanocomposite for visible-light-induced photocatalytic hydrogen evolution, Journal of Nanomaterials 2014 (2014) 4. https://doi.org/10.1155/2014/694073

[48] K.D. Shitole, R.K. Nainani, P. Thakur, Preparation, characterisation and photocatalytic applications of TiO_2-MWCNTs composite, Defence Science Journal 63 (2013) 435-441. https://doi.org/10.14429/dsj.63.4870

[49] Q. Yang, Z. Lu, J. Liu, X. Lei, Z. Chang, L. Luo, X. Sun, Metal oxide and hydroxide nanoarrays: Hydrothermal synthesis and applications as supercapacitors and nanocatalysts, Progress in Natural Science: Materials International 23 (2013) 351-366. https://doi.org/10.1016/j.pnsc.2013.06.015

[50] H. Köse, Ş. Karaal, A.O. Aydın, H. Akbulut, A facile synthesis of zinc oxide/multiwalled carbon nanotube nanocomposite lithium ion battery anodes by sol–gel method, Journal of Power Sources 295 (2015) 235-245. https://doi.org/10.1016/j.jpowsour.2015.06.135

[51] J.E. Ellis, U. Green, D.C. Sorescu, Y. Zhao, A. Star, Indium Oxide—Single-Walled Carbon Nanotube Composite for Ethanol Sensing at Room Temperature, The Journal of Physical Chemistry Letters 6 (2015) 712-717. https://doi.org/10.1021/jz502631a

[52] M. Wongaree, S. Chiarakorn, S. Chuangchote, Photocatalytic Improvement under Visible Light in Nanoparticles by Carbon Nanotube Incorporation, Journal of Nanomaterials 2015 (2015) 10. https://doi.org/10.1155/2015/689306

[53] C.-H. Wu, C.-Y. Kuo, S.-T. Chen, Synergistic effects between TiO_2 and carbon nanotubes (CNTs) in a TiO_2/CNTs system under visible light irradiation, Environmental technology 34 (2013) 2513-2519. https://doi.org/10.1080/09593330.2013.774058

[54] E. Soroodan Miandoab, S. Fatemi, Upgrading TiO_2 photoactivity under visible light by synthesis of MWCNT/TiO_2 nanocomposite, International Journal of Nanoscience and Nanotechnology 11 (2015) 1-12.

[55] N. Serpone, A. Emeline, Suggested terms and definitions in photocatalysis and radiocatalysis, International Journal of Photoenergy 4 (2002) 91-131. https://doi.org/10.1155/S1110662X02000144

[56] J.-M. Herrmann, Heterogeneous photocatalysis: fundamentals and applications to the removal of various types of aqueous pollutants, Catalysis Today 53 (1999) 115-129. https://doi.org/10.1016/S0920-5861(99)00107-8

[57] A.O. Ibhadon, P. Fitzpatrick, Heterogeneous photocatalysis: recent advances and applications, Catalysts 3 (2013) 189-218. https://doi.org/10.3390/catal3010189

[58] G. Sharma, Z.A. ALOthman, A. Kumar, S. Sharma, S.K. Ponnusamy, M. Naushad, Fabrication and characterization of a nanocomposite hydrogel for combined photocatalytic degradation of a mixture of malachite green and fast green dye, Nanotechnology for Environmental Engineering 2 (2017) 4. https://doi.org/10.1007/s41204-017-0014-y

[59] M.M. Khan, S.F. Adil, A. Al-Mayouf, Metal oxides as photocatalysts, Journal of Saudi Chemical Society 19 (2015) 462-464. https://doi.org/10.1016/j.jscs.2015.04.003

[60] S.H.S. Chan, T. Yeong Wu, J.C. Juan, C.Y. Teh, Recent developments of metal oxide semiconductors as photocatalysts in advanced oxidation processes (AOPs) for treatment of dye waste-water, Journal of Chemical Technology and Biotechnology 86 (2011) 1130-1158. https://doi.org/10.1002/jctb.2636

[61] A. Kumar, G. Sharma, M. Naushad, A.H. Al-Muhtaseb, A. Kumar, I. Hira, T. Ahamad, A.A. Ghfar, F.J. Stadler, Visible photodegradation of ibuprofen and 2,4-D in

simulated waste water using sustainable metal free-hybrids based on carbon nitride and biochar, Journal of Environmental Management 231 (2019) 1164-1175. https://doi.org/10.1016/j.jenvman.2018.11.015

[62] A. Fujishima, T.N. Rao, D.A. Tryk, Titanium dioxide photocatalysis, Journal of Photochemistry and Photobiology C: Photochemistry Reviews 1 (2000) 1-21. https://doi.org/10.1016/S1389-5567(00)00002-2

[63] G.G. Bessegato, T.T. Guaraldo, J.F. de Brito, M.F. Brugnera, M.V.B. Zanoni, Achievements and Trends in Photoelectrocatalysis: from Environmental to Energy Applications, Electrocatalysis 6 (2015) 415-441. https://doi.org/10.1007/s12678-015-0259-9

[64] S. Bagheri, N. Muhd Julkapli, S. Bee Abd Hamid, Titanium Dioxide as a Catalyst Support in Heterogeneous Catalysis, The Scientific World Journal 2014 (2014) 21. https://doi.org/10.1155/2014/727496

[65] J. Schneider, M. Matsuoka, M. Takeuchi, J. Zhang, Y. Horiuchi, M. Anpo, D.W. Bahnemann, Understanding TiO_2 Photocatalysis: Mechanisms and Materials, Chemical Reviews 114 (2014) 9919-9986. https://doi.org/10.1021/cr5001892

[66] D.K. Tiwari, J. Behari, P. Sen, Application of Nanoparticles in Waste Water Treatment 1, (2008).

[67] B.H. Nguyen, V.H. Nguyen, D.L. Vu, Photocatalytic composites based on titania nanoparticles and carbon nanomaterials, Advances in Natural Sciences: Nanoscience and Nanotechnology 6 (2015) 033001. https://doi.org/10.1088/2043-6262/6/3/033001

[68] K. Woan, G. Pyrgiotakis, W. Sigmund, Photocatalytic carbon-nanotube–TiO_2 composites, Advanced Materials 21 (2009) 2233-2239. https://doi.org/10.1002/adma.200802738

[69] S. Da Dalt, A.K. Alves, C.P. Bergmann, Photocatalytic degradation of methyl orange dye in water solutions in the presence of $MWCNT/TiO_2$ composites, Materials Research Bulletin 48 (2013) 1845-1850. https://doi.org/10.1016/j.materresbull.2013.01.022

[70] X. Zhang, D. Shi, J. Fan, One stone two birds: novel carbon nanotube/Bi_4VO_8Cl photocatalyst for simultaneous organic pollutants degradation and Cr (VI) reduction, Environmental Science and Pollution Research 24 (2017) 23309-23320. https://doi.org/10.1007/s11356-017-9969-2

[71] R. Zouzelka, Y. Kusumawati, M. Remzova, J. Rathousky, T. Pauporté, Photocatalytic activity of porous multiwalled carbon nanotube-TiO_2 composite layers

for pollutant degradation, Journal of Hazardous Materials 317 (2016) 52-59.
https://doi.org/10.1016/j.jhazmat.2016.05.056

[72] M. Ahmadi, H. Ramezani Motlagh, N. Jaafarzadeh, A. Mostoufi, R. Saeedi, G. Barzegar, S. Jorfi, Enhanced photocatalytic degradation of tetracycline and real pharmaceutical wastewater using MWCNT/TiO$_2$ nano-composite, Journal of Environmental Management 186 (2017) 55-63. https://doi.org/10.1016/j.jenvman.2016.09.088

[73] G. Zhu, H. Wang, G. Yang, L. Chen, P. Guo, L. Zhang, A facile synthesis of ZnO/CNT hierarchical microsphere composites with enhanced photocatalytic degradation of methylene blue, RSC Advances 5 (2015) 72476-72481. https://doi.org/10.1039/C5RA11873E

[74] W. Zhu, Z. Li, C. He, S. Faqian, Y. Zhou, Enhanced photodegradation of sulfamethoxazole by a novel WO$_3$-CNT composite under visible light irradiation, Journal of Alloys and Compounds 754 (2018) 153-162. https://doi.org/10.1016/j.jallcom.2018.04.286

[75] A. Payan, M. Fattahi, S. Jorfi, B. Roozbehani, S. Payan, Synthesis and characterization of titanate nanotube/single-walled carbon nanotube (TNT/SWCNT) porous nanocomposite and its photocatalytic activity on 4-chlorophenol degradation under UV and solar irradiation, Applied Surface Science 434 (2018) 336-350. https://doi.org/10.1016/j.apsusc.2017.10.149

[76] M. Oveisi, M. Alinia Asli, N.M. Mahmoodi, Carbon nanotube based metal-organic framework nanocomposites: Synthesis and their photocatalytic activity for decolorization of colored wastewater, Inorganica Chimica Acta 487 (2019) 169-176. https://doi.org/10.1016/j.ica.2018.12.021

[77] S. Zaman, K. Zhang, A. Karim, J. Xin, T. Sun, J.R. Gong, Sonocatalytic degradation of organic pollutant by SnO$_2$/MWCNT nanocomposite, Diamond and Related Materials 76 (2017) 177-183. https://doi.org/10.1016/j.diamond.2017.05.009

[78] A.M. Kamil, H.T. Mohammed, A.A. Balakit, F.H. Hussein, D.W. Bahnemann, G.A. El-Hiti, Synthesis, characterization and photocatalytic activity of carbon nanotube/titanium dioxide nanocomposites, Arabian Journal for Science and Engineering 43 (2018) 199-210. https://doi.org/10.1007/s13369-017-2861-z

[79] N.F. Khusnun, A.A. Jalil, S. Triwahyono, C.N.C. Hitam, N.S. Hassan, F. Jamian, W. Nabgan, T.A.T. Abdullah, M.J. Kamaruddin, D. Hartanto, Directing the amount of CNTs in CuO–CNT catalysts for enhanced adsorption-oriented visible-light-

responsive photodegradation of p-chloroaniline, Powder Technology 327 (2018) 170-178. https://doi.org/10.1016/j.powtec.2017.12.052

[80] Y. Wu, H. Liao, M. Li, CNTs modified graphitic C_3N_4 with enhanced visible-light photocatalytic activity for the degradation of organic pollutants, Micro & Nano Letters 13 (2018) 752-757. https://doi.org/10.1049/mnl.2017.0864

[81] C.H. Park, C.M. Lee, J.W. Choi, G.C. Park, J. Joo, Enhanced photocatalytic activity of porous single crystal TiO_2/CNT composites by annealing process, Ceramics International 44 (2018) 1641-1645. https://doi.org/10.1016/j.ceramint.2017.10.086

Photocatalysis
Materials Research Foundations **100** (2021) 113-160

Materials Research Forum LLC
https://doi.org/10.21741/9781644901359-4

Chapter 4

Photocatalytic and Adsorptional Removal of Heavy Metals from Contaminated Water using Nanohybrids

Ankita Guleria, Rohit Sharma, Pooja Shandilya*

School of Advanced Chemical Sciences, Shoolini University, Solan, 173229, Himachal Pradesh, India

*poojashandil03@gmail.com

Abstract

Water contaminated with heavy metals is a major menace for aquatic life and human health consequently its efficient removal remains a crucial challenge for researcher. The utilization of various photocatalytic nanohybrids to synergistically photo-reduce and adsorb heavy metals is a potent strategy to combat water pollution. This book chapter give an overview of the fundamental principle of photocatalysis and various single, binary, ternary and quaternary nanohybrids employed for simultaneous photoreduction and adsorption of heavy metals with its mechanistic insight. Further, conclusion and future prospective as well as limitation of available nanohybrids were addressed. We hope that this book chapter dispenses some noticeable information to heavy metal ions removal from polluted water.

Keywords

Photocatalysis, Nanohybrids, Water Purification, Organic Pollutants, Heavy Metals

Contents

1.   Introduction

Among the wide range of water pollutant, presence of heavy metals (HMs) poses a significant threat to the human society and has become a very prominent environmental issue across the world. HMs pollution can be geo-genic arises due to the natural processes such as weathering of rocks, leaching, wind erosion, glacial erosion and volcanism or it can be anthropogenic. The anthropogenic source is burning of fossil fuels, mining, smelting, automobile exhaust, nuclear power waste, electroplating processes, excessive use of fertilizers, pesticide and herbicide [1,2]. Commonly found heavy metals lead, zinc, cadmium, arsenic, copper, mercury, chromium, tin, and thorium entered directly or indirectly into the water bodies causing several health problems [3]. HMs is responsible for long-term toxic effect due to its non-biodegradable nature and high bio-accumulation rate in food chains [4-6].

The serious toxicological effects caused by HMIs are nausea, vomiting, diarrhoea, asthma, kidney and liver malfunctioning, weight loss, congenital abnormalities and cancer. Copper, above the permissible level damages gastrointestinal tract as well as anorexia and lethargy [7], cobalt induces kidney congestion and decline thyroid movement [8], zinc produces disruption of iron homeostasis along with neurological disorders [9], nickel is responsible for immunotoxicity, damage epithelial cells of respiratory system and also cause lungs or skin cancer [10] similarly, behavioural disorders, acute malfunction of kidneys, anemia were recognized by cause of lead poisoning [11], cadmium damages testicular tissues and bones, along with carcinogenic nature [12], mercury induces neurological disorders, retard intellectual ability, damage of pulmonary and also cause hyperactivity disorder [13]. The excessive concentration of HMs beyond its permissible limits in drinking water recommended by WHO[14] and USEPA[15], may leads to serious health issues efficient as shown in Table 1 [16-21].

Table 1. Permissible limits of HMs in drinking water as recommended by WHO and USEPA.

Heavy metal	Concentration (ppm)		Ill effect	Ref
	WHO	**USEPA**		
Hg	0.001	0.002	Impairment of pulmonary and kidney function, chest pain and dyspnea, nervous system damage	[16]
Pb	0.05	0.015	Dysfunction in the kidney, reduced neural development, impair liver and reproductive system, hypertension and brain damage	[17]
Cu	1.5	1.3	Vomiting, diarrhea, stomach cramps, and nausea, or even death.	[18]
Cd	0.005	0.05	High blood pressure, kidney damage, and destruction of testicular tissue, carcinogenic	[12]
Cr	0.05	0.1	Lung cancer, chromate ulcer, nasal septum perforation	[19]
As	0.05	0.01	Lung and skin damage, toxicological and carcinogenic effect, bladder cancer	[20]
Mn	0.5	0.05	Inhalation or contact damages to central system	[21]

Earlier, various physical, chemical and biological methods like chemical precipitation [22], ion-exchange [23], membrane separation [24], coagulation [25], adsorption [26], solvent extraction [27] and electrochemical treatment [28] etc. were already explored to detoxify water. These methods bear certain disadvantages such as consumption of costly chemical reagents, ion exchanger, semi-permeable membrane, large amount sludge formation, secondary pollutant generation and less efficient Table 2 [29-38].

Table 2. Advantages and disadvantages of conventional methods for HMs removal.

Conventional method	Advantages	Disadvantages	Ref.
Precipitation	Practicable to most metals, Low cost, Easy operation	Lower removal efficiency, Sludge generation, Secondary pollution, Concentration of heavy metals	[29]
Ions exchanges	Fast kinetics, High treatment capacity, Excellent selectivity of metal ions, Regenerable materials	High cost due to synthetic resins, Small range of action on metal ions.	[30]
Membrane separation	Resource recovery, High separation selectivity, Small space requirement	Requiring amount pretreatment, Low permeate flux, Pore blocking, Poor antibacterial property, Limited pH range,	[31]
Coagulation	Remove the turbidly in addition to heavy metal removal	Extra operational cost for sludge disposal	[32]
Adsorption	Low cost, Having wide pH range High efficiency, Possibility of regenerating adsorbents, Easy operation Effective and economical method	Limited pH range, Difficult to regenerate materials, Low selectivity, Long equilibrium time	[33]
Photocatalysis	Clean and efficient, Simultaneous removing metals and organic pollutant, Low-toxicity by-products.	High raw materials cost, Small range of action on metal ions, Long duration time	[34, 35]
Fenton like oxidation	Fast and efficient, Significant removal of residual organic chelates, Recovery metals	Low water volume treatment capacity, Vast Fe sludge, High cost, Secondary pollution from addition chemicals	[36]
Electrochemical	Metal selective, No consumption of chemicals	High capital and running cost, Initial solution pH, Current density	[37]
Replacements-co-precipitation	Easy operation, Relativity high efficiency, Make full use of iron-based materials	High chemical consumption, Large quantity of sludge and secondary pollution from release of organic chelates	[38]

Thusly, removal of HMs through simultaneous photocatalytic reduction and adsorption is more desirable due to cost effectiveness, easy operation, strong redox ability, green technologies and wide applicability which especially reduce the cost of water treatment. Firstly, TiO_2 photocatalyst a semiconducting material reported by Fujishima and Honda in 1972s utilized for water splitting [39]. Consequently, different semiconductors were originated and exploits for various photocatalytic activity because of eco-friendly nature. Photocatalysis is a promising advanced oxidation process displaying tremendous potential in reducing toxic heavy metal ions into less toxic metal ion and smashing various other pollutants that showed versatile nature of semiconducting material [40]. Generation of strong oxidative species can synergistically degrade organic pollutant and reduce noxious metal ions through photocatalytic reaction. Besides, water purification photocatalysis has been actively employed in water splitting, CO_2 reduction, disinfection process and organic synthesis.

Here, in this book chapter we have summarized the basic principle of photocatalysis with recent emerging direct Z-scheme migration pathway of charge carrier. Different type of single, dual, trinary and quaternary photocatalytic nanohybrids obeying conventional type-II, Z-scheme or dual Z-scheme migratory pathway of electron-hole pair were discussed. Subsequently, various nanocomposites exploited for adsorptional removal with its mechanism were also investigated. We believe that this extensive study on photocatalytic and adsorptional removal of HMs will be beneficial for the researchers and beginners to carry out their research a one step ahead from what reported previously in the literature. Lastly, conclusion and future perspectives were discussed in detail.

2. Fundamental principle of photocatalysis

The basic photocatalytic mechanism occurs in three steps where oxidation and reduction were induced on photocatalyst surface sensitized by solar light. Firstly, generation of charger carrier occurs on exposure to light possessing energy higher or equal to band gap of semiconductor. Secondly, migration of generated electron-hole pair arises on the surface of semiconductor where, electron jumps from valence band to the conduction band of photocatalyst. Thirdly, photogenerated holes oxidizes H_2O molecule to produce hydroxyl radicals ($^{\bullet}OH$) in the valence band and electrons reduces the O_2 molecule to form superoxide radical anion ($O_2^{\bullet -}$) in the conduction band, respectively as shown below in equation (1-4) [41].

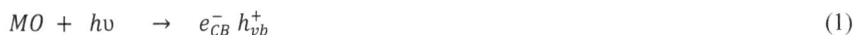

$$MO + h\upsilon \quad \rightarrow \quad e_{CB}^- \; h_{vb}^+ \tag{1}$$

$$MO(e_{cb}^-) \rightarrow {}^\circ O^{2-} \tag{2}$$

$$MO(h_{vb}^+) \rightarrow H^+ + {}^\circ OH \tag{3}$$

$$Pollutant + ROS \rightarrow CO_2 \;+\; H_2O + Degradation\,product \tag{4}$$

Among the various strategies applied to intensify photocatalytic efficiency, fabrication of heterojunction is a promising alternative magnetizing the interest of researchers across the world. Basically, heterojunction describe as the two semiconductors having different band alignment at the interface. Depending upon the band edge potential of semiconductor, conventional heterojunction was classified into three categories namely: straddling (type-I), staggered (type-II) and broken gap (type-III) as shown in (Fig. 1). In straddling type the CB and VB edges of SCI aligned higher and lower than the corresponding bands of SCII. Consequently, both charge carriers migrate towards SCII surface results in excessive accumulation of electron-hole pair which increases the probability of recombination. Also, both the oxidation and reduction reaction take place on SCII surface bearing lower redox potential will abate the potency of photocatalyst. For staggered heterojunction, CB and VB edges of SCI are higher than the SCII respectively. Thus, electrons will drift from SCI to SCII surface and holes form SCII to SCI due to suitable band alignment resulting in excellent charge separation. The reduction reaction occurs on the SCII with lower CB edge potential and oxidation reaction occur on SCI with lower VB edge potential which is analogous to straddling heterojunction. Therefore, type-II heterojunction also possess lower redox ability. However, in type-III heterojunction band edge potential become so extreme and impaired that the charge migration and separation altogether not possible. In these heterojunction staggered gap is more efficient though certain limitation such as occurrence of redox reaction on semiconductor with lower redox potential and electrostatic repulsion between electron-electron and holes-holes prevent its further migration which bounds its broader utilization.

To maximize the use of higher redox potential and to minimize the electrostatic repulsion in conventional type-II heterojunction, Bard et al. in 1979 come up with the concept of Z-scheme heterojunction where, path of charge migration is similar to letter "Z" [42]. The Z-scheme heterojunction evolves from 1st generation proposed in

1979 to 3rd generation in 2013. The 1st generation Z-scheme also called as conventional Z-scheme or liquid-phase Z-scheme in which two semiconductors are not in direct contact with each other.

Materials Research Forum LLC
https://doi.org/10.21741/9781644901359-4

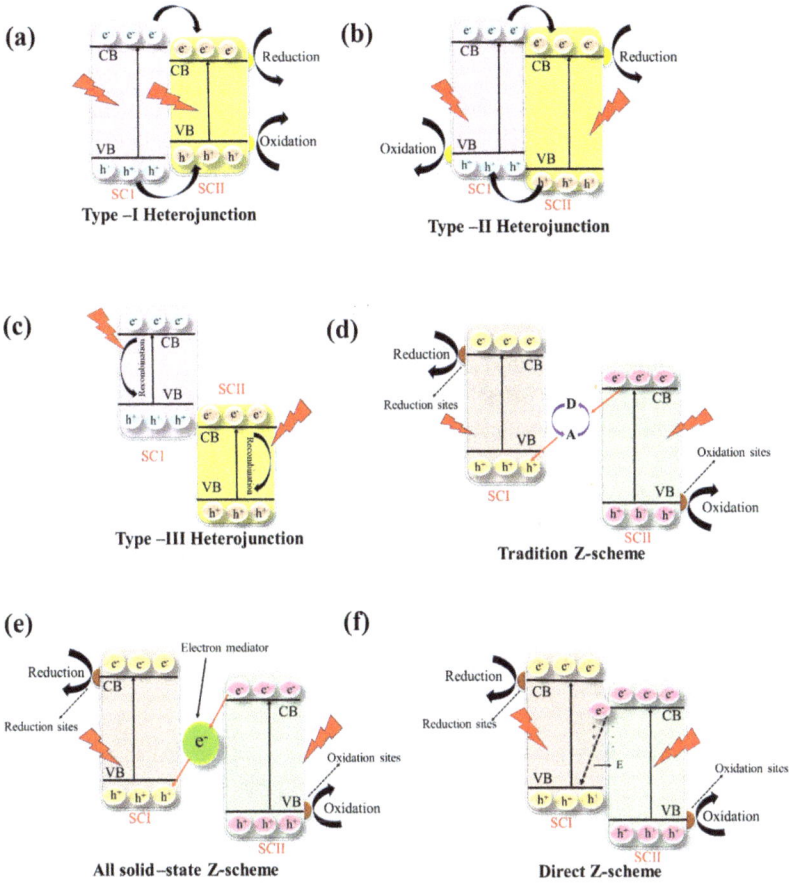

Figure 1: Schematic illustration of the (a) typical type-I heterojunction and (b) type-II (conventional heterojunction (c) type-III heterojunction (d) traditional Z-scheme (e) all solid-stat Z-scheme (f) direct Z-Scheme.

The electron migrates from CB of SCII to VB of SCI via acceptor-donor (A/D) pairs where, acceptor is reduced to donor by electron of SCII and donor is oxidized back to acceptor by photogenerated holes of SCI, respectively. Thus, redox reaction is carried out by SCI and SCII with higher reduction and oxidation potential. Such migration limits the rate of

Photocatalysis   Materials Research Forum LLC
Materials Research Foundations **100** (2021) 113-160   https://doi.org/10.21741/9781644901359-4

recombination of charge carrier and maintain the redox ability of semiconductor with high potential. However, their fabrication is limited to liquid phase which creates barrier on its wider application. The 2nd generation or all-solid-state Z-scheme was proposed in 2006 by Tada et al., where, the two photosystems are assembled with solid electron mediator such as Pt, Au and Ag [43]. All-solid-state Z-scheme can be applied to solid, liquid and gaseous phase however, high cost of electron mediator limits its usage. The 3rd generation or direct Z-scheme involves the combination of two photocatalyst without mediator was put forward in 2013 by Yu et al. which reduces the fabrication cost [44]. The band alignment in direct Z-scheme is similar to that of type-II heterojunction but differ in charge migration that is physically more favourable due to electrostatic attraction between electrons of SCII and holes of SCI whereas, in conventional heterojunction excessive accumulation of electron in the CB of SCII further cease the electron migration from CB of SCI due to electrostatic repulsion. Moreover, direct Z-scheme also maintain the redox ability of semiconductor with higher reduction and oxidation potential also the band potential can be reform easily according to different photocatalytic reactions. Further, trinary system displaying dual Z-scheme is another route offering immense charge separation with enhanced photo-efficiency and substantial amount of ROS generated on the surface. Since, great achievement in HMs removal have been attained by the various researchers therefore it is more significant to sum up recent achievement and challenges experienced in this filed.

3.   Nanohybrids for photocatalytic removal of heavy metals

3.1   Single system

The various conventional metal oxide based photocatalytic material such as TiO_2, ZnO, CuO, CdS, and V_2O_5 etc. were applied to remove inorganic and organic pollutant present in water. Among these TiO_2 and ZnO are largely explored for its photocatalytic activity due to low-cost, non-toxicity and abundance [45, 46]. Arsenate photo-catalytically reduced to arsenite with TiO_2 in the presence of hole scavenger so as to avoid recombination of photogenerated charge carrier [47]. Though, the direct reduction of arsenate with TiO_2 is thermodynamically not favourable because of the reduction potential value of As(V)/As(IV) is -1.2 V more negative than the CB potential -0.3 V of TiO_2. Therefore, indirect reduction using sacrificial electron donor methanol forming hydroxymethyl radical ($^\cdot CH_2OH$) a strong reducing agent possessing 1.4 V reduction potential is now capable of reducing arsenate. Consequently, the reduction process can be elaborated by following equations.

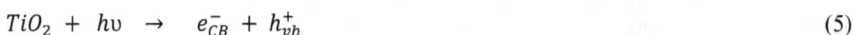

$$TiO_2 \ + \ h\upsilon \ \rightarrow \ e_{CB}^- \ + \ h_{vb}^+ \tag{5}$$

Materials Research Forum LLC
https://doi.org/10.21741/9781644901359-4

$$H_2O + h_{vb}^+ \rightarrow {}^\circ OH + H^+ \tag{6}$$

$$CH_3OH + {}^\circ OH \rightarrow {}^\circ CH_2OH + H_2O \tag{7}$$

$${}^\circ CH_2OH + As(V) \rightarrow CH_2O + As(IV) + H^+ \tag{8}$$

$$As(IV) + e_{CB}^- / e_{trap}^- ({}^\circ CH_2OH) \rightarrow As(III) + CH_2O + H^+ \tag{9}$$

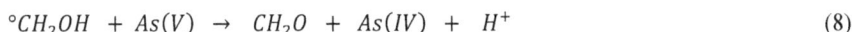

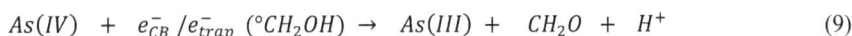

Nearly, complete elimination of Cu (II) were reported under UV light which is just comprise of 5% of solar spectrum. The Cr (II), Ni (II), Ag (I), Cd (II), Mn (II), Cu (II) and Pb (II) ions were eliminated in the presence UV or visible light using ZnO nanoparticle synthesized by solid precipitation method [46]. The removal efficiency in the presence of UV light is more than 85% for Cu (II), Pb (II) and Ag(I) metal ions. The wide band gap of conventional semiconductor restricts their practical implication.

Recently, innovative idea of converting toxic heavy metals into a photocatalyst employed for the photodegradation of methylene blue [48]. The steel slag derived calcium silicate hydrate with hierarchical structure and amorphous nature prepared by alkali activation method display outstanding adsorption capacity towards metals. On the surface of calcium silicate hydrate HMs transform into its hydroxide that is responsible for red shift in the absorption edge thus photocatalyst become visible light active. On performing scavenger experiment holes and hydroxyl radical appears to be main active species and the small degradation of organic pollutant were observed when EDTA or IPA is added (Fig 2a). The feasible photocatalytic mechanism of bare and metal coated calcium silicate hydrate is shown in Fig. 2b) indicating holes is responsible for the degradation.

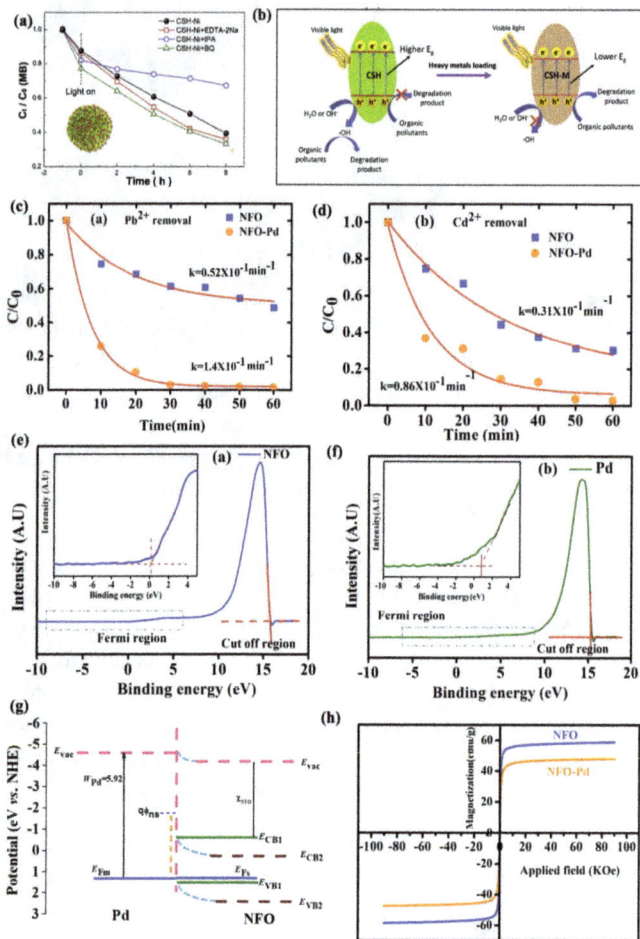

Figure 2. (a) Photocatalytic degradation of MB in the presence of various radical scavengers by CSH-Ni. (b) Possible photocatalytic mechanism of CSH and CSH-M (M = Cu, Ni, and Zn) under visible light. (c,d) Analysis of photocatalytic reduction of (c) Pb^{2+} (d) Cd^{2+} heavy metal impurity using NFO and NFO–Pd nanoparticles.(e,f) Ultraviolet photoemission spectroscopy (UPS) spectra of (e) NFO and (f) Pd nanoparticles. (g) The formation of Schottky barrier in the nanohybrid after contact between NFO semiconductor and Pd metal derived based on the UPS analysis. (h) Magnetic hysteresis plots suggesting superparamagnetic [48,49].

Photocatalysis   Materials Research Forum LLC
Materials Research Foundations **100** (2021) 113-160   https://doi.org/10.21741/9781644901359-4

The photocatalytic activity of bare photocatalysts can be enhanced by Fermi level modification. For instance, $NiFe_2O_4$ narrow band gap semiconductor with superior magnetic property synthesized by solvothermal method employed for Pb^{2+} and Cd^{2+} removal [49]. The rate constant of $Pd-NiFe_2O_4$ for Pb^{2+} and Cd^{2+} is 1.4×10^{-1} min^{-1} and 0.86×10^{-1} min^{-1}, respectively (Fig 2c and d). $NiFe_2O_4$ chemically bonded to Pd by surface modifier 3-aminopropyltriiethoxysilane undergo autocatalysis process due to the presence of amino groups. The band structure and alignment of $Pd-NiFe_2O_4$ were studied by ultraviolet photoemission spectroscopy Fig 2e and f. The work function calculated for $NiFe_2O_4$ and Pd using following equation is 5.28 eV and 5.92 eV, respectively.

$$W = h\upsilon - (E_{cut-off} - E_F) \tag{10}$$

From the Fermi edge region in UPS spectrum, the calculated ($E_{cut-off}$ - E_F) value for $NiFe_2O_4$ and Pd is 0.18 and 0.742 eV. The electron affinity of $NiFe_2O_4$ and Pd is 3.22 and 6.66 eV as calculated using below equation where, 0.187 and 0.742 eV were estimated value of (E_F - E_{VB}).

$$X = W + (E_F - E_{VB}) - E_g \tag{11}$$

Accordingly, schottky barrier establish between $NiFe_2O_4$ and Pd after contact is shown in Fig 2g. Further, magnetic hysteresis loop of $Pd-NiFe_2O_4$ suggest superparamagnetic nature of photocatalysts responsible for facile magnetic separation from aqueous suspension (Fig 2h). The bare semiconductor has wide band gap, undergo photo corrosion and agglomerate easily. Thus, the immobilization of nanoparticles onto certain stable, low-cost supportive material with large surface area and superior recycling efficiency is highly desirable to construct systematic nanohybrids. Though, photocatalysts can be immobilized over planar surface of glass and ceramics yet poor charge transfer restricts its wider application. Moreover, to enhance the photo-response various strategies of doping with metal, non-metal, noble metal and constructing heterojunction were put together. Among which heterojunction formation is highly in trend to increasing charge carrier separation.

3.2   Dual system

Keeping in view the various limitation of single photocatalytic system, various research groups have constructed dual system to promote photo-efficiency. Li and co-workers investigated TiO_2-graphene hydrogel with three-dimensional (3D) network structure which exhibited outstanding adsorption-photocatalysis activity and removed 100% Cr (VI) from wastewater within 30 min in the presence of UV light [50]. Nearly, 66.32 and 88.96% of Cd^{2+} and Pb^{2+} ions are removed using TiO_2/GO prepared by hummers, hydrothermal and freeze-drying method (Fig. 3a) [51]. From BET analysis 128.41 m^2g^{-1} surface area was calculated for TiO_2/GO as compared to bare TiO_2 which is just 59.51 m^2g^{-1}. The photocatalytic mechanism of e^-/h^+ pair's generation also reduction and oxidation process of heavy metals on CB and VB of TiO_2/GO surface are shown in (Fig. 3b). In the CB of nanohybrid O_2 is reduced to $°O_2^-$ anion and Cd^{2+} and Pb^{2+} reduced to Cd and Pb, respectively. On the same time holes in the VB produces $°OH$ radical which oxidizes Pb^{2+} to Pb^{4+} and precipitate out it in the form of PbO_2. Wang et al. prepared magnetically recyclable g-C_3N_4@Fe^0-rGO composite using vitamin C and KBH_4 as reducing agent (Fig. 3c) [52]. The mechanism of elimination of rhodamine-B dye and hexavalent chromium ion under visible light is shown in (Fig. 3d) and also explained by the equation 12 to 25. The higher surface area of composite provides large adsorption sites to capture e^- for elimination of heavy metal and dye. The quenching experiment reveals the major role of nano Fe^0 and $°OH$ for the removal of rhodamine-B and Cr (VI).

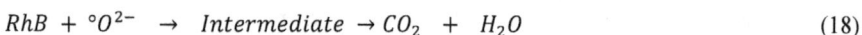

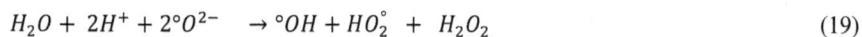

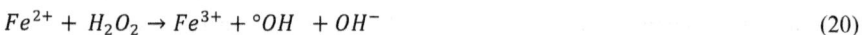

$$Fe^0 + Cr \rightarrow Fe^{2+} + Cr(III) \tag{12}$$

$$Fe + Cr(V) \rightarrow Fe^{3+} + Cr(III) \tag{13}$$

$$g-C_3N_4@Fe^0-rGO \rightarrow h^+ + e^- \tag{14}$$

$$Fe^{3+} + e_{CB}^- \rightarrow Fe^{2+} \tag{15}$$

$$Fe^{2+} + 2e_{CB}^- \rightarrow Fe^{2+} \tag{16}$$

$$O_2 + e^- \rightarrow °O^{2-} \tag{17}$$

$$RhB + °O^{2-} \rightarrow Intermediate \rightarrow CO_2 + H_2O \tag{18}$$

$$H_2O + 2H^+ + 2°O^{2-} \rightarrow °OH + HO_2° + H_2O_2 \tag{19}$$

$$Fe^{2+} + H_2O_2 \rightarrow Fe^{3+} + °OH + OH^- \tag{20}$$

$$H_2O_2 + e^- \rightarrow {}^{\circ}OH + OH^- \tag{22}$$

$$h^+ + H_2O \rightarrow H^+ + {}^{\circ}OH_{surf} \tag{22}$$

$$h^+ + OH^-_{surf} \rightarrow {}^{\circ}OH \tag{23}$$

$$h^+ + RhB \quad \rightarrow \quad Intermediate \rightarrow CO_2 + H_2O \tag{24}$$

$${}^{\circ}OH + RhB \quad \rightarrow \quad Intermediate \rightarrow CO_2 + H_2O \tag{25}$$

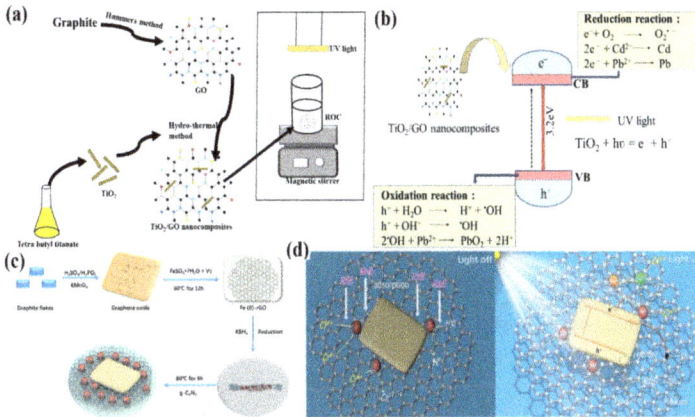

Figure 3. (a) The synthesis route of TiO₂/GO nanocomposites and photocatalytic experiments. (b) Schematic diagram of the charge transfer and separation in the TiO₂/GO nanocomposites under UV light irradiation and the main steps to reduce heavy metal ions. (c) Schematic illustration of synthesis of g-C₃N₄@Fe⁰ -rGO.(d) Proposed mechanism for the synergistic removal of RhB and Cr(VI) on g-C₃N₄@Fe⁰ -rGO photocatalyst [51,52].

Materials Research Forum LLC
https://doi.org/10.21741/9781644901359-4

The most acid stable CdS/TiO_2 heterojunction was reported capable of removing Cr (VI) in strong acidic conditions at pH=2.0–7.0 [53]. The detailed synthesis of CdS/TiO_2 is elaborated in (Fig 4a) where, HPC stands for hydroxypropyl cellulose, CAR-d represents CdS, anatase, rutile complex and d indicate ratio of Cd/Ti.

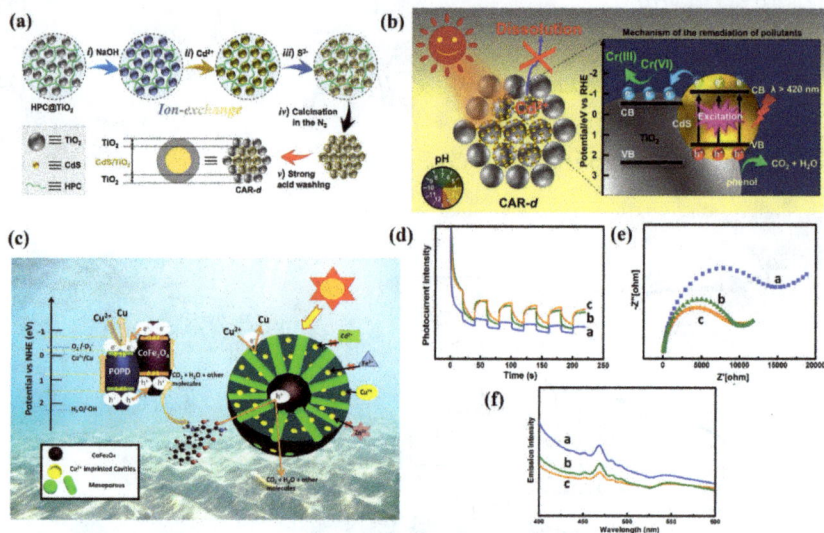

Figure 4. (a) Scheme of the synthetic process of the raspberry-like CAR-d using the ion-exchange strategy. (b) Mechanism of photocatalytic reduction of metal ions and oxidation of phenol under a strong acid condition. (c) Proposed selective reduction and photocatalytic degradation mechanism of the magnetic ion imprinted heterojunction photocatalyst. Transient photocurrent response (d), Nyquist plots of EIS (e) and PL (f) of $CoFe_2O_4$ [53,54].

Besides of expedite of charge separation, loading CdS to the bulk TiO_2 also prevents photochemical corrosion or dissolution of CdS nanoparticles (Fig 4b). The magnetically separable polyo-phenylenediamine/$CoFe_2O_4$ nanohybrids with good stability, active charge separation and recyclability were successfully prepared by microwave-assisted ion imprinted method [54]. The imprinted cavity of Cu^{2+} ion enhance the selective reduction of Cu^{2+} ion over Cd^{2+}, Fe^{2+} and Zn^{2+} on magnetic ion imprinted photocatalytic surface as displayed in (Fig 4c). According to the estimated position of band potential, migration pathway of electrons is from CB of $CoFe_2O_4$ to the CB of polyo-phenylenediamine and holes transfer from VB of polyo-phenylenediamine to VB of $CoFe_2O_4$. The separation of charge carrier further affirmed by better photocurrent response, minimum impedance in electrochemical impedance spectra, and lowest photoluminescence peak intensity for magnetic ion imprinted nanohybrids (Fig 4d, e and f). Gd_2O_3 modified CeO_2 nanoparticle creating p-n type heterojunction synthesized by solid state reaction eliminate Pb^{2+} ions by photoelectron-deposition method [55]. Similarly, 0.7 gL^{-1} CuO/ZnO eliminate 30 gL^{-1} of As (III) in 10 h irradiation of UV light at neutral pH [56].

In another work ZnO/Fe_2O_3 a type-II heterojunction reported in which photocatalytic mechanism involves the reduction of heavy metals at the CB of Fe_2O_3 (Fig 5a) [57]. The composites photo-reduces **88%** of Cr (VI) in 90 min whereas, 100 % reduction were observed in 45 min when tartaric acid is used a sacrificial agent (Fig 5b). Further, after reduction experiment the recovered sample were analyzed using X-ray photoelectron spectroscopy the spectra clearly indicate two peak of Cr (III) at binding energy 576.9 and 586.5 eV assigned to $2p_{3/2}$ and $2p_{1/2}$, respectively (Fig 5c). The various conventional nanohybrids functioning via type-II mechanism of charge migration bears limitation. Newly emerging Z-scheme heterojunction is highly in trend nowadays among scientist and continuously employed in various applications. Recently, Co_3O_4/g-C_3N_4 heterojunction prepared by growing nano-cubes of Co_3O_4 on 2D g-C_3N_4 nanosheet via hydrothermal method almost reduces 81% of Cr^{+6} into Cr^{+3} [58]. The valence band XPS spectra indicate the position of Fermi level of g-C_3N_4 and Co_3O_4 at 1.2 eV and 0.6 eV, respectively (Fig 5d and e). Additionally, Fermi level of g-C_3N_4 and Co_3O_4 is 5.0 and -.4 eV/AVS as explicit by CASTEP code (Fig 5f and g). Based on both the results VB edges of g-C_3N_4 and Co_3O_4 is -6.2 and -7.0 eV whereas, the calculated band gap is 2.8 and 2.1 eV, respectively. Apparently, the suggest photocatalytic mechanism for Co_3O_4/g-C_3N_4 heterojunction operating via Z-scheme is given in Fig 5h.

Constructing Z-scheme nanohybrids is an ideal way to develop highly efficient photocatalyst to treat water pollution. $Ag_xH_{3-x}PMo_{12}O_{40}$/Ag/g-$C_3N_4$ a 1D//2D Z-scheme nanomaterial prepared by self-assembly method is highly efficient for the reduction of Cr (VI) and photodegradation of methylene orange and tetracycline under visible light [59]. The distinctive band structure, appropriate band gap of 2.7 eV, superior chemical and thermal stability and eco-friendly nature of g-C_3N_4 make it suitable candidate for constructing Z-scheme heterojunction. Further, Ag loading between the two assemblies direct the electron to recombine from CB of $Ag_xH_{3-x}PMo_{12}O_{40}$ to VB of g-C_3N_4 thereby maintains the charge separation process more accurately. Similarly, BiOCl-Ag-AgBr running through Z-scheme mechanism where Ag displayed surface plasmon resonance effect producing negatively charged electrons and positively charged holes on its surface. On photo-excitation electron in the CB of BiOCl migrates to Ag and recombine with surface plasmon resonance induced holes. On the same time holes rom VB of AgBr recombine with surface plasmon resonance induced electron on Ag. Apart from this oxygen vacancy, local surface plasmon resonance and presence of iodine ions are also some off the effective strategy to enhance photo-efficiency [60]. The oxygen vacancy generated by loss of oxygen atom assist in generating more ROS by activating molecular oxygen and also acts as an electron capturing cites. Not only this, but it also creates local surface plasmon resonance that can inject electrons to promote charge transfer.

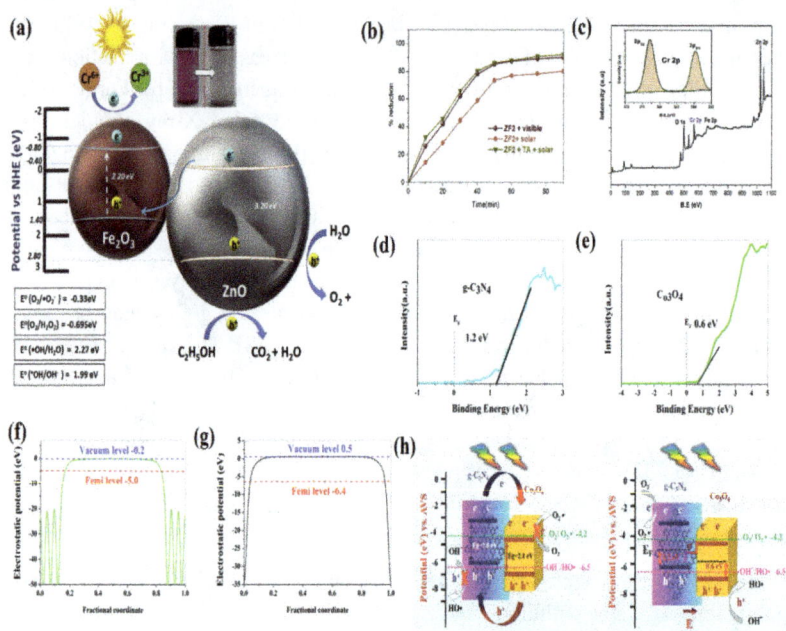

Figure 5. (a) ZF2 heterojunction showing movement of charge carriers and mechanism for Cr(VI) reduction. (b) The extent of photo-reduction of Cr(VI) under solar light with ZF2 (c) XPS survey scan and Cr 2p spectrum (Inset) of unused ZF2 before final recovery (ZF2 + visible + TA system). VB-XPS spectra of the (d) g-C₃N₄ and (e) Co₃O₄. The Fermi level EF is located at E = 0 eV, as marked by the vertical dotted line. The work function of (f) gC3N4, and (g) Co₃O₄. (h) Proposed type-II staggered band alignment of Co₃O₄/g-C₃N₄ and photocatalysis enhancement mechanism of the Z-scheme Co₃O₄/g-C₃N₄ [57,58].

3.3   Ternary system

By combining different photocatalysts with individual inherent advantages and to achieve much higher photo efficient, stability, and larger surface area trinary system were constructed. Hou and co-workers reported photocatalytic reduction of Cr (VI) in the presence of $Ag/TiO_2/rGO$ [61]. Ag loading on TiO_2 reduces the rate of recombination and also assist in the reduction of Cr (VI) to Cr (III) ion. On the same time superoxide radical anion and hydroxyl radical generated at the CB and VB of TiO_2 oxidizes tetracycline into CO_2 and H_2O. The pH of reaction solutions plays a vital role in adsorption and degradation as it not only affects the surface charge on photocatalysts but also affect the ionic state of organic pollutant. The zero-point charge of $Ag/TiO_2/rGO$ is in the range of $pH_{pzc} = 5\text{-}6$ therefore when pH < 5 photocatalysts surface is positively charged whereas pH > 6 surface becomes negatively charges. Moreover, adsorption of Cr (VI) increases with the decrease in pH as positively charged surface of photocatalyst can easily attract negatively charged $Cr_2O_7^{2-}$ ions via electrostatic interaction. Li and co-worker reported $Ag@TiON/CoAl\text{-}LDH$ prepared by dark-deposition method exhibit catalytic memory effect with increased performance [62]. The phenomena of retaining catalytic activity for a short period of time even when the light is turned off are called catalytic memory effect. Even certain noble metals such as Ag and Au also demonstrate electron storage properties. Also, broad range absorption, presence of surface hydroxyl group, tunable chemical compositions etc. are some of the advantages associated with 2D layered materials. Further, nitrogen doping in TiO_2 is done to generate oxygen vacancies and to extend visible light absorption. Keeping in view the above advantages $Ag@TiON/CoAl\text{-}LDH$ is constructed that reduces 61% of Cr (VI) and degraded 94% methyl orange. To check the catalytic memory effect, photodegradation experiment performed in dark where, excess of electrons entrapped by Ag under light illumination was discharged in dark to various electron acceptors to subsequently generate active free radicals thus maintaining round-the-clock photodegradation. Further, new strategy by combining homogeneous and heterogeneous a multiphase catalytic system was devised to simultaneously remove organic pollutant and heavy metals. In hierarchical 3D $Ag/ZnO@carbon$ foam porous photocatalysts where, Ag used as co-catalysts to smoothen charge separation process indicative of heterogeneous catalysis [63].

Besides, the formation of phenol-chromate (VI) ester intermediates via ligand to metal i.e. benzene ring to Cr (VI) charge transfer process represent homogeneous catalysis as shown in (Fig 6a). DFT calculation were performed to understand the intermolecular interaction between highest occupied molecular orbital (HOMO) and lowest unoccupied molecular orbital (LUMO) of ester intermediate formed. As shown in (Fig 6b and c) benzene ring and chromium ion contributes for HOMO and LUMO of phenol-chromate (VI) ester,

Photocatalysis   Materials Research Forum LLC
Materials Research Foundations **100** (2021) 113-160   https://doi.org/10.21741/9781644901359-4

respectively where the central Cr (VI) reduced to Cr (III) and phenol oxidizes to innoxious substances. Apart from electrons, superoxide radical also reduces Cr (VI) ion and the generation of OH° and °O_2^- were further confirmed by ESR technique (Fig 6d and e).

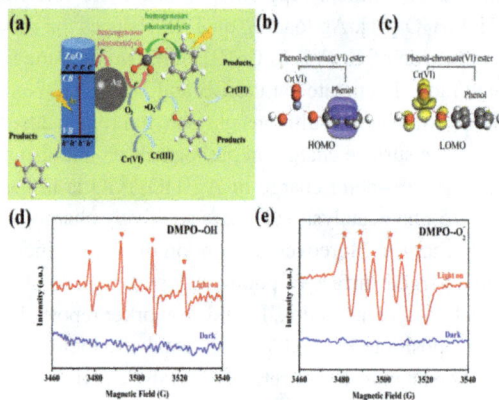

Figure 6. (a) Schematic diagrams for the possible mechanism of simultaneous heterogeneous and homogeneous catalysis for efficient removal of phenol and Cr (VI) over 2.0Ag/ZnO@CF under full arc light irradiation. The charge densities of the (b) HOMO, (c) LUMO of the phenol-chromate (VI) ester. ESR spectra of (d) DMPO-˙OH and (e) DMPO ˙O₂ adducts in aqueous and methanol solution [63].

TiO_2/g-C_3N_4/rGO ternary nanohybrids assembled by hydrothermal methods own large surface area, huge photoactive cites, low fluorescence intensity; enhance visible light absorption and photocurrent intensity [64]. The superior photo-catalytic activity of nanocomposites was witnessed reducing 90% of Cr (VI) to Cr (III) in 4h. TiO_2 nanoparticles evenly dispersed over g-C_3N_4 and rGO surface preventing its leaching and agglomeration (Fig 7a). MoS_2/ZnS/ZnO fabricated by hydrothermal method possesses 27 m^2g^{-1} of surface area and ample of active cites exhibit remarkable reduction and adsorption ability for Cr(VI) and rhodamine B [65]. The dual-II heterojunction degraded 98.7% of Cr(VI) in 90 min and scavenging experiment reveals h^+ and °O_2^- are the major active species for rhodamine B degradation. The photocatalytic mechanism involves the shifting of electron from CB of MoS_2 and ZnS to the CB of ZnO after irradiation this electron reduces Cr (VI) as well as assist for the generation of °O_2^- (Fig 7b). Another ternary $BiVO_4$/$FeVO_4$@rGO heterojunction creating nano-channels at interfacial unction was constructed with suitable band alignment for directional charge transfer (Fig 7c) [66]. The

2D rGO provides ample of active sites, large surface area and superb electrical conductivity whereas, $BiVO_4$ and $FeVO_4$ visible light driven narrow band gap semiconductor with excellent stability. Ternary heterojunction is employed for the oxidation of tetracycline and reduction of Cr (VI) and ESR measurement clearly depicts h^+ and $°O_2^-$ are the main active species for the photodegradation of tetracycline and electron for the reduction of Cr (VI). Movement of electron-hole pair completely depend upon band alignment, the flat potential of $BiVO_4$ and $FeVO_4$ calculated by Mott-Schottky diagram (Fig 7d and e). The CB potential of $BiVO_4$ and $FeVO_4$ are -0.57 and –0.64 eV relative to Ag/AgCl which is equivalent to -0.37 and -0.44 eV of NHE at pH 7. Using equation $E_{VB} = E_{CB} + Eg$, the VB potential comes out to be 2.09 and 1.53 eV for $BiVO_4$ and $FeVO_4$, respectively.

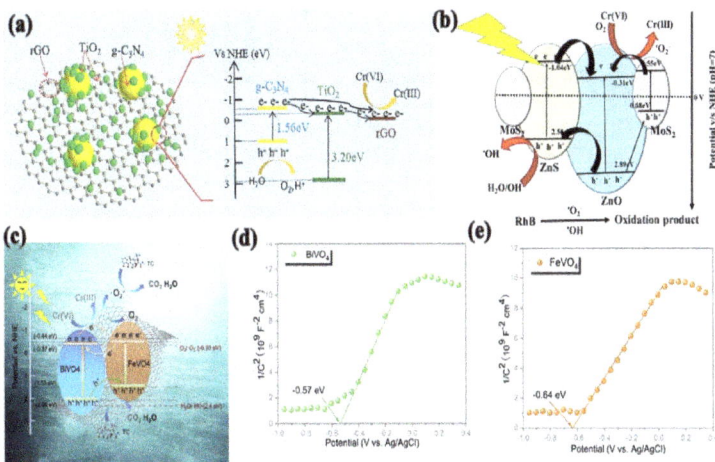

Figure 7. (a) Mechanism diagram of photocatalytic removal of Cr(VI) in T-CNS-G system.(b) Schematic illustration of photodegradation RhB and Cr (VI) over the MZ2 photocatalyst under Xe lamp irradiation. (c) Possible photocatalytic mechanism scheme of BVO/FVO@rGO under visible-light irradiation. Mott-Schottky plots of (d) BiVO₄ and (e) FeVO₄ [64,65,66].

$HPW_{12}/CN@Bi_2WO_6$ organic-inorganic nanocomposite synthesized via one-step hydrothermal method exhibit high photocatalytic reduction and oxidation capability towards Cr^{6+} and tetracycline under visible-light irradiation as compared to bare component [67]. The large surface area and pore size of $HPW_{12}/CN@Bi_2WO_6$ responsible for higher adsorption of Cr^{6+} and tetracycline. Nearly, 98.7% and 97.5% of metal ion and antibiotic were removed under 90 and 100 min of irradiation (Fig 8a and b). The degradation kinetic for of Cr^{6+} and tetracycline reveals pseudo first-order reactions (Fig. 8c and d). The possible photocatalytic mechanism where the electron-hole pair's generated on Bi_2WO_6 and HPW_{12} surface under visible light shown in (Fig. 8e). Since the CB edges of both Bi_2WO_6 and HPW_{12} is more positive than $O_2/°O_2^-$ (−0.33 V) it will not generate $°O_2^-$ anion. The conjugated organic CN is an excellent electron capturing materials and immediately transfer electrons from CB of Bi_2WO_6 to VB of HPW_{12} enhancing the charge separation process. On the same time electrons in the CB of HPW_{12} also migrates to CN which is thus participating in reduction process while holes in the VB of Bi_2WO_6 contributes for oxidation.

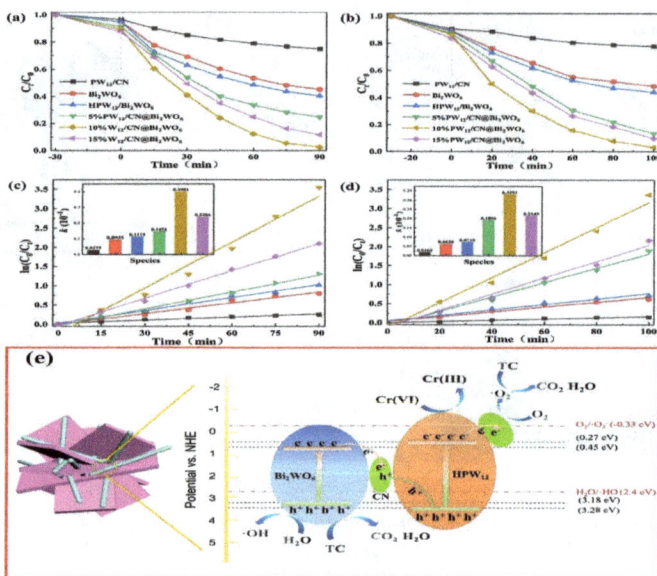

Figure 8. (a) Photodegradation performance of Cr (VI) (b) Tetracycline under visible light irradiation; (c) Kinetic linear fitting curves of Cr (VI) (d) TC and corresponding apparent rate constants (k) as insets; (e) Possible photocatalytic mechanism scheme of PW12/CN@Bi2WO6 under visible-light irradiation (Reproduced from Ref. 204, with permission from Separation and Purification Technology) [67].

Photocatalysis   Materials Research Forum LLC
Materials Research Foundations **100** (2021) 113-160   https://doi.org/10.21741/9781644901359-4

$BiOIO_3$/C500/MoS_2 all solid-state Z-scheme heterostructure synthesized by sol-gel and hydrothermal method where C500 provide channels for the electron in the CB of $BiOIO_3$ to recombine with the holes in the VB of MoS_2 [68]. The mechanism of charge migration on the surface of heterojunction is shown in (Fig 9a). The band bending of $BiOIO_3$ and MoS_2 at the interface is displayed in (Fig 9b). Firstly, electron diffuses from $BiOIO_3$ surface with larger work function of 5.8 eV towards MoS_2 having smaller work function of 5.2 eV, respectively. The electron diffusion creates surface charge layer generating internal electric field at the interface. The diffusion of electrons continues till the Fermi level equilibrates by shifting upward in $BiOIO_3$ and downward MoS_2 forming Z-scheme heterojunction after contact. This causes downward bending of CB and VB edges in $BiOIO_3$ and upward bending of CB and VB edges in MoS_2. (In Fig 9c), the CB of MoS_2 -0.29 eV is more negative as compared to $\dot{O}_2^-$ potential -0.28 eV whereas, VB potential of $BiOIO_3$ is 3.51 eV more positive than $\dot{O}H$ potential of 2.2 eV, respectively. Both $\dot{O}H$ and $\dot{O}_2^-$ generated in the VB of $BiOIO_3$ and CB MoS_2 oxidizes 78.32% of Hg^0 to HgO. The mechanism was further assisted by performing scavenger experiments and ESR techniques. In comparison to conventional type-II heterojunction, Z-scheme based heterojunction is more efficient in terms of accelerating the charge carrier separation and also maintaining the strong redox ability of individual photocatalysts [69].

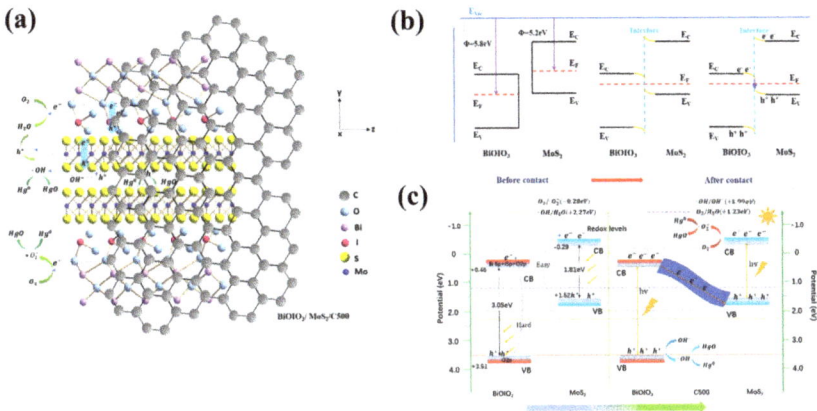

Figure 9. (a) A schematic diagram of $BiOIO_3$/MoS_2/C500; (b) The band bending diagram and electron transfer at interface of $BiOIO_3$ and MoS_2 (c) Photocatalytic mechanism diagram [68].

A multi-component layered double hydroxide based heterostructure $Ag@Ag_3PO_4/g$-$C_3N_4/NiFe$ were formed by in-situ photoreduction and self-assembly by electrostatic attraction [70]. The quasi-type-II heterojunction photo-catalytically reduced 97% hexavalent chromium in 2 h. The surface plasmon resonance effect of Ag nanoparticles and surface defects on layered material created by oxygen vacancy hastens the charge transfer. Further, the formation of Schottky barrier at the interface of quasi-type-II heterojunction was evident by much spectroscopic technique. Another multicomponent system synthesized by hydrothermal method in which carbon quantum dot is decorated over CeO_2/g-C_3N_4/V_2O_5 operating through dual Z-scheme mechanism (Fig 10a) [71]. The multicomponent system avails up-conversion phenomena of carbon quantum dots, broad absorption range, large charge separation, operates via two-channel charge transfer. Nearly 99% of Cr (VI) is reduced in 100 min without using sacrificial agent whereas complete reduction is accomplished in 30 min when 10^{-4} mol of tartaric acid as hole scavenger is used as shown in (Fig 10b and c). Tartaric acid retains all the electron free and oxidized itself by scavenging holes. The complete reduction of Cr (VI) into Cr (III) is confirmed by XPS showing the presence of trivalent chromium in the survey scan performed after reduction test (Fig 10 d and e).

On further simplification two peak of Cr (III) at 576.9 and 586.5 eV were obtained assigned to $2p_{3/2}$ and $2p_{1/2}$, respectively. The broad spectrum response of carbon quantum dot decorated CeO_2/g-C_3N_4/V_2O_5 is due to the red shift accounts for lowering of band gap. Further, two-channel charge transfer mechanism, oxygen vacancies on carbon quantum dots and induced defects of photocatalyst surface by carbon quantum dots arise and their up-conversion phenomena accounts for high photo efficiency. The surface defects entrap the charge carrier and minimizes the process of recombination. Through the bridging material migration of photogenerated electrons from the CB of g-C_3N_4 to the VB of CeO_2 and V_2O_5 becomes more facile. Such type of electron migration renders the electron with high reduction ability in CB of CeO_2 and V_2O_5 for reduction process (Fig 10f). Different bare, binary, trinary and multicomponent nanohybrids employed for photocatalytic reduction of HMs are summarized in Table 3 (72-85).

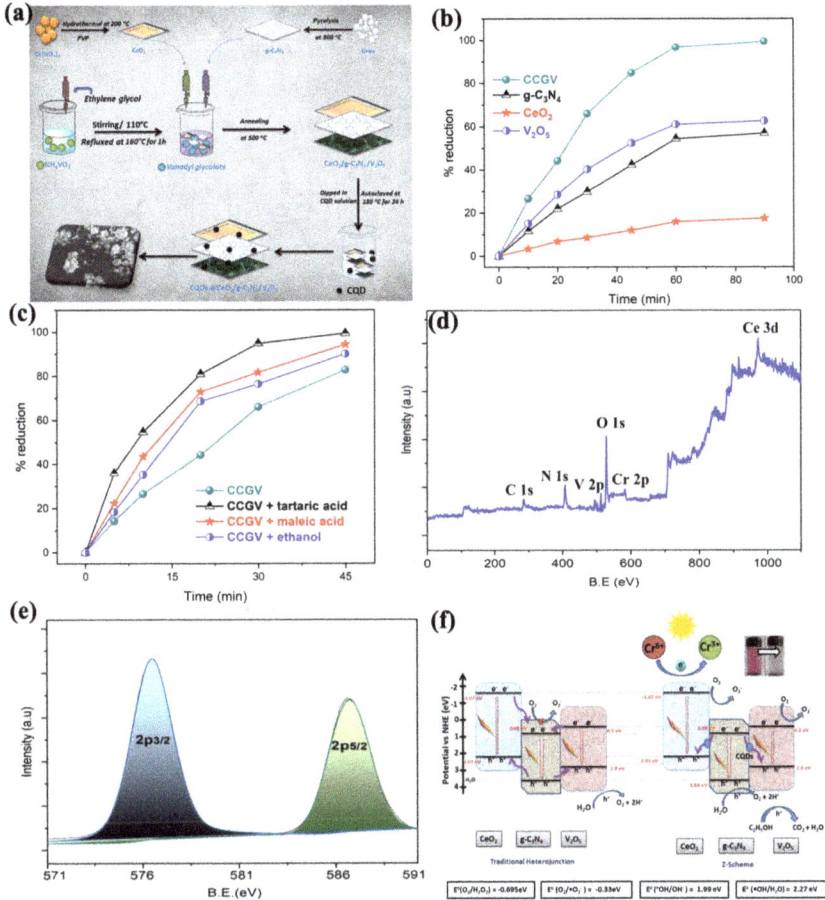

Figure 10. (a) Synthesis Scheme. (b) The extent of photo-reduction of Cr(VI) under visible light(c) Effect of reaction parameters onto Cr(VI) photo-reduction with CCGV (f) Initial co-catalysts or sacrificial agents; [catalyst] = 0.3 mg/mL [Cr⁶⁺] = 20 mg L-1, pH=2, temperature = 30 ± 0.5 °C, Visible light intensity 480 mW cm −2. XPS survey scan (d) Cr²⁺ spectrum (Inset) of unused CCGV before final recovery (CCGV + visible + TA system) (e) Reusability studies for Cr(VI) reduction with CCGV under visible light; (f)Traditional and Z-scheme heterojunction CCGV with possible charge flow [71].

Table 3. Summary of various photocatalytic nanohybrids for the removal of HMs

Nanohybrids	S_{BET} (m^{2}g^{-1})	Pore volume (cm^3 g^{-1})	Pore Diameter (nm)	Active species	Targeted pollutant	Photodegradation efficiency	Reusability	Ref
Cr-SrTiO$_3$	CrSTO=118.8 m^{2}g^{-1} STO=5.0 m^{2}g^{-1}	CrSTO = 0.24 m^{2}g^{-1} STO= 0.23 m^{2}g^{-1}	CrSTO = 3.0 nm STO= 2.5 nm	-	Cr (VI)	Photoreduction of of Cr (VI) (up to ~92%) under visible light	Up to ~6 photocatalytic cycle	[72]
B-Sn$_3$O$_4$	Sn$_3$O$_4$=31.35 m^{2}g^{-1} B$_{0.3}$-Sn$_3$O$_4$= 69.14m^{2}g^{-1},	Sn$_3$O$_4$= 0.126 cm^{3}g^{-1} B$_{0.3}$-Sn$_3$O$_4$ = 0.078cm^{3}g^{-1}	Sn$_3$O$_4$= 16.13nm B$_{0.3}$-Sn$_3$O$_4$ = 5.77nm	h$^+$, ·O$_2^-$	Cr (VI) Phenol	Photoreduction of of Cr (VI) (up to ~80%) under visible light	Up to ~5 photocatalytic cycle	[73]
CdS/TiO$_2$	TiO$_2$= 241.32 m^2 g^{-1} CAR$_{0.35}$=15.88 m^2 g^{-1}	TiO$_2$ = 0.18cm^3 g^{-1} CAR$_{0.35}$ = 0.12cm^3 g^{-1}	TiO$_2$ = 2.99nm CAR$_{0.35}$ = 41.07nm	-	Cr (VI) Phenol	73.1 % reduction of Cr(VI) and oxidation of phenol under visible light.	Up to ~5 photocatalytic cycle	[53]
Ag/Ag$_3$PO$_4$/MIL-125-NH$_2$	Ag/Ag$_3$PO$_4$ = 12.1 m^2/g. AAMN-120= 478.95 m^2/g, MIL-125-NH=1000 m^2/g	-	-	·O$_2^-$	Cr (VI)	-	Up to ~5 photocatalytic cycle	[74]
Fe-TNTs	TNTs=272.3 m^2 g^{-1} Fe-TNT=162.8 m^2 g^{-1}	TNTs =1.26cm^3 g^{-1} Fe-TNT = 0.38 cm^3 g^{-1}	TNTs= 18.5nm Fe-TNT = 9.3nm	-	As(III)	Photoreduction of of As(III) (up to ~99.6%) under visible light	Up to~ 5 photocatalytic cycle	[75]
Bi$_2$MoO$_6$/Ti$_3$C$_2$	Bi$_2$MoO$_6$/Ti$_3$C$_2$ =33.24 m^{2}g^{-1}	2 to 45 nm	-	·O$_2^-$ h$^+$	Cr(VI) Tetracycline hydrochloride	Photoreduction of of Cr(VI) (up to ~100 %) under visible light	Up to ~ 4 photocatalytic cycle	[76]
BiOBr/BiOIO$_3$	BiOBr=31.83 m^2 g^{-1}, BiOIO$_3$=22.26 m^2 g^{-1}, BI-10=21.22m^2 g^{-1}	BiOBr = 0.0675cm^3 g^{-1} BiOIO$_3$= 0.1498cm^3 g^{-1} BI-10= 0.1679cm^3 g^{-1}	BiOBr = 8.49nm BiOIO3 = 26.93nm BI-10=31.66 nm	·O$_2^-$ h$^+$	Hg^0	Photoreduction of of Hg^0 (up to ~ 90.25%) under visible light	Up to ~ 5 photocatalytic cycle	[77]

Cd/Co-based MOFs	-	-	-	$\cdot O_2^-$	Cr(VI)	Photoreduction of (up to Cr(VI) ~100%) under visible light under UV light	Up to ~ 5 photocatalytic cycle	[78]
Ag@TiON/Co Al	TiO_2= 0.228 (m²/g), TiON= 0.178(m²/g), , TiON/LDH= 0.182 (m²/g), Ag@TiON/Co Al-LDH= 0.132 (m2/g), ,	TiO_2=106. 486nm TiON=79.2 48nm TiON/LDH = 160.554nm Ag@TiON /CoAl-LDH=168. 832nm	TiO_2= 3.471 (m³) TiON= 9.293m³ TiON/LDH =6.253m³ Ag@TiON /CoAl-LDH= 5.860m³	-	Cr(VI) Methyl orange, Methyle ne blue	Photoreduction of (up to Cr(VI) ~94%) under visible light under UV light	Up to ~ 5 photocatalytic cycle	[79]
SnIn₄S₈-CdS	CdS=14.8060 m²/g SnIn₄S₈=9.522 6 m²/g SnIn₄S₈– CdS=34.6320 m²/g	0-10 nm	-	$\cdot O_2^-$ h⁺	Cr(VI), Methyl orange (MO)	Photoreduction of (up to Cr(VI) ~82% and 100% of MO under visible light	Up to ~5 photocatalytic cycle	[80]
TiO₂-2-naphthol	P25 =46.1 m²g⁻¹ TiO_2 = 133 m²g⁻¹ TiO2-2-NAP = 130 m²g⁻¹	-	TiO_2 =8.79nm TiO₂-2-NAP = 8.77 nm	-	Cr(VI)	Photoreduction of (up to Cr (VI) ~82%)	-	[81]
Bi₂S₃/β-Bi₂O 3/ZnIn₂S₄	-	-	15 and 25 nm	$\cdot O_2^-$	Cr (VI)	96.3% Tetracycline (TCN) degradation and Cr(VI) reduction	-	[82]
Functional amorphous phosphate titanium oxide	-	446 m² /g	3-7nm	-	Pb (II) Cd (II) Cr (III) Fe (III)	90% degradation of Pb (II), Cd (II), Cr (III) and Fe (III)	Up to 5 photocatalytic cycle	[83]
CoFe-LDH/ g-C₃N₄	g-C₃N₄= 15.16 CoFe-LDH = 67.08 50%-CoFeLDH/g-C3N4= 101.61	g-C₃N₄= 2.58 nm CoFe-LDH =0.61 50%-CoFeLDH/ g-C3N4= 0.36	g-C₃N₄= 4.12 CoFe-LDH =3.01 50%-CoFeLDH/ g-C3N4= 2.58	$\cdot O_2^-$	Cr (VI)	90% reduction of Cr(VI) under	Up to 5 photocatalytic cycle	[84]

P/SnS₂	SnS₂ = $35.38\ m^2\ g^{-1}$ RP= 36.79 $m^2\ g^{-1}$ 50%- RP/SnS2 = $40.96\ m^2\ g^{-1}$	SnS₂ = $3.05\ cm^3$ g^{-1} RP=3.74 $cm^3\ g^{-1}$ 50%- RP/SnS2 =3.76 cm^3 g^{-1}	SnS₂=0.20 nm RP=0.21n m 50%- RP/SnS₂= 0.23nm	$\dot{O}_2^{-}$	Cr(VI)	removal of Cr(VI) and RhB by 50%	Up to 5 photocat alytic cycle	[85]

4.   Nanohybrids for adsorptional removal of organic pollutants and heavy metals

Adsorption technique is widely used in industrial applications to decontaminate waste water due to its simple operation, high efficiency and low-cost. Physisorption or chemisorption are the two important phenomena through which adsorbent holds a molecule of adsorbate onto it. Adsorption process depends on different parameters such as adsorbent dosage, pH, temperature, contact time, concentration and electrostatic force between the adsorbate and adsorbent. This method is more flexible as adsorption site on the surface of adsorbent can be easily regenerated by reversible adsorption mechanism [86]. Metal, zeolite and carbon based nanosorbent are commonly applied adsorbent for water purification. For the adsorptional removal of HMIs different types of adsorbent like humic acid [87], activated carbons [88], bio-chars, sawdust [89], coal ash [90], resin [91], carbon nanotubes [92], graphene oxide [93], zeolites [94], bio-chars [95], sepiolites [96], palygorskite [97], mesoporous materials [98], and layered double hydroxides (LDHs) [99] were reported for heavy metal ion removal. A broad range of nano-adsorbents based on polyaniline (PANI) and its derivatives also appeared as promising candidate for the removal of heavy metals form polluted water [100].

Apart from bare photocatalytic system, adsorption capacity of simple metal oxide, inorganic bimetallic Mo-Fe-S clusters, graphene and graphene oxide were also reported for heavy metals. The highest adsorption capacity up to 81.2 mg/g at pH 5 for Pb^{+2} and lowest about 24.6 mg/g at pH 5 for Cr^{+3} ion was estimated by the iron oxide as adsorbent [101]. Whereas, in another report maximum adsorption of 319.35 mg g^{-1} for Pb^{2+} were reported by inorganic bimetallic Mo-Fe-S clusters prepared by chemical fixation of tetra thio-molybdate on Fe₃O₄ nanoparticles at room temperature [102]. Further, hierarchical nanoporous material synthesized using trimercaptos--triazine-trisodium salt and trimercaptos-triazine-monosodium salt display efficient adsorption capacity of 735.3 and 3775.9 mgg⁻¹for Hg^{2+} and Pb^{2+} ions, respectively [103]. The outstanding adsorption capacity and mechanical properties of graphene and graphene oxide imparts broader application in water purification [104,105]. About 98.46% and 99.99% is elimination efficiency acquired by graphene and graphene oxide towards Pb^{2+} ion. The various inorganic NPs were immobilized on certain organic support to avoid agglomeration, enhance surface area, improve recyclability and inhibit leaching of baser nanoparticles [106,107]. Numerous fibrous polymeric materials are also utilized as supportive systems for nanoparticles. The fabrication of polymeric semiconducting materials via polymeric matrices can be achieved by chemical, thermal and photo-induced reaction. The polystyrene based core/shell gel adsorbent display 526.3, 434.8, 555.6 and 476.2 mgg^{-1}, adsorption capacity for Pb^{2+}, Cu^{2+}, Cd^{2+} and Cr^{2+}, respectively [108].

Chitosan is a most abundant alkaline polymer of polysaccharides exhibits excellent hydrophilicity, biocompatibility, biodegradability and renewable capacity. Besides, presence of hydroxyl and amine group serves as adsorption site and make it more beneficial for adsorption of anionic pollutants and heavy metal ions. Guo et al. reported poly-dopamine modified chitosan aerogel and determine their adsorption capacity of 374.4 and 441.2 mg g^{-1}were observed towards Cr (VI) and Pb (II) ions [109]. The optimal pH for Cr (VI) and Pb (II) ions removal is 2.0 and 5.5, respectively and pH$_{ZPC}$ of nanocomposites is 7.5. At lower pH hydroxyl and amine group on nanocomposites get protonated thus possess positively charged surface.

PANI/rGO nanohybrid having surface area of 36.70 m^{2}g^{-1} adsorb nearly 1000 mg/g of Hg^{2+} ions at pH 4.0 obeying second order kinetic model [110]. The surface charge on PANI, rGO, and PANI/rGO is studied by zeta potential analysis (Fig 11a) indicating pH$_{ZPC}$ 5.8, 2.5, and 3.0, respectively. If the pH of reaction solution is below pH$_{ZPC}$ surface become positively charged whereas higher pH makes the surface negatively charged. Therefore, pH of the reaction is extremely important in deciding the optimum pH range for adsorption of pollutant. At low pH adsorption process is greatly reduced due to the electrostatic repulsion between Hg^{2+} ions and positively charged PANI/rGO although higher pH increases adsorption process hence 3-4.0 is the optimal pH range displaying maximum adsorption. Gusain and co-workers reported effective adsorption mechanism for eliminating Pb (II) and Cd (II) by MoS$_2$/SH-MWCNT nanocomposite synthesize by hydrothermal method [111]. Presence of sulphur cite on MoS$_2$ form metal−sulphur complex on the inner layer of MoS$_2$ evident by increased interlayer spacing. Nearly, 90.0 and .66.6 mgg^{-1} of Pb (II) and Cd (II) were adsorbed obeying pseudo second order kinetics. The electrostatic interaction, complex formation and ion-exchange process blend together for effective adsorptional removal of metals (Fig 11b).

New emerging class of porous organic materials, metal organic framework such as porphyrins were also exploited for heavy metal ion removal [112,113]. The mesoporous zirconium-based metal-organic framework displayed 277 mgg^{-1} adsorption capacity for Hg^{2+} ions at 7.1 pH where, pyrrolic functional group is an activity for the adsorption of metal ions [114]. In another work sulphonic acid modified metal organic framework demonstrated 88.7 mgg^{-1} adsorption capacity of cadmium ions [115] (Fig 11c). Heavy metal removal via adsorption mechanism commonly takes place through electrostatic adsorption, surface bonding, hydrogen bonding, ion exchange etc. The adsorption behaviour can be analyzing using advanced spectroscopic techniques, theoretical calculations and surface complexation models. Another metal organic framework-based nanocomposite possessing high surface area of 268.7 m^{2}g^{-1} adsorb 200.6 and 152.1 mgg^{-1} of Cu^{2+} and Cr^{6+}, respectively [116]. The XPS spectra displaying binding energy peak at 934 eV and 955 eV of 2p$_{3/2}$ and 2p$_{1/2}$ orbital in Cu^{2+} ion whereas, 579.2 and 588.1 eV binding energy represent Cr^{+6} (Fig 11d and e).

Thiosalicylhydrazide-modified magnetic nano-adsorbent employed for Pb^{2+}, Cd^{2+}, Cu^{2+} Zn^{2+}, and Co^{2+}, ions removal from industrial wastes with adsorption capacity 188.7, 107.5, 7.9, 51.3, and 27.7 mgg^{-1}, respectively [117]. Fe$_2$O$_3$ nanoparticle surface is modified with thiosalicylhydrazide ligating agent so as to minimize the involvement of alkaline earth metal on adsorption process. Moreover, immobilization of Fe$_2$O$_3$ nanoparticle onto poly-acrylic acid polymer shell reduces its aggregation and maintains dispersion. The schematic preparation of magnetic nano-adsorbent with possible adsorption mechanism of metal ion shown in (Fig 11f).

Cellulose-based amphoteric adsorbent owning amino and carboxyl functional group on the surface beneficial for the removal of Cr^{2+}, Cd^{2+}, Cu^{2+}, Zn^{2+} and Pb^{2+} ions [118]. The percentage removal and adsorption ability are decided by density of functional group which should be at least 10.0 mmolg^{-1} that prove absolute removal of metal ions and can be calculated using below mention equations.

$$Amino\ group\ contents = \frac{m\ X\ W_t}{m\ X\ M}\ X\ 1000 \tag{26}$$

$$Carboxyl\ group\ contents = \frac{m\ X\ W_t}{m\ X\ M}\ X\ 1000\ X\ W_t' \tag{27}$$

Where, m, M, Wt and W_t' represents sample weight used in EA (g), relative atomic mass (molg^{-1}), element proportion (wt%), and percentage of carboxyl group measured by XPS, respectively. Owing to the high density of amino group having great affinity for anionic Cr (VI) responsible for its complete removal as compared to cationic Cd^{2+}, Cu^{2+}, Zn^{2+} and Pb^{2+} ions. The electrostatic interactions, chelating reactions, and oxidation–reduction reactions explained the adsorption mechanism elaborated in (Fig 11g). Further, 3D hydrogel having great potential to undergo structure deformation arises due to swelling property by simply absorbing water that facilitate penetration of ions. EDTA cross-linked chitosan and polyacrylamide based hydrogel employed for Cd^{2+}, Cu^{2+} and Pb^{2+} ions with sorption capacity of 86.0, 99.4 and 138.4 mg g^{-1}, respectively [119].

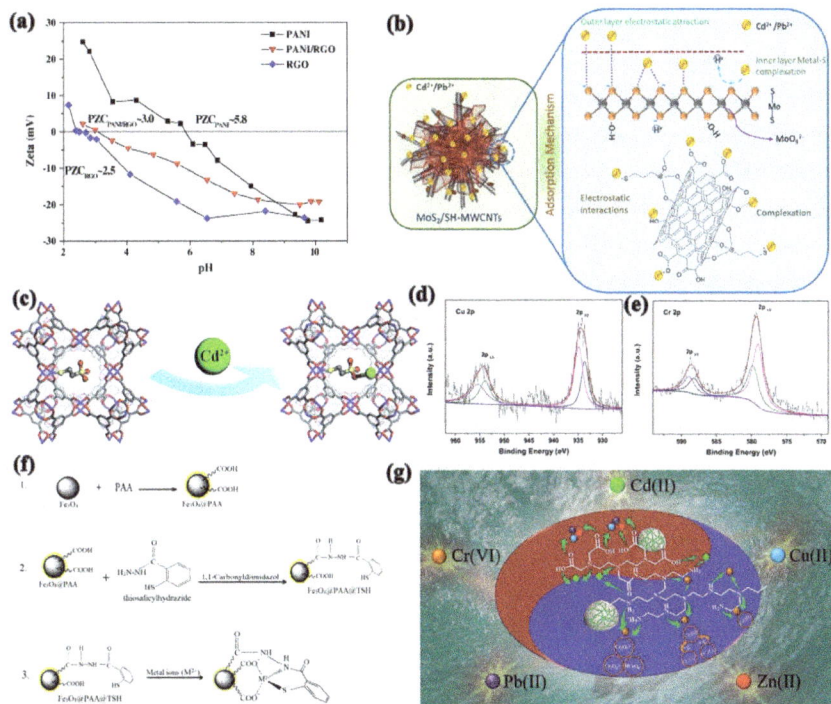

Figure 11(a)Zeta potential of PANI, RGO and PANI/RGO composites in a 0.02M NaNO₃ solution [110]; (b) Diagrammatic illustration of Pb(II) and Cd(II) adsorption mechanism on MoS₂/SH-MWCNTs [112]; (c) Proposed mechanism for Cd(II) adsorption in Cu₃(BTC)₂-SO₃H [115]; XPS spectra of (d) Cu²⁺ signal region (e)Cr²⁺ signal region [116]; (f) Scheme for the preparation of Fe₃O₄@PAA@TSH MNPs and a possible mechanism for adsorption of metal ions on them [117] (g) Specialization and cooperation mechanism of MCC/DTPA-PEIA for heavy metal ions [119].

V_2O_5-WO_3/TiO_2 oxidizes 65% of Hg^0 to Hg^{+2} at 250-350 °C in CO_2 enriched atmosphere [120]. Several parallel mechanisms occur for the oxidation of Hg^0 elaborated in successive

Materials Research Forum LLC
https://doi.org/10.21741/9781644901359-4

equations. Different multicomponent nanohybrids employed for adsorptional removal of HMs are summarized in Table 4 (121-140).
(1) Oxidation by V_2O_5

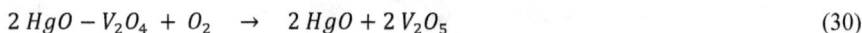

$$Hg_{(g)} \rightarrow Hg_{(ad)} \tag{28}$$

$$Hg_{(ad)} + V_2O_5 \rightarrow HgO - V_2O_4 \tag{29}$$

$$2\,HgO - V_2O_4 + O_2 \rightarrow 2\,HgO + 2\,V_2O_5 \tag{30}$$

(2) Oxidation by chemisorbed oxygen

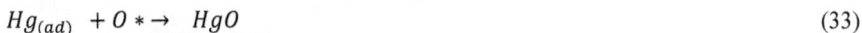

$$Hg_{(g)} \rightarrow Hg_{(ad)} \tag{31}$$

$$O_2 \rightarrow O * (\text{chemisorbed oxygen}) \tag{32}$$

$$Hg_{(ad)} + O * \rightarrow HgO \tag{33}$$

(3) Oxidation by CO_2

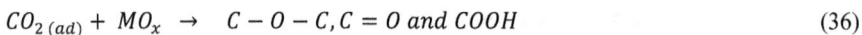

$$Hg_{(g)} \rightarrow Hg_{(ad)} \tag{34}$$

$$CO_{2\,(g)} \rightarrow CO_{2\,(ad)} \tag{35}$$

$$CO_{2\,(ad)} + MO_x \rightarrow C-O-C, C=O \text{ and } COOH \tag{36}$$

$$Hg_{(ad)} + 2\,C=O \rightarrow C-O-C + HgO \tag{37}$$

Table 4. Summary of different nanohybrids with distinct adsorption capacity

Adsorbent	Targeted pollutant	Concent ration	Dosages	pH	Surface area (m^2/g^{-1})	Pore size (nm)	Adsorption capacity (mg/g)	Ref.
EDTA/Nanoporous starch-based nanomaterial (EDTA/3D-PSN)	Cd (II) Hg (II) Pb (II) Cu (II)	2.00 mg/L	0.25g	2.4	83.75 m^2/g^{-1}.	-	Cd (II)= 532.28 mg/g Hg (II)= 381.47 mg/g Cu (II)=354.1 5 mg/g Pb (II)=238.3 9 mg/g	[121]
Sodium alginate grafted polyacrylamide / graphene oxide hydrogel (SA-PAM/GO)	Cu (II) Pb (II)	25 mg/L 50 mg/L	-	5	SA-PAM/G O= 190.89 dm^2/g 9.451 dm^3/g	SA-PAM/G O=6.794 9 nm to 19.8052 nm	68.76 mg/g 240.69 mg/g,	[122]
N-[4-morpholinecarboximidam idoyl] carboximidamidoylmethy latedpolyphenylenesulfide (MCMPPS)	Pb (II) Cr (III) Cu (II) Cd (II) Ni (II)	-	-	4.5 4.5 4.5 4.0 4.5	4.49m^2 g^{-1}.	-	Pb (II)= 186.92 Cd (II)= 189.75 Cr (III) = 120.04 Cu (II)= 119.33 Ni (II)= 134.05	[123]
Fe$_2$O$_3$@CaCO$_3$	As (V) Cr (VI) Pb (II)	1.0-30 ppm	0.5 mg/mL	2-12	53.4 $m^2.g^{-1}$	13.2 nm	As (V)=184.1 Cr (VI)= 251.6, Pb (II) = 1041.9	[124]
Inorganic polymer microspheres	Pb (II) Cu (II) Cd (II)	200 mg/L		5	87.74m^2 /g	8.9 nm	Pb =310.84 mg/g Cu =47.71 mg/g Cd = 36.26 mg/g	[125]

Gum ghatti – *graft* – poly(4-acryloylmorpholine) hydrogel (Ggh-*g*-PAcM hydrogel polymer)	Cu (II) Hg (II)	20-1000 mgL^{-1}		1.2	0.11 to 0.21 m^{2}g^{-1}	90.6, 103.6, 224.8, 213.8 mgg-1	Cu(II) =249.9 mg/g, Hg(II) =235.1 mg/g,	[126]
PANI/JF (jute fiber) composite	Cd (II) Cr (VI)	50 mg·L^{-1}	0.05 to 1.0 g	7	1.0 m^2 g^{-1}	140 and 50mg.g^{-1}	8.43 mg·g^{-1}	[127]
Ganguemicrosphere/geop olymer (GM/KGP)	Cu (II) Cd (II) Zn (II) Pb (VI)	100 mg/L	6 g/L	5	KGP=1 7.59m^2/ g GM=6.0 3m^2/g	KGP =45.18n m GM= 28.39nm	~30 mg/g.	[128]
Polysulfide/MgAl LDH	Hg (II) Pb (II) Ag (II) Zn (II)	10 mM	0.02 g	3-5	-	-	Hg (II) =686 Pb (II) =483, Ag (II) =383, Zn (II)=145m g/g	[129]
Amino functionalized silica (Ni (II), Cd (II) and Pb (II)	Ni (II), Cd (II) Pb (II)	50 mg/L	5 g/L	5	966 m^2/g	4.85 nm	Ni (II)= 12.36 Cd (II)=18.25 Pb (II)= 57.74 mg/g	[130]
Nano-hydroxyapatite/chitosan composite	Cd (II)	100-500 mg/L	5 g/L	5.6	-	-	122 mg/g	[131]
Fe$_3$O$_4$/PANI/MnO$_2$ Core-Shell Hybrids	Cd (II) Zn (II) Pb (II) Cu (II)	20 mg/L	1 g/L	6.2 - 6.4	63.96 m^2/g	4.12 nm	158 (428) mg/g	[132]
Fe$_3$O$_4$/ MnO$_2$	Cd (II)	10 mg/L	1 g/L (0.02 g)	6.2 - 6.4	118 m^2/g	3.3 nm	53.2 mg/g	[133]
Fe$_3$O$_4$@Mesoporous Silica-Graphene Oxide	Pb (II) Cu (II)	20 mg/L	100 mg/L	7.1	-	-	Pb (II)= 333 mg/g Cu (II)= 167 mg/g	[134]
Fe$_3$O$_4$ magnetic polypyrrole–graphene oxide Hg (II)	Hg (II)	20-100 mg/L	0.01-0.09 g/L	7	1737.6 m^2/g	4.2 nm	400 mg/g	[135]

Polyaniline/maghemite (PANI/γ-Fe$_2$O$_3$)	Cr (VI) Cu (II)	100 mg/L	2 mg	Cr (VI) =2. 0 Cu (II) =5. 5	60 m^2/g	5.1 nm	Cr (VI)=195. 7 mg/g Cu (II)= 106.8 mg/g	[136]
Fe$_3$O$_4$/xanthate GO	Hg (II)	20 mg/L	2-14 mg	7	30.13 m^2/g	-	118.55 mg/g	[137]
MnO$_2$/carbon nanotubes	Hg (II)	1-50 mg/L	0.02 g	6	110.38 m^2/g	2.70 nm	58.8 mg/g	[138]
Multiwalled carbon nanotubes	Hg (II)	0.1-10 mg/L	0.5 g	7	270 m^2/g	-	25.64 mg/g	[139]
Magnetic graphene oxide	Hg (II)	100 mg/L	0.5g/L	6	58.6 m^2/g	-	71.3 mg/g	[140]

Conclusions and future prospects

Water pollution is the major challenge faced by the humans of 21st century and the speedy growth in industrialization and population poses considerable threat on water resources. Due to the toxicity, recalcitrance persistence and large scale discharge of heavy metals into the environment, photocatalysis is a coherent approach. Though various adsorbent were also reported for their removal yet construction of multitasking nanohybrids that only reduces HMs but also work upon other toxic pollutant present in aqueous solution. The incomplete removal of HMs by implementing bare adsorption process is their major limitation due to which the treated water could left out with certain amount of toxic metal ion. Optimizing reaction condition to design adsorbing material retaining high density of functional group that accomplishes complete removal.

The use of photocatalytic system is cost effective technique providing large specific surface area, enhance photostability, less toxicity, less energy consumption and broad solar absorption range. Many heterojunction formations of semiconductors have been explored operating through conventional type-II charge migration however, recently Z-scheme type of migration of charge carrier is broadly explored by scientist. Z-scheme based photocatalyst is a potential candidate for different photocatalytic reactions as it retains the strong redox ability of individual semiconductor and reforming their band potential is also more facile. Fast rate of recombination of charge carrier, narrow absorption range and small surface area are some of the major limitation of photocatalysts that should be addressed while constructing a next generation photocatalysts. Thus, band gap engineering of semiconductors will remain open and challenging field for researchers to work upon in the future.

Rather than using individual photocatalytic reduction and adsorption method for HMs removal, constructing and implementing such nanohybrids that can simultaneously photo-catalytically reduce metal ion into less toxic ions followed by their complete adsorptional removal is most recommendable. Therefore, nanohybrid should be such that assure complete removal of HMs so that its concentration should met the standard of safe drinking level as per US-EPA protocols. And if somehow there is the chance of incomplete removal from the water system then at least the metal ion is photo-catalytically reduced to innocuous metal ion. Thusly, considering the economic benefit such photocatalytic system needs to come-up with simultaneous and efficient elimination of coexisting hazardous organic contaminants and heavy metals from the polluted water. The aim of integrating different photocatalysts is to synergistically avail the independent property of semiconductors and to construct ideal systems with enhanced photo-efficiency by regenerating the active cites.

References

[1] A. Azimi, A. Azari, M. Rezakazemi, M. Ansarpour, Removal of heavy metals from industrial wastewaters: a review, Chemical and Biochemical Engineering Review, 4 (2017) 37–59. https://doi.org/10.1002/cben.201600010

[2] S. Xu, Y. Lv, X. Zeng, D. Cao, ZIF-derived nitrogen-doped porous carbons as highly efficient adsorbents for removal of organic compounds from wastewater, Chemical Engineering Journal, 323 (2017) 502–511. https://doi.org/10.1016/j.cej.2017.04.093

[3] I. Díaz, Environmental uses of zeolites in Ethiopia, Catalysis Today, 285 (2017) 29–38. https://doi.org/10.1016/j.cattod.2017.01.045

[4] Y. Wang, G. Ye, H. Chen, X. Hu, Z. Niu, S. Ma, Functionalized Metal−Organic Framework as a New Platform for Efficient and Selective Removal of Cadmium (II) from Aqueous Solution, Journal of Materials Chemistry A, 3 (2015) 15292−15298. https://doi.org/10.1039/C5TA03201F

[5] X. Li, C. Bian, X. Meng, F.S. Xiao Design and Synthesis of an Efficient Nanoporous Adsorbent for Hg^{2+} and Pb^{2+} Ions in Water. Journal of Materials Chemistry A, 4 (2016) 5999−6005. https://doi.org/10.1039/C6TA00987E

[6] H. C. J. Godfray, J. R. Beddington, I. R. Crute, L. Haddad, D. Lawrence, J. F. Muir, Food Security: The Challenge of Feeding 9 Billion People, Science, 327 (2010) 812−818. https://doi.org/10.1126/science.1185383

[7] T. Theophanides, J. Anastassopoulou, Copper and carcinogenesis. Critical reviews in oncology/hematology, 42 (2002) 57–64. https://doi.org/10.1016/S1040-8428(02)00007-0

[8] S. Strachan, Trace Elements, Current Anaesthesia & Critical Care, 21 (2010) 44–48. https://doi.org/10.1016/j.cacc.2009.08.004

[9] J. Lemire, R. Mailloux, V.D. Appanna, Zinc Toxicity Alters Mitochondrial Metabolismand Leads to Decreased ATP Production in Hepatocytes, Journal of Applied Toxicology: An International Journal, 28 (2008) 175–182. https://doi.org/10.1002/jat.1263

[10] G.D. Leikauf, S.A. McDowell, S.C. Wesselkamper, W.D. Hardie, J.E. Leikauf, T.R. Korfhagen, D.R Prows. Acute Lung Injury: Functional Genomics and Genetic Susceptibility, Chest, 121 (2002) 70S– 75S. https://doi.org/10.1378/chest.121.3_suppl.70S

[11] M.R. Awual, M.M Hasan, Novel conjugate adsorbent for visual detection and removal of toxic lead (II) ions from water, Microporous and mesoporous materials, 196 (2014) 261-269. https://doi.org/10.1016/j.micromeso.2014.05.021

[12] M. Dinari, M. Hatami, Novel N-riched crystalline covalent organic framework as a highly porous adsorbent for effective cadmium removal, Journal of Environmental Chemical Engineering, 7 (2019) 102-907. https://doi.org/10.1016/j.jece.2019.102907

[13] B. Weiss, P.J. Landrigan, The Developing Brain and the Environment: An Introduction, Environmental health perspectives, 108 (2000) 373– 374. https://doi.org/10.1289/ehp.00108s3373

[14] W.H. Organization, Guidelines for drinking-water quality: recommendations, World Health Organization, 2004

[15] C. Griffiths, H. Klemick, M. Massey, C. Moore, S. Newbold and D. Simpson, et al., US Environmental Protection Agency valuation of surface water quality improvements, Review of Environmental Economics and Policy, 6 (2012) 130-146. https://doi.org/10.1093/reep/rer025

[16] J.P.F. D'Mello Food Safety: Contaminants and Toxins, Scottish Agricultural College, UK (2003) 206. https://doi.org/10.1079/9780851996073.0000

[17] M.R. Awual, M.M. Hasan, Novel conjugate adsorbent for visual detection and removal of toxic lead (II) ions from water, Microporous and mesoporous materials, 196 (2014) 261-269. https://doi.org/10.1016/j.micromeso.2014.05.021

[18] R.A.K. Rao, S. Ikram, Sorption studies of Cu (II) on gooseberry fruit (emblica officinalis) and its removal from electroplating wastewater. Desalination, 277 (2011) 390–398. https://doi.org/10.1016/j.desal.2011.04.065

Materials Research Forum LLC
https://doi.org/10.21741/9781644901359-4

[19] X. He, P. Li Surface water pollution in the middle Chinese Loess Plateau with special focus on hexavalent chromium (Cr $^{6+}$): occurrence, sources and health risks, Exposure and Health, (2020) 1-7. https://doi.org/10.1007/s12403-020-00344-x

[20] T. Joseph, B. Dubey, E.A. Mc Bean, Human health risk assessment from arsenic exposures in Bangladesh, Science of the Total Environment, 527 (2015) 552-560. https://doi.org/10.1016/j.scitotenv.2015.05.053

[21] A.Galadima, Z.N. Garba, L. Leke, M.N.Almustapha, I.K. Adam, Domestic water pollution among local communities in Nigeria-causes and consequences, European Journal of Scientific Research, 52(4) (2011) 592-603. https://doi.org/10.5539/jsd.v4n4p22

[22] M. Kobya, E. Demirbas, E. Senturk, M. Ince, Adsorption of heavy metal ions from aqueous solutions by activated carbon prepared from apricot stone, Bioresource technology, 96 (2005) 1518-1521. https://doi.org/10.1016/j.biortech.2004.12.005

[23] A. Oehmen, R. Viegas, S. Velizarov, M.A.M. Reis, J.G. Crespo. Removal of heavy metals from drinking water supplies through the ion exchange membrane bioreactor, Desalination, 199 (2006) 405-407. https://doi.org/10.1016/j.desal.2006.03.091

[24] E. Salazar, M.I. Ortiz, A.M. Urtiaga, Irabien JA. Kinetics of the separation–concentration of chromium (VI) with emulsion liquid membranes, Industrial & engineering chemistry research, 31 (1992)1523. https://doi.org/10.1021/ie00006a015

[25] F.M. Pang, P. Kumar, T.T. Teng, A. K. M Omar, K.L. Wasewar Removal of lead, zinc and iron by coagulation–flocculation. Journal of the Taiwan Institute of Chemical Engineers, 42 (2011) 809–815. https://doi.org/10.1016/j.jtice.2011.01.009

[26] W. Maatar, S. Boufi, Poly (methacylic acid-co-maleic acid) grafted nanofibrillated cellulose as a reusable novel heavy metal ions adsorbent, Carbohydrate Polymers, 126 (2015) 199–207. https://doi.org/10.1016/j.carbpol.2015.03.015

[27] S.K. Sahu, V.K. Verma, Bagchi D., V. Kumar, B.D. Pandey, Recovery of chromium (VI) from electroplating effluent by solvent extraction with tri-n-butyl phosphate. Indian J Chem Technol 126 (2008) 397.

[28] N. Meunier., P. Drogui, C. Montane, R. Hausler, G. Mercier, J.F. Blais, Comparison between electrocoagulation and chemical precipitation for metals removal from acidic soil leachate, Journal of hazardous materials, 137 (2006) 581-590. https://doi.org/10.1016/j.jhazmat.2006.02.050

[29] T.A. Kurniawan, G.Y. Chan, W.H. Lo, S. Babel Physico–chemical treatment techniques for wastewater laden with heavy metals, Chemical engineering journal, 118 (2006) 83-98. https://doi.org/10.1016/j.cej.2006.01.015

[30] A. Dabrowski, Z. Hubicki, P. Podkoscielny, E. Roben, Selective removal of the heavy metal ions from waters and industrial wastewaters by ion-exchange method, Chemosphere 56, 91 (2004) 106. https://doi.org/10.1016/j.chemosphere.2004.03.006

[31] J. Gao, S.P. Sun, W.P. Zhu, T.S. Chung, Chelating polymer modified P84 nanofiltration (NF) hollow fiber membranes for highly efficient heavy metal removal, Water Research, 63 (2014) 252-261. https://doi.org/10.1016/j.watres.2014.06.006

[32] D. Kong, B. Liang, H. Yun, H. Cheng, J. Ma, et al. Cathodic degradation of antibiotics: characterization and pathway analysis, Water Research, 72 (2015)281–292. https://doi.org/10.1016/j.watres.2015.01.025

[33] M. Dakiky, M. Khamis, A. Manassra, M. Mereb. Selective adsorption of chromium (VI) in industrial wastewater using low-cost abundantly available adsorbents, Advances in environmental research, 6 (2002) 533–540. https://doi.org/10.1016/S1093-0191(01)00079-X

[34] M. Brienza, I.A. Katsoyiannis. Sulfate radical technologies as tertiary treatment for the removal of emerging contaminants from wastewater, Sustainability, 9(9) (2017) 1604. https://doi.org/10.3390/su9091604

[35] K. Villa, X. Domènech, U.M. García-Pérez, J. Peral Photocatalytic Hydrogen Production Under Visible Light by Using a CdS/WO$_3$ Composite, Catalysis Letters,146 (2016) 100−108. https://doi.org/10.1007/s10562-015-1612-6

[36] L. Soler, V. Magdanz, V.M. Fomin, S. Sanchez, O.G. Schmidt. Self-Propelled Micromotors for Cleaning Polluted Water, ACS Nano, 7 (2013) 9611−9620. https://doi.org/10.1021/nn405075d

[37] Liu C, Wu T, Hsu PC, Xie J, Zhao J, Liu K, et al. Direct/Alternating Current Electrochemical Method for Removing and Recovering Heavy Metal from Water Using Graphene Oxide Electrode, ACS nano, 2019. https://doi.org/10.1021/acsnano.8b09301

[38] He HP, Wu DL, Zhao LH. Luo C. Sequestration of chelated copper by structural Fe (II): reductive decomplexation and transformation of Cu (II)-EDTA, Journal of hazardous materials, 309 (2016) 116–125. https://doi.org/10.1016/j.jhazmat.2016.02.009

[39] A. Fujishima, K. Honda, Electrochemical Photolysis of Water at a Semiconductor Electrode, Nature, 238 (1972) 37-38. https://doi.org/10.1038/238037a0

[40] J. Efome, D. Rana, T. Matsuura, C. Lan, Metal organic frameworks supported on nanofibers to remove heavy metals, Journal of Materials Chemistry A, 22(2018) 24577-24583.

[41] J. Low, C. Jiang, B. Cheng, S. Wageh, A.A. Al-Ghamdi, J. Yu, A review of direct Z-scheme photocatalysts, Small Methods, 5 (2017) 1700080. https://doi.org/10.1002/smtd.201700080

[42] A.J. Bard, Photoelectrochemistry and heterogeneous photo-catalysis at semiconductors, Journal of Photochemistry, 10(1) (1979) 59-75. https://doi.org/10.1016/0047-2670(79)80037-4

[43] H. Tada, T.Mitsui, T. Kiyonaga, T. Akita, K. Tanaka, All-solid-state Z-scheme in CdS–Au–$_{TiO2}$ three-component nanojunction system, Nature materials, 5(10) (2006) 782-6. https://doi.org/10.1038/nmat1734

[44] J. Yu, S. Wang, J. Low, W. Xiao, Enhanced photocatalytic performance of direct Z-scheme gC 3 N 4–TiO 2 photocatalysts for the decomposition of formaldehyde in air, Physical Chemistry Chemical Physics, 15(39) (2013) 16883-90. https://doi.org/10.1039/c3cp53131g

[45] M.A. Barakat, Y.T. Chen, C.P. Huang, Removal of toxic cyanide and Cu (II) Ions from water by illuminated TiO_2 catalyst, Applied Catalysis B: Environmental, 53 (2004) 13-20. https://doi.org/10.1016/j.apcatb.2004.05.003

[46] A.T. Le, S.Y. Pung, S. Sreekantan, A. Matsuda, Mechanisms of removal of heavy metal ions by ZnO particles. Heliyon, 5 (2019) 01440. https://doi.org/10.1016/j.heliyon.2019.e01440

[47] A. Samad, S. Ahsan, I. Tateishi, M. Furukawa, H. Katsumata, T. Suzuki, S. Kaneco, Indirect photocatalytic reduction of arsenate to arsenite in aqueous solution with TiO_2 in the presence of hole scavengers, Chinese Journal of Chemical Engineering, 26(3) (2018) 529-33. https://doi.org/10.1016/j.cjche.2017.05.019

[48] N. Shao, S. Li, F. Yan, Y. Su, F. Liu, Z. Zhang, An all-in-one strategy for the adsorption of heavy metal ions and photodegradation of organic pollutants using steel slag-derived calcium silicate hydrate. Journal of hazardous materials, 382 (2020) 121120. https://doi.org/10.1016/j.jhazmat.2019.121120

[49] B. Thomas, L.K. Alexander Removal of Pb^{2+} and Cd^{2+} toxic heavy metal ions driven by Fermi level modification in $NiFe_2O_4$–Pd nano hybrids, Journal of Solid-State Chemistry, (2020) 121417. https://doi.org/10.1016/j.jssc.2020.121417

[50] Y. Li, W. Cui, L. Liu, R. Zong, W. Yao, Removal of Cr (VI) by 3D TiO_2-graphene hydrogel via adsorption enriched with photocatalytic reduction, Applied Catalysis B: Environmental, 199 (2016) 412-423. https://doi.org/10.1016/j.apcatb.2016.06.053

[51] H. Zhang, X. Wang, N. Li, J. Xia, Q. Meng, Synthesis and characterization of TiO_2/graphene oxide nanocomposites for photoreduction of heavy metal ions in

reverse osmosis concentrate, RSC advances, 8 (2018) 34241-34251.
https://doi.org/10.1039/C8RA06681G

[52] X. Wang, M. Lu, J. Ma, P. Ning, Preparation of air-stable magnetic g-C_3N_4@ Fe^0-graphene composite by new reduction method for simultaneous and synergistic conversion of organic dyes and heavy metal ions in aqueous solution, Separation and Purification Technology, 1 (2019) 586-96.
https://doi.org/10.1016/j.seppur.2018.11.052

[53] Y. Deng, Y. Xiao, Y. Zhou, T. Zeng, M. Xing, J. Zhang, A structural engineering-inspired CdS based composite for photocatalytic remediation of organic pollutant and hexavalent chromium, Catalysis Today, 335 (2019) 101-109.
https://doi.org/10.1016/j.cattod.2018.09.012

[54] F. He, Z. Lu, M. Song, X. Liu, H. Tang, P. Huo, W. Fan, H. Dong, X. Wu, S. Han, Selective reduction of Cu2+ with simultaneous degradation of tetracycline by the dual channels ion imprinted POPD-$CoFe_2O_4$ heterojunction photocatalyst, Chemical Engineering Journal, 369 (2019) 750-61. https://doi.org/10.1016/j.cej.2018.12.034

[55] J. Ayawanna, W. Teoh, S. Niratisairak, K. Sato, Gadolinia-modified ceria photocatalyst for removal of lead (II) ions from aqueous solutions, Materials Science in Semiconductor Processing, 40 (2015) 136-9.
https://doi.org/10.1016/j.mssp.2015.06.058

[56] A. Samad, M. Furukawa, H. Katsumata, T. Suzuki, S. Kaneco, Photocatalytic oxidation and simultaneous removal of arsenite with CuO/ZnO photocatalyst, Journal of Photochemistry and Photobiology A: Chemistry, 325 (2016) 97-103.
https://doi.org/10.1016/j.jphotochem.2016.03.035

[57] P. Dhiman, S. Sharma, A. Kumar, M. Shekh, G. Sharma, M. Naushad, Rapid visible and solar photocatalytic Cr (VI) reduction and electrochemical sensing of dopamine using solution combustion synthesized ZnO–Fe_2O_3 nano heterojunctions: mechanism Elucidation, Ceramics International, 2020.
https://doi.org/10.1016/j.ceramint.2020.01.275

[58] W. Zhao, J. Li, T. She, S. Ma, Z. Cheng, G. Wang, P. Zhao, W. Wei, D. Xia, D.Y. Leung, Study on the Photocatalysis Mechanism of the Z-Scheme Cobalt Oxide Nanocubes/Carbon Nitride Nanosheets Heterojunction Photocatalyst with High Photocatalytic Performances, Journal of Hazardous Materials, 402 (2020) 123839.
https://doi.org/10.1016/j.jhazmat.2020.123839

[59] X. Zhao, Y. Zhang, Y. Zhao, H. Tan, Z. Zhao, H. Shi, E. Wang, Y. Li, $Ag_x H_{3-x}$ $PMo_{12} O_{40}$/Ag nanorods/gC_3N_4 1D/2D Z-scheme heterojunction for highly efficient

visible-light photocatalysis, Dalton Transactions, 48(19) (2019) 6484-91.
https://doi.org/10.1039/C9DT00744J

[60] T. Jia, J. Wu, J. Song, Q. Liu, J. Wang, Y. Qi, P. He, X. Qi, L. Yang, P. Zhao, In situ
self-growing 3D hierarchical BiOBr/BiOIO$_3$ Z-scheme heterojunction with rich
oxygen vacancies and iodine ions as carriers transfer dual-channels for enhanced
photocatalytic activity, Chemical Engineering Journal, (2020) 125258.
https://doi.org/10.1016/j.cej.2020.125258

[61] Y. Hou, S. Pu, Q. Shi, S. Mandal, H. Ma, S. Xue, G. Cai, Y. Bai, Ultrasonic
impregnation assisted in-situ photoreduction deposition synthesis of Ag/TiO$_2$/rGO
ternary composites with synergistic enhanced photocatalytic activity, Journal of the
Taiwan Institute of Chemical Engineers, 104 (2019) 139-150.
https://doi.org/10.1016/j.jtice.2019.08.023

[62] C. Li, G. Zhao, T. Zhang, T. Yan, C. Zhang, A novel Ag@ TiON/CoAl-layered
double hydroxide photocatalyst with enhanced catalytic memory activity for removal
of organic pollutants and Cr (VI), Applied Surface Science, 504 (2020) 144352.
https://doi.org/10.1016/j.apsusc.2019.144352

[63] H. Liang, T. Li, J. Zhang, D. Zhou, C. Hu, X. An, R. Liu, Liu H., 3-D hierarchical
Ag/ZnO@ CF for synergistically removing phenol and Cr (VI): Heterogeneous vs.
homogeneous photocatalysis, Journal of colloid and interface science, 558 (2020) 85-
94. https://doi.org/10.1016/j.jcis.2019.09.105

[64] G. Li, Y. Wu, M. Zhang, B. Chu, W. Huang, Enhanced Removal of Toxic Cr (VI) in
Wastewater by Synthetic TiO$_2$/g-C$_3$N$_4$ Microspheres/rGO Photocatalyst under
Irradiation of Visible Light, Industrial & Engineering Chemistry Research, 58 (2019)
8979-8989. https://doi.org/10.1021/acs.iecr.8b05990

[65] D. Zhao, T. Wu, Y. Zhou, Dual II heterojunctions metallic phase MoS$_2$/ZnS/ZnO
ternary composite with superior photocatalytic performance for removing
contaminants, Chemistry–A European Journal, 41 (2019) 9710-20.
https://doi.org/10.1002/chem.201901715

[66] R. Yang, Z. Zhu, C. Hu, S. Zhong, L. Zhang, B. Liu, W. Wang, One-step preparation
(3D/2D/2D) BiVO$_4$/FeVO$_4$@rGO heterojunction composite photocatalyst for the
removal of tetracycline and hexavalent chromium ions in water, Chemical Engineering
Journal, 390 (2020) 124522. https://doi.org/10.1016/j.cej.2020.124522

[67] R. Yang, S. Zhong, L. Zhang, Liu B., PW$_{12}$/CN@ Bi$_2$WO$_6$ composite photocatalyst
prepared based on organic-inorganic hybrid system for removing pollutants in water,
Separation and Purification Technology, 235 (2020) 116270.
https://doi.org/10.1016/j.seppur.2019.116270

[68] T. Jia, K. Xu, J. Wu, Q. Liu, Y. Lin, M. Gu, F. Tian, W. Pan, J. Wu, Y. Xiao, Constructing 2D BiOIO$_3$/MoS$_2$ Z-scheme heterojunction wrapped by C500 as charge carriers transfer channel: enhanced photocatalytic activity on gas-phase heavy metal oxidation, Journal of Colloid and Interface Science (2019). https://doi.org/10.1016/j.jcis.2019.11.100

[69] F. Chen, C. Yu, L. Wei, Q. Fan, F. Ma, J. Zeng, J. Yi, K. Yang, H. Ji, Fabrication and characterization of ZnTiO$_3$/Zn$_2$Ti$_3$O$_8$/ZnO ternary photocatalyst for synergetic removal of aqueous organic pollutants and Cr (VI) ions, Science of The Total Environment, 706 (2020) 136026. https://doi.org/10.1016/j.scitotenv.2019.136026

[70] S. Nayak, K.M. Parida, Dynamics of charge-transfer behavior in a plasmon-induced quasi-type-II p–n/n–n dual heterojunction in Ag@Ag$_3$PO$_4$/g-C$_3$N$_4$/NiFe LDH nanocomposites for photocatalytic Cr (VI) reduction and phenol oxidation, ACS omega, 3 (2018) 7324-7343. https://doi.org/10.1021/acsomega.8b00847

[71] A. Kumar, S.K. Sharma, G. Sharma, M. Naushad, F.J. Stadler, CeO$_2$/g-C$_3$N$_4$/V$_2$O$_5$ ternary nano hetero-structures decorated with CQDs for enhanced photo-reduction capabilities under different light sources: Dual Z-scheme mechanism, Journal of Alloys and Compounds, 2020 155692. https://doi.org/10.1016/j.jallcom.2020.155692

[72] D. Yang, X. Zhao, X. Zou, Z. Z. Zhou, Jiang. Removing Cr (VI) in water via visible-light photocatalytic reduction over Cr-doped SrTiO$_3$ nanoplates. Chemosphere 215 (2019) 586-595. https://doi.org/10.1016/j.chemosphere.2018.10.068

[73] C. Yu, D. Zeng, O. Fan, K. Yang, J. Zeng, L. Wei, J. Yi, H. Ji, The distinct role of boron doping in Sn$_3$O$_4$ microspheres for synergistic removal of phenols and Cr (vi) in simulated wastewater. Environmental Science: Nano, 1 (2020) 286-303. https://doi.org/10.1039/C9EN00899C

[74] Y.C. Zhou, P. Wang, H. Fu, C. Zhao, C.C. Wang., Ternary Ag/Ag$_3$PO$_4$/MIL-125-NH2 Z-scheme heterojunction for boosted photocatalytic Cr (VI) cleanup under visible light. Chinese Chemical Letters. (2020)25. https://doi.org/10.1016/j.cclet.2020.02.048

[75] Liu W, Zhao X, Borthwick AG, Wang Y, Ni J. Dual-enhanced photocatalytic activity of Fe-deposited titanate nanotubes used for simultaneous removal of As (III) and As (V). ACS Appl. Mater Interfaces 2015; 7: 19726-19735. https://doi.org/10.1021/acsami.5b05263

[76] Zhao, D. and Cai, C., 2020. Preparation of Bi2MoO6/Ti3C2 MXene Heterojunction Photocatalyst for Fast Tetracycline Degradation and Cr (VI) Reduction. Inorganic Chemistry Frontiers. https://doi.org/10.1039/D0QI00540A

[77] Jia, T., Wu, J., Song, J., Liu, Q., Wang, J., Qi, Y., He, P., Qi, X., Yang, L. and Zhao, P., 2020. In situ self-growing 3D hierarchical BiOBr/BiOIO3 Z-scheme heterojunction

with rich oxygen vacancies and iodine ions as carriers transfer dual-channels for enhanced photocatalytic activity. Chemical Engineering Journal, p.125258. https://doi.org/10.1016/j.cej.2020.125258

[78] Yi XH, Wang FX, Du XD, Fu H, Wang CC. Highly efficient photocatalytic Cr (VI) reduction and organic pollutants degradation of two new bifunctional 2D Cd/Co-based MOFs. Polyhedron 2018; 152: 216-24. https://doi.org/10.1016/j.poly.2018.06.041

[79] Li C, Zhao G, Zhang T, Yan T, Zhang C et al. A novel Ag@TiON/CoAl-layered double hydroxide photocatalyst with enhanced catalytic memory activity for removal of organic pollutants and Cr (VI). Appl Surf Sci 2020; 504: 144352. https://doi.org/10.1016/j.apsusc.2019.144352

[80] Zhang, B. Zhang, Y. Jiang, Y. Xiao, W. Zhang, H. Xu, X. Yang, Z. Liu, Zhang J. In-situ constructing of one-dimensional SnIn4S8-CdS core-shell heterostructure as a direct Z-scheme photocatalyst with enhanced photocatalytic oxidation and reduction capabilities, App Surf Sci 2021; 148618. https://doi.org/10.1016/j.apsusc.2020.148618

[81] P. Karthik, R. Vinoth, S.G. Babu, M. Wen, T. Kamegawa, H. Yamashita, B. Neppolian, Synthesis of highly visible light active TiO_2-2-naphthol surface complex and its application in photocatalytic chromium (vi) reduction, RSC Advances, 50 (2015)39752-9. https://doi.org/10.1039/C5RA03831F

[82] D. Majhi, K. Das, R. Bariki, S. Padhan, A. Mishra, R. Dhiman, P. Dash, B. Nayak, B.G. Mishra, A facile reflux method for in situ fabrication of a non-cytotoxic Bi_2S_3/β-Bi_2O_3/$ZnIn_2S_4$ ternary photocatalyst: a novel dual Z-scheme system with enhanced multifunctional photocatalytic activity, Journal of Materials Chemistry A, 41 (2020) 21729-43. https://doi.org/10.1039/D0TA06129H

[83] P. Wang, D. Sun, M. Deng, S. Zhang, Q. Bi, W. Zhao, F. Huang, Amorphous phosphated titanium oxide with amino and hydroxyl bifunctional groups for highly efficient heavy metal removal, Environmental Science: Nano, 4 (2020)1266-74. https://doi.org/10.1039/C9EN01466G

[84] B. Ou, J. Wang, Y. Wu, S. Zhao, Z., Wang Efficient removal of Cr (VI) by magnetic and recyclable calcined CoFe-LDH/g-C_3N_4 via the synergy of adsorption and photocatalysis under visible light. Chemical Engineering Journal, 380 (2020) 122600. https://doi.org/10.1016/j.cej.2019.122600

[85] X. Bai, Y. Du, W. Xue, X. Hu, J. Fan, J. Li, E. Liu, Enhancement of the photocatalytic synchronous removal of Cr (vi) and RhB over RP-modified flower-like SnS_2. Nanoscale Advances, 9 (2020) 4220-8. https://doi.org/10.1039/D0NA00489H

[86] X. Liu, M. Liu, L. Zhang, Co-adsorption and sequential adsorption of the co-existence four heavy metal ions and three fluoroquinolones on the functionalized

ferromagnetic 3D $NiFe_2O_4$ porous hollow microsphere, Journal of colloid and interface science, 511 (2018)135-44. https://doi.org/10.1016/j.jcis.2017.09.105

[87] D.P. Sounthararajah, P. Loganathan, J. Kandasamy, S. Vigneswaran, Effects of humic acid and suspended solids on the removal of heavy metals from water by adsorption onto granular activated carbon, International Journal of Environmental Research and Public Health, (2015) 10475-89. https://doi.org/10.3390/ijerph120910475

[88] M.M. Rahman, M. Adil, A.M. Yusof, Y.B..Kamaruzzaman, R.H. Ansary, Removal of heavy metal ions with acid activated carbons derived from oil palm and coconut shells, Materials, 7(5) 2014 3634-50. https://doi.org/10.3390/ma7053634

[89] M.N. Sahmoune, A.R. Yeddou, Potential of sawdust materials for the removal of dyes and heavy metals: examination of isotherms and kinetics, Desalination and Water Treatment, 57(50) (2016) 24019-34. https://doi.org/10.1080/19443994.2015.1135824

[90] J. Hizal, E. Tutem, K. Guclu, M. Hugul, S. Ayhan, R. Apak, F.Kilinckale, Heavy metal removal from water by red mud and coal fly ash: an integrated adsorption–solidification/stabilization process, Desalination and water treatment, 51(37-39) (2013) 7181-93. https://doi.org/10.1080/19443994.2013.771289

[91] Z. Yang, X. Huang, X. Yao, H. Ji, Thiourea modified hyper-crosslinked polystyrene resin for heavy metal ions removal from aqueous solutions, Journal of Applied Polymer Science, 135(1) (2018) 45568. https://doi.org/10.1002/app.45568

[92] C. Zhang, J. Sui, J. Li, Y. Tang, W. Cai, Efficient removal of heavy metal ions by thiol-functionalized superparamagnetic carbon nanotubes, Chemical Engineering Journal, 210 (2012) 45-52. https://doi.org/10.1016/j.cej.2012.08.062

[93] Zhang Y., Zhang S., Chung T.S. Nanometric graphene oxide framework membranes with enhanced heavy metal removal via nanofiltration, Environmental science & technology, 49(16) (2015) 10235-42. https://doi.org/10.1021/acs.est.5b02086

[94] S. Wang, Y. Peng, Natural zeolites as effective adsorbents in water and wastewater treatment, Chemical engineering journal, 156(1) (2010) 11-24. https://doi.org/10.1016/j.cej.2009.10.029

[95] E.B. Son, K.M. Poo, J.S. Chang, K.J. Chae, Heavy metal removal from aqueous solutions using engineered magnetic biochars derived from waste marine macro-algal biomass, Science of the Total Environment, 615 (2018) 161-168. https://doi.org/10.1016/j.scitotenv.2017.09.171

[96] M.F. Brigatti, C. Lugli, L. Poppi, Kinetics of heavy-metal removal and recovery in sepiolite, Applied Clay Science, 16 (2000) 45-57. https://doi.org/10.1016/S0169-1317(99)00046-0

[97] X. Zhang, Y. Qin, G. Zhang, Y. Zhao, C. Lv, X. Liu, L. Chen, Preparation of PVDF/hyperbranched-nano-palygorskite composite membrane for efficient removal of heavy metal ions, Polymers, 11 (2019) 156. https://doi.org/10.3390/polym11010156

[98] Y. Kim, J. Yi, Advances in environmental technologies via the application of mesoporous materials, Journal of Industrial and Engineering Chemistry, 10 (2004) 41-51.

[99] J. Wang, P. Wang, H. Wang, J. Dong, W. Chen, Preparation of molybdenum disulfide coated Mg/Al layered double hydroxide composites for efficient removal of chromium (VI), ACS Sustainable Chemistry & Engineering, 5 (2017) 7165-7174. https://doi.org/10.1021/acssuschemeng.7b01347

[100] J. Chen, N. Wang, Y. Liu, J. Zhu, J. Feng, W. Yan, Synergetic effect in a self-doping polyaniline/TiO$_2$ composite for selective adsorption of heavy metal ions, Synthetic Metals, 245 (2018) 32-41. https://doi.org/10.1016/j.synthmet.2018.08.006

[101] M. Maiti, M. Sarkar, M.A. Malik, S. Xu, Q. Li, S. Mandal, Iron oxide NPs facilitated a smart building composite for heavy-metal removal and dye degradation, ACS Omega, 3 (2018) 1081-1089. https://doi.org/10.1021/acsomega.7b01545

[102] W. Zhang, S. Shi, W. Zhu, C. Yang, S. Li, In-situ fixation of all-inorganic Mo–Fe–S clusters for the highly selective removal of lead (II), ACS Applied Materials & Interfaces 9 (2017) 32720-32726. https://doi.org/10.1021/acsami.7b08967

[103] X. Li, C. Bian, X. Meng, F.S. Xiao, Design and synthesis of an efficient nanoporous adsorbent for Hg^{2+} and Pb^{2+} ions in water, Journal of Materials Chemistry A, 4(16) (2016) 5999-6005. https://doi.org/10.1039/C6TA00987E

[104] M. Musielak, A. Gagor, B. Zawisza, E. Talik, R. Sitko, Graphene Oxide/Carbon Nanotube Membranes for Highly Efficient Removal of Metal Ions from Water, ACS Applied Materials & Interfaces, 11.31 (2019) 28582-28590. https://doi.org/10.1021/acsami.9b11214

[105] J. Hao, L. Ji, C. Li, C. Hu, K. Wu, Rapid, efficient and economic removal of organic dyes and heavy metals from wastewater by zinc-induced in-situ reduction and precipitation of graphene oxide, Journal of the Taiwan Institute of Chemical Engineers, 88 (2018) 137-145. https://doi.org/10.1016/j.jtice.2018.03.045

[106] G. Ognibene, D.A. Cristaldi, R. Fiorenza, I. Blanco, G. Cicala, Photoactivity of hierarchically nanostructured ZnO-PES fibre mats for water treatments, RSC advances, 6 (2016) 42778–42785. https://doi.org/10.1039/C6RA06854E

[107] D. Mauc̆ec, A. Šuligoj, A. Ristic, G. Dražic, A. Pintar, N.N. Tušar, Titania versus zinc oxide nanoparticles on mesoporous silica supports as photocatalysts for removal

of dyes from wastewater at neutral pH, Catalysis Today, 310 (2018) 32–41. https://doi.org/10.1016/j.cattod.2017.05.061

[108] K. Naseem, R. Begum, W. Wu, M. Usman, A. Irfan, Adsorptive removal of heavy metal ions using polystyrene-poly (N-isopropylmethacrylamide-acrylic acid) core/shell gel particles: adsorption isotherms and kinetic study. Journal of Molecular Liquids, 277 (2019) 522-531. https://doi.org/10.1016/j.molliq.2018.12.054

[109] D.M. Guo, Q.D. An, Z.Y. Xiao, S.R. Zhai, D.J. Yang, Efficient removal of Pb (II), Cr (VI) and organic dyes by polydopamine modified chitosan aerogels, Carbohydrate polymers, 202 (2018) 306–314. https://doi.org/10.1016/j.carbpol.2018.08.140

[110] R. Li, L. Liu, F. Yang, Preparation of polyaniline/reduced graphene oxide nanocomposite and its application in adsorption of aqueous Hg (II), Chemical Engineering Journal, 229 (2013) 460-468. https://doi.org/10.1016/j.cej.2013.05.089

[111] R. Gusain, N. Kumar, E. Fosso-Kankeu, S.S. Ray. Efficient removal of Pb (II) and Cd (II) from industrial mine water by a hierarchical MoS_2/SH-MWCNT nanocomposite. ACS omega 4 (9) (2019) 13922-13935. https://doi.org/10.1021/acsomega.9b01603

[112] A. Fateeva, P.A. Chater, C.P. Ireland, A.A. Tahir, Y.Z. Khimyak, A Water-Stable Porphyrin-Based Metal–Organic Framework Active for Visible-Light Photocatalysis, Angewandte Chemie International Edition,124 (2012) 7558-7562. https://doi.org/10.1002/ange.201202471

[113] X. Wang, C. Yang, S. Zhu, M. Yan, S. Ge, J. Yu, 3D origami electrochemical device for sensitive Pb^{2+} testing based on DNA functionalized iron-porphyrinic metal-organic framework, Biosensors and Bioelectronics, 87 (2017) 108-115. https://doi.org/10.1016/j.bios.2016.08.016

[114] Z.S. Hasankola, R. Rahimi, H. Shayegan, E. Moradi, V., Safarifard, Removal of Hg^{2+} heavy metal ion using a highly stable mesoporous porphyrinic zirconium metal-organic framework, Inorganica Chimica Acta 501, (2020) 1192649. https://doi.org/10.1016/j.ica.2019.119264

[115] Y. Wang, G. Ye, H. Chen, X. Hu, Z. Niu, S. Ma, Functionalized metal–organic framework as a new platform for efficient and selective removal of cadmium (II) from aqueous solution, Journal of Materials Chemistry A, 3 (2015) 15292-8. https://doi.org/10.1039/C5TA03201F

[116] D. Li, X. Tian, Z. Wang, Z. Guan, X. Li, H. Qiao, H. Ke, L. Luo, Q. Wei, Multifunctional adsorbent based on metal-organic framework modified bacterial cellulose/chitosan composite aerogel for high efficient removal of heavy metal ion and

organic pollutant, Chemical Engineering Journal, 383 (2020) 123127.
https://doi.org/10.1016/j.cej.2019.123127

[117] K. Zargoosh, H. Abedini, A. Abdolmaleki, M.R. Molavian, Effective removal of heavy metal ions from industrial wastes using thiosalicylhydrazide-modified magnetic nanoparticles, Industrial & Engineering Chemistry Research 52, (2013) 14944-14954. https://doi.org/10.1021/ie401971w

[118] H. Zhou, H. Zhu, F. Xue, H. He, Wang S., Cellulose-based amphoteric adsorbent for the complete removal of low-level heavy metal ions via a specialization and cooperation mechanism, Chemical Engineering Journal, (2019) 123879. https://doi.org/10.1016/j.cej.2019.123879

[119] J. Ma, G. Zhou, L. Chu, Y. Liu, C. Liu, S. Luo, Y. Wei, Efficient removal of heavy metal ions with an EDTA functionalized chitosan/polyacrylamide double network hydrogel, ACS Sustainable Chemistry & Engineering, 5 (2016) 843-851. https://doi.org/10.1021/acssuschemeng.6b02181

[120] F. Wang, G. Li, B. Shen, Y. Wang C. He, Mercury removal over the vanadia– titania catalyst in CO_2-enriched conditions, Chemical Engineering Journal, 263 (2015) 356-63. https://doi.org/10.1016/j.cej.2014.10.091

[121] Y. Fang, X. Lv, X. Xu, J. Zhu, P. Liu, L. Guo, C. Yuan, B. Cui., Three-dimensional nanoporous starch-based material for fast and highly efficient removal of heavy metal ions from wastewater, International Journal of Biological Macromolecules, 164 (2020) 415-26. https://doi.org/10.1016/j.ijbiomac.2020.07.017

[122] H. Jiang, Y. Yang, Z. Lin, B. Zhao, J. Wang, J. Xie, A. Zhang., Preparation of a novel bio-adsorbent of sodium alginate grafted polyacrylamide/graphene oxide hydrogel for the adsorption of heavy metal ion, Science of The Total Environment, 744 (2020) 140653. https://doi.org/10.1016/j.scitotenv.2020.140653

[123] C Z. Zhang, H. Sheng, Y X. Su, J Q. Xu., An efficient and health-friendly adsorbent N-[4-morpholinecarboximidamidoyl] carboximidamidoylmethylated polyphenylene sulfide for removing heavy metal ions from water, Journal of Molecular Liquids, 296 (2019) 111860. https://doi.org/10.1016/j.molliq.2019.111860

[124] MS. Islam, W. San Choi, B. Nam, C. Yoon, H J. Lee., Needle-like iron oxide@ $CaCO_3$ adsorbents for ultrafast removal of anionic and cationic heavy metal ions, Chemical Engineering Journal, 307 (2017) 208-19. https://doi.org/10.1016/j.cej.2016.08.079

[125] Q. Tang, K. Wang, J. Su, Y. Shen, S. Yang, Y. Ge, X. Cui., Facile fabrication of inorganic polymer microspheres as adsorbents for removing heavy metal ions,

Materials Research Bulletin, 113 (2019) 202-8.
https://doi.org/10.1016/j.materresbull.2019.02.009

[126] P. Kulal, V. Badalamoole. Efficient removal of dyes and heavy metal ions from waste water using Gum ghatti–graft–poly (4-acryloylmorpholine) hydrogel incorporated with magnetite nanoparticles, Journal of Environmental Chemical Engineering, 8 (2020) 104207. https://doi.org/10.1016/j.jece.2020.104207

[127] Q. Huang, D. Hu, M. Chen, C. Bao, X. Jin., Sequential removal of aniline and heavy metal ions by jute fiber biosorbents: A practical design of modifying adsorbent with reactive adsorbate, Journal of Molecular Liquids, 285 (2019) 288-98.
https://doi.org/10.1016/j.molliq.2019.04.115

[128] S. Yan, F. Zhang, Wang. L, Y. Rong, P. He, D. Jia, J.Yang., A green and low-cost hollow gangue microsphere/geopolymer adsorbent for the effective removal of heavy metals from wastewaters. Journal of environmental management, 246 (2019) 174-83.
https://doi.org/10.1016/j.jenvman.2019.05.120

[129] S. Ma, Q. Chen, H. Li, P. Wang, S.M. Islam, Q. Gu, X. Yang., M.G Kanatzidis. Highly selective and efficient heavy metal capture with polysulfide intercalated layered double hydroxides, Journal of Materials Chemistry A, 2(2014) 10280-9.
https://doi.org/10.1039/C4TA01203H

[130] A. Heidari, H. Younesi, Z. Mehraban., Removal of Ni (II), Cd (II), and Pb (II) from a ternary aqueous solution by amino functionalized mesoporous and nano mesoporous silica, Chemical Engineering Journal, 153 (2009) 70-9.
https://doi.org/10.1016/j.cej.2009.06.016

[131] T.A. Salah, A.M. Mohammad, M.A. Hassan, B.E. E.l-Anadouli, Development of nano-hydroxyapatite/chitosan composite for cadmium ions removal in wastewater treatment, Journal of the Taiwan Institute of Chemical Engineers, 4 (2014) 1571-7.
https://doi.org/10.1016/j.jtice.2013.10.008

[132] J. Zhang, J. Han, M. Wang, R. Guo., Fe_3O_4/PANI/MnO_2 core–shell hybrids as advanced adsorbents for heavy metal ions, Journal of Materials Chemistry A, 5(8) (2017) 4058-66. https://doi.org/10.1039/C6TA10499A

[133] EJ. Kim, CS. Lee, YY. Chang, YS. Chang., Hierarchically structured manganese oxide-coated magnetic nanocomposites for the efficient removal of heavy metal ions from aqueous systems, ACS applied materials & interfaces, 5(19) (2013) 9628-34.
https://doi.org/10.1021/am402615m

[134] Y. Wang, S. Liang, B. Chen, F. Guo, S. Yu, Y. Tang., Synergistic removal of Pb (II), Cd (II) and humic acid by Fe_3O_4@ mesoporous silica-graphene oxide composites, PloS one, 8(6) (2013) 65634. https://doi.org/10.1371/journal.pone.0065634

Materials Research Forum LLC
https://doi.org/10.21741/9781644901359-4

[135] C. Zhou, H. Zhu, Q. Wang, J. Wang, J. Cheng, Y. Guo, X. Zhou, R. Bai., Adsorption of mercury (II) with an Fe_3O_4 magnetic polypyrrole–graphene oxide nanocomposite, RSC advances, 7(30) (2017) 18466-79. https://doi.org/10.1039/C7RA01147D

[136] A.E. Chávez-Guajardo, J.C. Medina-Llamas, L. Maqueira, C.A. Andrade, K.G. Alves, C.P. de Melo., Efficient removal of Cr (VI) and Cu (II) ions from aqueous media by use of polypyrrole/maghemite and polyaniline/maghemite magnetic nanocomposites, Chemical Engineering Journal, 281 (2015) 826-36. https://doi.org/10.1016/j.cej.2015.07.008

[137] L. Cui, X. Guo, Q. Wei, Y. Wang, L. Gao, L. Yan, T. Yan, B. Du., Removal of mercury and methylene blue from aqueous solution by xanthate functionalized magnetic graphene oxide: sorption kinetic and uptake mechanism, Journal of colloid and interface science, 439 (2015)112-20. https://doi.org/10.1016/j.jcis.2014.10.019

[138] HK. Moghaddam, M. Pakizeh., Experimental study on mercury ions removal from aqueous solution by MnO2/CNTs nanocomposite adsorbent, Journal of Industrial and Engineering Chemistry, 21 (2015) 221-9. https://doi.org/10.1016/j.jiec.2014.02.028

[139] K. Yaghmaeian, RK. Mashizi, S. Nasseri, AH. Mahvi, M. Alimohammadi, S. Nazmara., Removal of inorganic mercury from aquatic environments by multi-walled carbon nanotubes, Journal of Environmental Health Science and Engineering, 13(1) (2015) 55. https://doi.org/10.1186/s40201-015-0209-8

[140] Y. Guo, J. Deng, J. Zhu, X. Zhou, R. Bai., Removal of mercury (II) and methylene blue from a wastewater environment with magnetic graphene oxide: adsorption kinetics, isotherms and mechanism. RSC Advances, 6 (2016) 82523-36. https://doi.org/10.1039/C6RA14651A

Photocatalysis
Materials Research Foundations 100 (2021) 161-192

Materials Research Forum LLC
https://doi.org/10.21741/9781644901359-5

Chapter 5

Graphitic Carbon Nitride based Photocatalytic Systems for High Performance Hydrogen Production: A Review

Shweta Sharma[1,2], Amit Kumar[1,2,3**], Gaurav Sharma[1,2,3], Anu Kumari[2], Mu. Naushad[3*]

[1]International Research Centre of Nanotechnology school of Advanced Chemical Sciences, Shoolini University, Solan, Himachal Pradesh, India-173229

[2]School of Advanced Chemical Sciences, Shoolini University, Solan, Himachal Pradesh, India-173229

[3]College of Materials Science and Engineering, Shenzhen Key Laboratory of Polymer Science and Technology, Guangdong Research Center for Interfacial Engineering of Functional Materials, Nanshan District Key Laboratory for Biopolymers and Safety Evaluation, Shenzhen University, Shenzhen, 518055, PR China

[4]Department of Chemistry, College of Science, Building#5, King Saud University, Riyadh, Saudi Arabia-11451

* shad123@gmail.com (M. Naushad), mittuchem83@gmail.com (A. Kumar)

Abstract

The hydrogen (H_2) production using solar-power is a novel way of harvesting and meeting the ever-increasing energy demands. It also helps in alleviating the global warming issues by decreasing the carbon footprints. The photocatalyst required for H_2 production should possess some specific properties such as high charge separation, appreciable H_2 production activity, chemical and thermal stability, abundant availability, and cost effectiveness, etc. The metal free graphitic- carbon nitride (g-C_3N_4)-based photocatalysts have grabbed considerably large interest in the area of visible-light photocatalytic H_2 production. It has extraordinary chemical stability, earth-abundant, attractive band structure, and easy fabrication. Herein, various approaches such as band gap engineering, nanostructure design, and heterojunction construction using various examples presented in literature are described. We have also discussed various nano heterojunctions of g-C_3N_4 which have been used effectively for the hydrogen production.

Keywords

g-C_3N_4, Visible Light, Hydrogen Generation, Heterojunction, Photocatalytic

Contents

1.   Introduction

The rapid enhancement in the consumption of fossil fuels as an energy source has led to their fast depletion since they are limited and have high exhaustibility. Presently, 85% of the H_2 production is achieved by the combustion of fossil fuels, that leads to the annual production of ~500 metric tonnes of carbon dioxide (CO_2) as a by-product [1]. Solar energy on the other hand is an inexhaustible, pure and sustainable source. So, we require highly efficient materials that can convert the solar energy into some other form that could be utilized by the human race [2, 3]. In recent years, conversion of solar energy to chemical energy via water splitting has fascinated many researchers because about 72% of the earth's surface is covered with water [4, 5]. Until now, multiple techniques have been formulated for the production of H_2 and that includes coal gasification, renewable liquid, pyrolysis of biomass, and bio-oil processing, water splitting using energy resources, water splitting by electrochemical and biological materials [6].

The water splitting using TiO_2 catalyst under the influence of UV radiations was performed firstly by Fujishima and Honda in 1972 [7]. Following that, various studies were performed using variety of semiconductors. Process of water splitting for H_2 generation occurs in three steps; (i) production of electrons and holes upon light illumination; (ii) separation of

Materials Research Forum LLC
https://doi.org/10.21741/9781644901359-5

generated charges and their proper movement over the catalyst surface and (iii) reduction of H^+ to H_2 by the electrons and oxidation of H_2O to O_2 by the holes [8]. During water splitting, there occurs two processes, namely, oxygen evolution reaction (OER) by holes and hydrogen evolution reactions (HER) by electrons. The material or catalyst performance during water splitting is determined by solar to hydrogen conversion efficiency (STH) and apparent quantum efficiency (AQE). Their characteristic equations are [9]:

$$STH = \frac{Output\ energy\ as\ H_2}{Energy\ of\ incident\ solar\ light} \tag{i}$$

$$AQE = \frac{n* H_2\ generation}{Incident\ photon} \tag{ii}$$

Where n represents the number of electrons required for the process (2 in case of H_2 production)

2. Metal free photocatalyst: Graphitic carbon nitride

As mentioned in the above section, the semiconductor materials can be used for water splitting. It should meet some of the characteristic features such as (a) the valence and conduction band or both should match with the reduction and oxidation potential of water; (b) the photocatalyst should be active in the visible range and (c) it should have high separation and easy transportation of charges [10, 11]. Metal- free photocatalysts have high utilization in the past few decades as a substitute to metal-based photocatalysts [12]. Some of the metal free photocatalysts illustrates even similar and enhanced activity as compared to the metal- based photocatalysts. In the recent decades, the utilization of carbon-based metal-free photocatalysts for water splitting have experienced much significance due to their low cost, easy availability and ecofriendly nature. Among these, g-C_3N_4 is the most popular photocatalyst [13, 14].

2.1 Chemical structure of g-C_3N_4

The molecular formula of this family is near to C_3N_4, however, it does contain some amount of hydrogen too. These structures are based on s- heptazine and s-triazine units which exhibits different properties under different reaction conditions [15]. The final structure formed using these two different units are shown in Figure 1. The s-triazine based C_3N_4 units exhibits some special electronic properties due to the presence of delocalized electrons and lone pair of electrons on nitrogen. In addition, its thermal and chemical permanence marks it a favourable candidate for several applications. Because of its

Photocatalysis   Materials Research Forum LLC
Materials Research Foundations **100** (2021) 161-192   https://doi.org/10.21741/9781644901359-5

distinctive semiconductor properties, it can display high catalytic activity for multiple reactions [16].

Various methodologies have been reported in literature for the synthesis of g-C_3N_4 such as solvothermal, chemical vapour deposition (CVD), thermal nitration, and solid state reaction [10, 17, 18]. However, recently employed method, thermal condensation is quite a facile technique as compared to the above mentioned. In this method, the nitrogen-rich species such as melamine, urea, cyanamide, and dicyandiamide are used that experience self-condensation with deammoniation process via thermal reaction. Figure 2 shows the different fabrication methods with the probable precursors for the synthesis of g-C_3N_4.

In spite of the above advancements, the complete crystal structure of g-C_3N_4 has not been fully determined. To the best of our knowledge the most recent work on the 3D structure of g-C_3N_4 is carried out by Tyborski [19]. They used XRD to solve the structure of g-C_3N_4. However, due to the lack of long range order in the material which prevents the proper structure refinement, conventional XRD alone is not sufficient. On the other hand, total scattering data, which associates information from the diffuse scattering regarding short range order, can provide additional information useful for its structure determination [19].

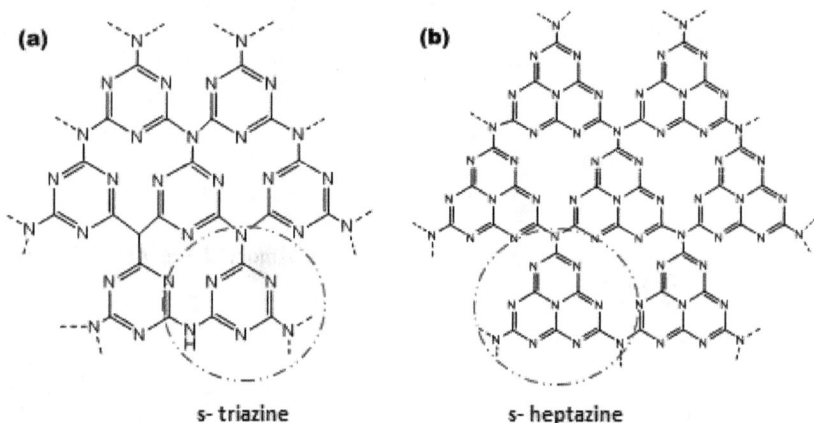

s- triazine   s- heptazine

Figure 1: Chemical structure of triazine.

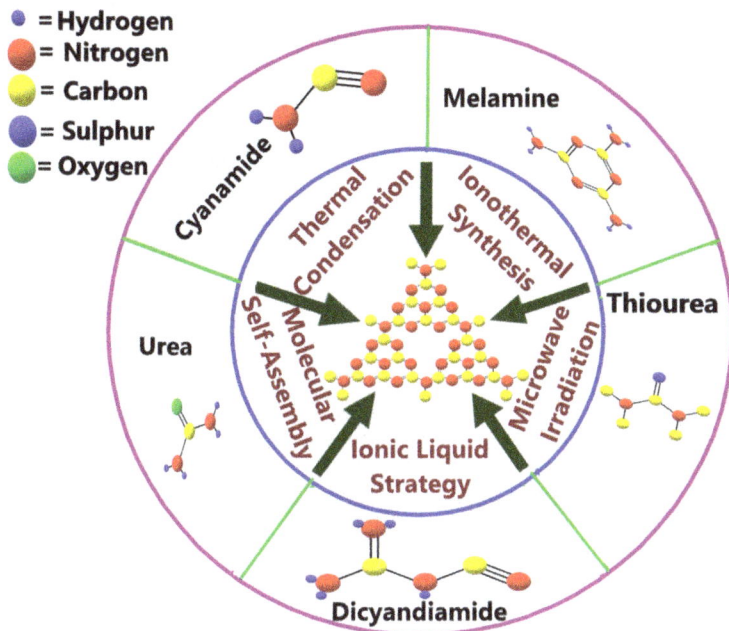

Figure 2. Different fabrication methods with the probable precursors for the synthesis of g-C$_3$N$_4$.

2.2 Electronic properties of g-C$_3$N$_4$

At ambient environment, g-C$_3$N$_4$ is considered to be the most stable allotrope [20, 21]. They are polymeric organic materials that have five different allotropes such as α, β, cubic, pseudocubic, and graphitic [22]. The g-C$_3$N$_4$ is a unique semiconductor photocatalyst that has visible-light response up to 455 nm, equivalent to the narrow band gap of 2.7 eV [23, 24]. Its valence band is formed by the N2p orbitals and the conduction band displays the hybridized C2p and N2p orbitals [25, 26].

The g-C$_3$N$_4$ is a poly conjugated semiconductors made of carbon and nitrogen atoms and it is characterized by a layered graphitic-like structure [27]. The fully polymerized form of g-C$_3$N$_4$, characterized by a C: N of 0.75, cannot be practically attained. It can be easily prepared via solid state synthesis from cheap materials such as melamine. It is insoluble in most solvents and shows great stability in superlative conditions (pH=0 and pH=14). Due

Photocatalysis   Materials Research Forum LLC
Materials Research Foundations **100** (2021) 161-192   https://doi.org/10.21741/9781644901359-5

to these properties, it has currently attracted great attention for various catalytic applications, in specific photo catalysis for hydrogen evolution from water [27].

2.3   Strategies for improving the properties of g-C_3N_4

The g-C_3N_4 with a layered structure has shown brilliant applications, like as in the field of hydrogen evolution reaction and oxygen reduction [28]. However, there is still limited electron transfer ability of perfect g-C_3N_4 due to its chemical stability. Proof of concept studies has been focused on promoting electron transfer and enhancing the concentration of active site [15, 29], among which several strategies has been reported such as (a) hetero atoms doping is a successful strategy to modify the materials with various morphologies and electronic structures [30, 31]. Zhu *et al.*, 2015 synthesized the doped g-C_3N_4 by a condensation method, as a result, doping not only significantly changed its morphology and its surface property of g-C_3N_4 but also changed its electronic structure and increased its ionic conductivity [32]. (b) The structural modification has also been reported for increasing the activity of pristine g-C_3N_4. By modifying the structure to 0-D, 1-D, 2-D and 3-D structures, the optical, physiochemical, diffusion distance of charge carriers and number of redox active sites can be properly optimized [33]. (c) In addition, copolymerization of pristine g-C_3N_4 helps in narrowing the band gap that aids in utilizing the entire visible range. (d) Formation of heterojunction with other semiconductors enhances the separation of charge carriers and transfer to different sites [34, 35].

2.3.1   Structural modification

By monitoring the nanostructure, the physical, chemical, and optical properties of g-C_3N_4 based photocatalysts can be easily controlled or modified. Various factors can be easily tuned by just controlling the nanostructure such as number of redox sites, and diffusion distance of e^- and h^+ to reach these sites. Different dimensional structures of g-C_3N_4 have been reported in literature such as 1-D (nanorods), 2-D (sheets) and 3-D (various porous structures). The transformation of structures into 1-D nanorods or tubes, decreases the charge diffusion length and helps in improving the light harvesting by increasing the light interactions. In case of 2-D nanosheets, the charge transport along the sheet got enhanced and the recombination rate decreases drastically. It also increases the availability of abundant reactive sites for various activities. Single layered g-C_3N_4 nanosheets can be easily prepared by exfoliation process using H_2SO_4 which led to its intercalation between the g-C_3N_4 nanosheets [36]. However, the 3-D nanostructures acquired by accumulating 1-D nanostructural units possess high specific surface area, less mass transport resistance, high light absorption ability, and large number of active sites. However, the construction of 3-D nanostructures by using the 2-D nanosheets is tough because of the randomness in their sizes and orientations [37].

2.3.2 Doping

Doping is quite an effective way for enhancing the properties of g-C$_3$N$_4$ based photocatalysts by narrowing the band gap. Addition of dopant increases the localized states and also lifts the valence band maximum that helps in the narrowing of band gap. Doping of non-metals increase the delocalization of conjugated electrons, which helps in enhancing the conductance, mobility and separation of photo-generated charges [38]. Instead of a single atom doping, co-doping and more than one heteroatom doping is also preferable since it tune the band gap to a much greater extent. Among the halogens, bromine (Br), fluorine (F) and iodine (I) have been used as dopants for g-C$_3$N$_4$. Instead of non-metals, if the doping of g-C$_3$N$_4$ is done with organic molecules, then it may result in delocalization of π-conjugated electrons and red shift. However, the doping of aromatic molecules enhances the charge separation on the g-C$_3$N$_4$ surface [39].

2.3.3 Heterojunction formation

Designing of a heterojunction of g- C$_3$N$_4$ with other semiconductors helps in enhancing the optical properties, thus improving the charge separation [40, 41]. The coupling of g-C$_3$N$_4$ with other semiconductors with diverse properties prompts the band bending of the heterojunction at interface with the aid of inner electric field. This helps in improving the transfer and separation of photo-generated charges. In general, heterojunction systems can be categorized into; Type II, Z- scheme and S- scheme [42, 43].

The type II heterojunction is typically composed of two semiconductors, SI and SII in which one semiconductor possesses higher valence and conduction band potential. When the heterojunction is irradiated with sufficient energy, this generates electron-hole pairs. The electrons then move from the higher conduction band position of one semiconductor to the lower conduction band position of the other. The holes in the other hand follow the reverse path. The photoinduced electrons and holes get accumulated on the different semiconductors and thus helps in promoting the separation of charges and suppressing their recombination. This significantly enhances the energy conversion ability of synthesized heterojunction.

On the other hand, the Z- scheme mimics the natural photosynthesis process. This system is similar to type II heterojunction but the difference occurs in its charge-carrier migration mechanism. The charge-carrier migration pathway of this system looks like the letter "Z" [44]. This system comprises of a unique charge transfer mechanism in which the oxidation photocatalyst with strong oxidation ability exhibits low VB position and the reduction photocatalyst with robust reduction capability own high CB value [45]. The presence of charge carrier mediator and the type which is used, distinguishes the Z-scheme photocatalysts into three types: (i) traditional Z-scheme photocatalyst, (ii) all-solid state Z-

Photocatalysis   Materials Research Forum LLC
Materials Research Foundations **100** (2021) 161-192   https://doi.org/10.21741/9781644901359-5

scheme photocatalyst, and (iii) direct Z-scheme photocatalyst [46]. The traditional Z-scheme photocatalyst consists of two different photocatalysts, in which an appropriate shuttle redox ion mediator is used to couple these two photocatalysts. Fe^{3+}/Fe^{2+} and IO^{3-}/I^- are the most common used shuttle redox ion pairs. When these photocatalysts are irradiated with light, the electron acceptor species consumes the photogenerated electrons of CB of photocatalyst having lower potential value while the electron donor species consumes the photogenerated holes of VB of other photocatalyst. The other remained photogenerated holes and electrons take part in separate oxidation and reduction reactions that have the required potential for carrying out effective reactions [47]. The all-solid state Z-scheme photocatalysts contains an electronic solid conductor which is used as charge carrier transfer-bridge. In the first reported all-solid state Z-scheme photocatalysts, Au as electronic conductor was used to create a bridge between CdS and TiO_2 photocatalysts [48]. In this system, the selection of suitable electron conductor is most important because it is not only limited to efficient transfer of photogenerated charge carriers but also helps in improving the stability of the photocatalyst. Various nanoparticles such as Au, Ag, and Cu have been reported as excellent electron mediators to construct all-solid state Z-scheme photocatalysts [49]. Apart from these two, the direct Z-scheme photocatalyst consists of two semiconductors which are in close contact with one another without any charge carrier mediator. These photocatalyst systems are resistant to corrosion. In contrast to traditional heterojunction, the back reactions in direct Z- scheme are considerably suppressed due to the absence of redox mediators [50].

S- scheme or "Step- scheme" is a recently formulated explanation for the mechanism of charge transfer in heterojunction. In this scheme, the two photocatalysts are arranged in similar way as in case of type-II heterojunction but follow a different charge transfer route. with staggered band structures. Here, an electric field is generated between the two photocatalysts that results in band bending. As a result of band bending, the electrons from the low potential conduction band moves to the higher potential valence band. Thus, the higher oxidation and reduction potential bands remain conserved for the generation of reactive radicals [43].

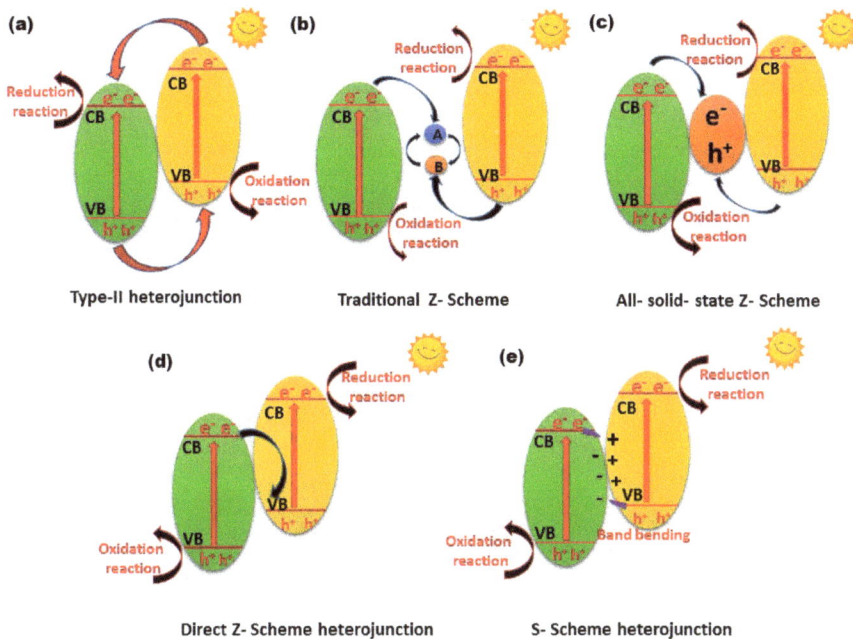

Figure 2. Different types of heterojunctions possible with their possible charge transfer mechanism.

3.   C₃N₄ based heterojunctions for H₂ production

H_2 is the most abundant element in the universe available in water and organic compounds. It is the lightest element which contains only one electron and one proton. It is a colorless, odorless, and a flammable gas. It is non- toxic and spread rapidly when released. It is transportable, storable, and utilizable and has the ability to act as alternate energy storage system. By using solar energy, water splitting for the production of H_2 is considered as one of the most proficient approaches to deal with the current global energy issues [44, 51-53].

Considering the above discussion, numerous work has been done on the utilization of g-C_3N_4 based photocatalyst for solving the issue of energy crisis. In the same context, Bi_2MoO_6/g-C_3N_4 based heterojunction was designed by in-situ solvothermal method using 2-D g-C_3N_4 nanosheets. The H_2 production ability of Bi_2MoO_6/g-C_3N_4 from water splitting was performed. They also studied the bacterial disinfection under visible light. The complete mechanism of H_2 production and bacterial disinfection is presented in Figure 3. They exfoliated the bulk g-C_3N_4 to g-C_3N_4 nanosheets that considerably increased the surface area and reduced the diffusion distance for photogenerated charges. This helped in promoting not only the photocatalytic activity [54]. The photocatalytic H_2 production by Bi_2MoO_6/g-C_3N_4 was approximately 3.84 time higher than bulk g-C_3N_4 and 32.19 times higher than Bi_2MoO_6.

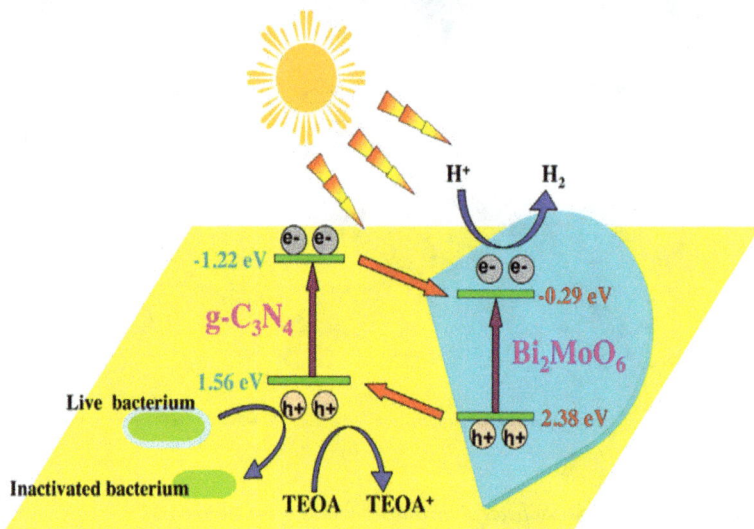

Figure 3: (a) Photocatalytic mechanism for H_2 production by Bi_2MoO_6/g-C_3N_4 heterojunction [54].

In another study, novel C_{60}/graphene/g-C_3N_4 composite was utilized as a photocatalytic material for H_2 production by water splitting. The superior electron conduction ability of graphene and strong electron-attracting tendency of C_{60} made the composite quite effective for the H_2 production. The synergistic effect between graphene and C_{60} helped in expanding the movement and utilization proficiency of photo-generated electrons and enhanced the separation of photo-generated charges. The H_2 production amount and rate of C_{60}/graphene/g-C_3N_4 (10 mg/L C_{60} and graphene) was found to be about 539.6 times greater than that of pure g-C_3N_4 under the same experimental conditions. Also, the rate obtained was 50.8 and 4.24 times higher than that of graphene/g-C_3N_4 and C_{60}/g-C_3N_4, respectively. The quantum yield of the synthesized composite in 97 h was about 7.2% [55].

Materials Research Forum LLC
https://doi.org/10.21741/9781644901359-5

Due to higher redox capability, ample oxygen defects and cost effectiveness, the CeO_2 and its various derivatives have high potential as photocatalysts. Based on this, CeO_2/g-C_3N_4 was designed with tuneable CeO_2 crystal planes [56, 57]. Characterization techniques such as photoelectrochemical, Raman, XPS, and ESR results proposed that the charge separation efficacy, oxygen defects and Ce^{3+}/Ce^{4+} reversibility pairs were dependent on the CeO_2 and g-C_3N_4 contact. The photocatalytic activity of H_2 production under visible light was of the order: $CeO_2(110)/g$-$C_3N_4 > CeO_2(100)/g$- $C_3N_4 > CeO_2(111)/g$-$C_3N_4 > g$-C_3N_4 [58].

Aiming for the synthesis of highly efficient photocatalyst, Chen and his co-workers, synthesized 2D/2D $MnIn_2S_4/g$-C_3N_4 photocatalyst by hydrothermal method. They carried out the in-situ loading of $MnIn_2S_4$ nanoflakes on the nanosheets of g-C_3N_4. The synthesized photocatalyst formed a direct Z-scheme heterojunction that showed higher activity than that of the individual components under the visible light irradiation. Complete transfer mechanism of the photogenerated charges for the H_2 production and degradation of pharmaceutical wastewater is shown in Figure 4. The formation of direct Z-scheme between the two components significantly increased the photocatalytic activity due to the efficient charge separation and high charge transfer. They even reported the excellent stability of the synthesized photocatalyst [59].

In one another very interesting study, g-$C_3N_4/KNbO_3$ heterojunction and Pt-loaded g-$C_3N_4/KNbO_3$ heterojunctions were synthesized for studying its H_2 production ability under simulated sunlight using water splitting. In the Pt-loaded g-$C_3N_4/KNbO_3$ heterojunction, the Pt acted as the co-catalyst that helped in enhancing the light absorption ability and in reducing the charge recombination. The H_2 production efficiency of the heterojunction was even compared with the individual components and the results proved its high ability. The schematic presentation of g-$C_3N_4/KNbO_3$ and Pt-g-$C_3N_4/KNbO_3$ for H_2 production is shown in Figure 5 [60].

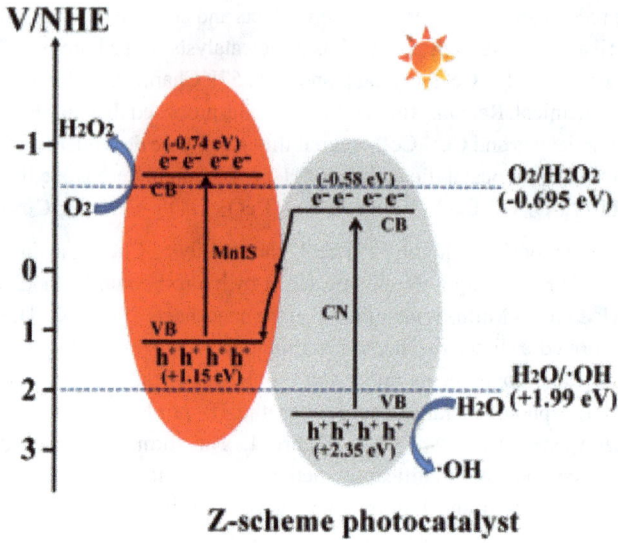

Figure 4: Charge transfer mechanism in direct Z-scheme heterojunction formed between MnIn$_2$S$_4$ and g-C$_3$N$_4$ [59].

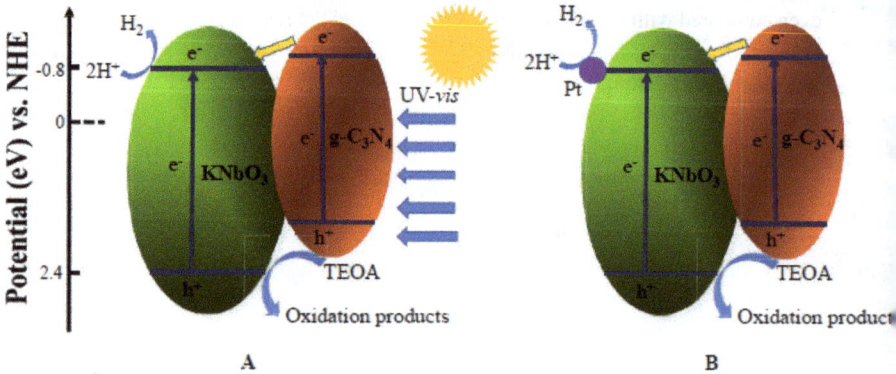

Figure 5: Schematic presentation of g-C$_3$N$_4$/KNbO$_3$ and Pt-g-C$_3$N$_4$/KNbO$_3$ for H$_2$ production [60].

Photocatalysis   Materials Research Forum LLC
Materials Research Foundations **100** (2021) 161-192   https://doi.org/10.21741/9781644901359-5

Pure g-C_3N_4 shows negligible activity for H_2 production due to its high recombination rate. So, as a modification, a heterojunction of g-C_3N_4 and NiS_2 was synthesized and studied for H_2 generation. The production was found to be dependent upon the amount of NiS_2. The H2 production rate varied significantly with variation in the content of NiS_2. For 3%, 5%, 7%, and 10% NiS_2 content, the corresponding H_2 production rates were 365.72, 550.81, 715.83, and 540.1 lmol/h/g respectively. They studied the effect of co-catalyst loading with 2% Pt. They found that the H_2 production rate was 2.75 times as much as that of 2% Pt as a co-catalyst. They suggested that the enhanced activity was due to a wide range of photoresponse and efficient separation of photogenerated charges [61].

The g-C_3N_4, rGO and TiO_2 are some of the most capable substance for photocatalytic hydrogen production activity. However, the H_2 evolution rate reported for bare g- C_3N_4, P25 (degussa), TiO_2, g-C_3N_4/TiO_2, and TiO_2/rGO [62, 63] nanocomposites in the literature are comparatively low. In 2017, Hafeez with his co-workers fabricated rGO supported g-C_3N_4-TiO_2 using simple ultrasound assisted wet impregnation. Various concentrations of photocatalyst were synthesized by varying the concentration of TiO_2 and g-C_3N_4 and they found that 70:30 (g-C_3N_4: TiO_2) exhibited the highest activity of 23,143 μmol/g/h for H_2 production. High light absorption and efficient charge transfer resulted in the excellent activity of the synthesized photocatalyst [64]. A probable mechanism for H_2 production using g-C_3N_4-TiO_2/rGO was also proposed and is presented in Figure 6.

In 2016, Liu and his co-workers reported the designing of complete metal- free photocatalyst in which they synthesized isotype triazine/heptazine- g-C_3N_4 based heterojunction by microwave-assisted molten-salt process. They used eutectic mixture of KCl/LiCl as the molten salt and melamine as the nitrogen- rich precursor and treated them in the microwave. A complete synthesis procedure is presented in Figure 7 (a). A huge improvement in HER proficiency was attained that was linked to the isotype triazine/heptazine based g-C_3N_4 heterojunctions. It had an appropriate arrangement of the electronic band structures that provided the proficient transport path and high separation of photo-generated charges. Furthermore, the synergistic effect of microwave reaction and molten-salt liquid polycondensation condition afford an improved way for the fabrication of g-C_3N_4 based photocatalysts [65]. A probable charge transfer mechanism involved in the synthesized photocatalyst for HER is shown in Figure 7 (c).

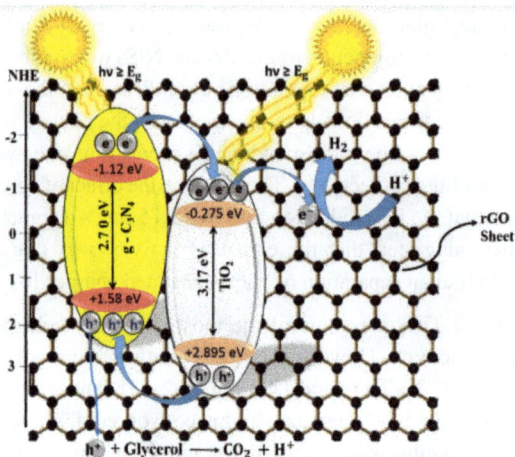

Figure 6: A possible mechanism of H₂ production using g-C₃N₄-TiO₂/rGO under UV-light [64].

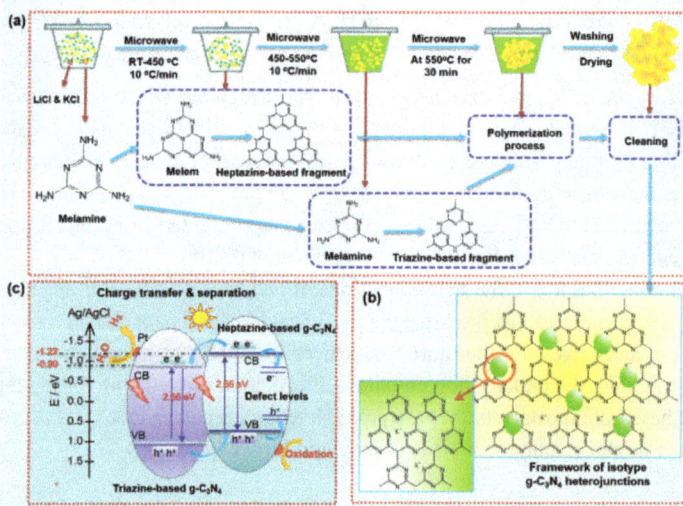

Figure 7: (a) Microwave-assisted molten-salt route for designing of isotype triazine-/heptazine based g-C₃N₄ heterojunctions; (b) Structure of the isotype g-C₃N₄ heterojunction; and (c) Probable charge transfer paths in the isotype g-C₃N₄ heterojunction [65].

H_2 evolution by water splitting using a semiconductor photocatalyst is deliberated as quite an interesting and a supreme procedure for obtaining a clean and sustainable energy [66-68]. The photocatalytic performance and stability of any photocatalyst can be proficiently amended by coupling with other semiconductors and doping as explained in Section 2.3 [69]. Liu and his co-workers [70] revealed that the coupling of g-C3N4 with $ZnIn_2S_4$ formed a highly proficient photocatalyst that showed advanced H_2 production and degradation ability than that of the individual components [71].

The sunlight-driven water splitting at semiconductor's boundaries has fascinated extensive attention due to its cost and energy saving nature [72, 73]. In the same context, Cheng and Jiang fabricated the g-C3N4/nanocarbon/$ZnIn_2S_4$ nanocomposite. The resulting nanocomposite showed artificial Z-scheme that was active under visible light for the photocatalytic H_2 evolution and maximum rate obtained was 50.32 $\mu mol \cdot h^{-1}$ [74].

Using in-situ photo- deposition method, a novel Co–Pi/g-C3N4 was synthesized. The photocatalytic activity of Co–Pi/g-C3N4 for oxygen and hydrogen production via water splitting was analyzed under visible light irradiation. The Co–Pi modified g-C3N4 exhibited considerably improved photocatalytic activity, and the g-C3N4 with deposition time of 20 min (CP-20) displayed the maximum activity. The H_2 production performance reported was about 9.6 times than that of bulk g-C3N4. The photo- deposition of Co- Pi onto g-C3N4 significantly improved the separation of photogenerated carriers and improved the H_2 and O_2 production rate [75]. A possible mechanism of charge transfer in the Co–Pi/g-C3N4 is shown in Figure 8.

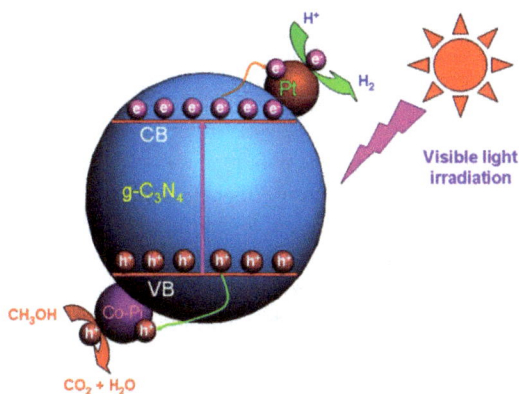

Figure 8: The hydrogen generation by Co–Pi/g-C3N4 composite photocatalysts [75].

A novel $CuS-ZnS_{1-x}O_x/g-C_3N_4$ heterostructure based photocatalyst was synthesized by combination of thermal decomposition and hydrothermal method. The introduction of CuS into the nanocomposite significantly improved the absorption of the resulting photocatalyst. The decoration of CuS highly improved the H_2 production rate from 9200 to 10,900 µmol/h/g and $g-C_3N_4$ provided the efficient separation of photogenerated charges. The $CuS-ZnS_{1-x}O_x/g-C_3N_4$ nanocomposite displayed superior photocatalytic activity for H_2 production. The composite photocatalyst $CuS-ZnS_{1-x}O_x$ with 5 wt% $g-C_3N_4$ showed the maximal H_2 production rate of 12,200 µmol/g/h [76].

Semiconductor hybridization is a proficient substitute practice for enhancing the utilisation of $g-C_3N_4$ for H_2 production under visible light. This approach is established on the alignment of bands between $g-C_3N_4$ and other semiconductors, forcing the photogenerated charges to migrate in the opposite directions. It ultimately results in the spatial separation of the charges on the heterojunction and to decrease the charge recombination to advance the photocatalytic proficiency. In the same context, $CdS/g-C_3N_4$ heterojunction was synthesized. In this, the CdS quantum dots were grown on the surface of $g-C_3N_4$ [77]. In another study, the $CdS/g-C_3N_4$ core/shell nanowires were prepared [69]. The formation of a heterojunction between CdS and $g-C_3N_4$ resulted in efficient charge separation and the transfer of corrosive holes from CdS to $g-C_3N_4$ (as shown in Figure 9). As a result, highlighting the permanence of CdS and attained a much higher H_2 production rate under visible light irradiation as compared to that for pure CdS and pure $g-C_3N_4$ [77].

Similarly, in another study, the in- situ growth of In_2O_3 nanocrystals on the $g-C_3N_4$ nanosheets was carried out. The synthesized $In_2O_3-g-C_3N_4$ photocatalyst showed substantial advancement in the photocatalytic activity for H_2 production. The improved rate was due to the efficient transfer of photogenerated charges between the two constituting components [78].

At the same time, $g-C_3N_4$ can also be excited by absorbing photons with energy exceeding its band gap and subsequently transfer the photo induced electrons directly to the Pt co-catalyst to generate H_2. As a result, the sensitized photocatalyst shows a high H_2 evolution rate. Activity under irradiation of visible light longer than the absorption edge of $g-C_3N_4$ and achieves an apparent quantum efficiency of 19.4% even at 550 nm. Wang and his co-workers also attained a high deceptive quantum efficiency of 33.4% at 460 nm by alerting the $g-C_3N_4$ layer with low-cost Erythrosin B dye, which is greater than that of CdS-cluster-decorated graphene nanosheets [79] and comparable to that of CdS quantum dots/ graphene/ZnIn2S4 heterostructures [80].

Photocatalysis   Materials Research Forum LLC
Materials Research Foundations **100** (2021) 161-192   https://doi.org/10.21741/9781644901359-5

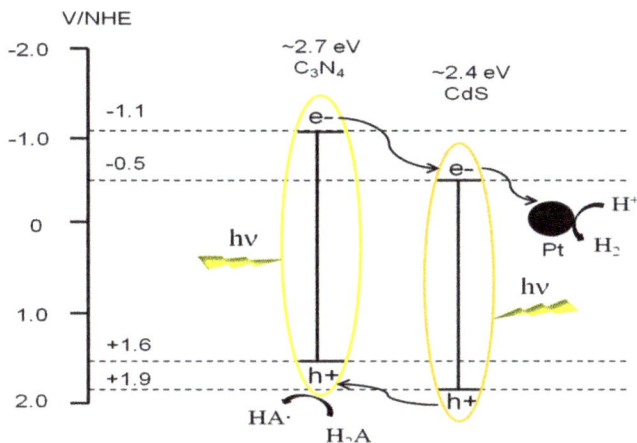

Figure 9: The hydrogen generation by CdS/g-C₃N₄ photocatalysts [69].

Delamination of layered materials into 2-D single sheets prompt some extraordinary properties in it, such as, high surface area, extreme charge carrier mobility, and distinct variations in the band energy system. With the exclusive structural benefits of aligned energy levels, charge transfer, increased light absorption range, and reaction site availability, the as-prepared mesoporous g-C₃N₄ displayed an AQE of 5.1% at 420 nm and photocatalytic H_2 evolution rate of 8510 μmol/h/g, maximum among all the metal-free photocatalyst [81].

InVO₄ was a capable photocatalyst for H_2 production under visible light irradiation. The conduction band of InVO₄ was appropriate for H_2 production [82]. Additionally, many data reported that the preferred morphology and size of photocatalysts could regular the position of energy band for achieving higher redox ability [67, 83]. Yan et al reported that the nano-sized InVO₄ nanoparticles with the size of 20 nm showed higher photocatalytic activity of H_2 production than InVO₄ microspheres [84, 85]. In another study, the g-C₃N₄/nano-InVO₄ nanocomposites were designed by hydrothermal technique. The g-C₃N₄ sheets performed a very important role in the establishment of g-C₃N₄/InVO₄ nanocomposites, which revealed that g-C₃N₄ was an auspicious support for in-situ growth of other nano-sized components. The construction of interfaces enhanced the charge transfer and introverted the recombination of charges, which considerably enriched the photocatalytic H_2 evolution [86].

Materials Research Forum LLC
https://doi.org/10.21741/9781644901359-5

Table 1: g-C$_3$N$_4$ based photocatalysts for H$_2$ production with their synthesis method.

Sr. No.	g-C$_3$N$_4$ based photocatalyst	Synthesis method	Amount of H$_2$ produced in µmol/g/h	References
1.	g-C$_3$N$_4$/MoS$_2$	Photodeposition	660	[87]
2	g-C$_3$N$_4$ nanodots	Templateless infrared heating	57.8	[88]
3	g-C$_3$N$_4$/TiO$_2$	Vapour deposition method	513	[89]
4	3-D g-C$_3$N$_4$ nanobelts	Hydrothermal	27.2	[37]
5.	ZnO/ZnS/g-C$_3$N$_4$	Chemical conversion	1205	[90]
6	Br-doped g-C$_3$N$_4$	Heating urea with NH4Br	48.00	[91]
7	O self- doped g-C$_3$N$_4$	Hydrothermal treatment	158.7	[92]
8	N self- doped g-C$_3$N$_4$	Thermal treatment	44.28	[93]
9	Graphene/g-C$_3$N$_4$	Impregnation-chemical method	36.08	[94]
10	N-GQDs/g-C$_3$N$_4$	Chemical	43.6	[95]
11	C-ZIF/g-C$_3$N$_4$	Calcination	32.6	[96]
12	Tapered C-PAN/g-C$_3$N$_4$ nanotubes	Thermal treatment	177.5	[97]
13	g-C$_3$N$_4$ /SrTiO$_3$	Calcination	440	[98]
14	Pt-Pd/g-C$_3$N$_4$	Chemical deposition precipitation	1600.8	[99]
15	Oxygen-doped g-C$_3$N$_4$	Heating treatment	189.3	[100]
16	g-C$_3$N$_4$/SiC	In- situ precipitation	182	[101]
17	CoP/g-C$_3$N$_4$	Thermal decomposition + Electroless plating	936	[102]
18	Zn$_x$Cd$_{1-x}$In$_2$S$_4$/g-C$_3$N$_4$	Hydrothermal	170.3	[103]
19	Carbon fibre/g-C$_3$N$_4$	Electrospinning + Calcination	1080	[104]
20	Cu$_2$O/g-C$_3$N$_4$	In-situ technique	33.2	[105]

21	CdSe Quantum Dots/g-C_3N_4	Thermal polymerizatinon + chemical method	615	[106]
22	Organic-Organic Hybrid g-C3N4/ethanediamine	Solvothermal reaction	3055	[107]
23	g-C_3N_4/CoO	Calcination + Chemical reaction	651.3	[108]
24	g-C_3N_4/WS_2	Gas- solid reaction	101	[109]
25	0D-ZnO/g-C_3N_4	Solution conversion and calcination	322	[110]
26	SnS_2/g-C_3N_4	In-situ growth method	928	[111]
27	N-ZnO/g-C_3N_4	Hydrothermal and calcination	152.7	[112]
28	$SnFe_2O_4$/g-C_3N_4	In-situ precipitation	132.5	[113]
29	2D WO_3/g-C_3N_4	Hydrothermal	1853	[114]
30	Ta_2O_5/g-C_3N_4	One- step heating	36.4	[115]
31	Sulfur-doped holey g-C_3N_4 nanosheets	Self- templating approach	6225.4	[116]
32	Benzene ring modified g-C_3N_4	Thermal approach	8.9	[117]
33	Black phosphorus/MXene/Ultrathin-g-C_3N_4 composite	Chemical method	18.42	[118]
34	Nb_2O_5/2D g-C_3N_4	Subtractive manufacturing method	22,600	[119]
35	S,N-co-doped carbon dots (S,N-CDs)/g-C_3N_4 nanosheet	Hydrothermal	832	[120]
36	Hollow tubular g-C_3N_4	Hydrothermal calcination	235.68	[121]

Conclusions

In conclusion, the advancements in the utilization of g–C_3N_4 based photocatalysts have perceived an epitome of novel research work in the utilization of abundantly available solar light for H_2 production. From this perspective, the H_2 generation via water splitting have well- exploited the exclusive features of g-C_3N_4, which comprises the abundant availability of its constituting elements; carbon and nitrogen, visible-light absorbance, required band-energy gap, metal-free 2-D structure, and high chemical stability. A tremendous work has been done in this field since its first utilization in 2009. Various methodologies have been discussed for enhancing the H_2 production ability of g-C_3N_4 based photocatalysts such as band gap engineering, introduction of various dopants, and nanostructure designing, etc. Finally, it is supposed that this chapter will be helpful in understanding the utilization of g-C_3N_4 based photocatalysts for H_2 production and their modification with nanoscale features for substantial advances in accomplishing an effective, eco-friendly and economical energy production system.

References

[1] I. Staffell, D. Scamman, A.V. Abad, P. Balcombe, P.E. Dodds, P. Ekins, N. Shah, K.R. Ward, The role of hydrogen and fuel cells in the global energy system, Energy & Environmental Science 12 (2019) 463-491. https://doi.org/10.1039/C8EE01157E

[2] N.S. Lewis, D.G. Nocera, Powering the planet: Chemical challenges in solar energy utilization, Proceedings of the National Academy of Sciences 103 (2006) 15729-15735. https://doi.org/10.1073/pnas.0603395103

[3] I. Dincer, C. Acar, Review and evaluation of hydrogen production methods for better sustainability, International Journal of Hydrogen Energy 40 (2015) 11094-11111. https://doi.org/10.1016/j.ijhydene.2014.12.035

[4] C. Acar, I. Dincer, Review and evaluation of hydrogen production options for better environment, Journal of Cleaner Production 218 (2019) 835-849. https://doi.org/10.1016/j.jclepro.2019.02.046

[5] D. Pathania, G. Sharma, A. Kumar, N. Kothiyal, Fabrication of nanocomposite polyaniline zirconium (IV) silicophosphate for photocatalytic and antimicrobial activity, Journal of alloys and compounds 588 (2014) 668-675. https://doi.org/10.1016/j.jallcom.2013.11.133

[6] R. Malik, V.K. Tomer, State-of-the-art review of morphological advancements in graphitic carbon nitride (g-CN) for sustainable hydrogen production, Renewable and Sustainable Energy Reviews 135 (2017) 110235. https://doi.org/10.1016/j.rser.2020.110235

[7] A. Fujishima, K. Honda, Electrochemical photolysis of water at a semiconductor electrode, nature 238 (1972) 37. https://doi.org/10.1038/238037a0

[8] D.Y. Leung, X. Fu, C. Wang, M. Ni, M.K. Leung, X. Wang, X. Fu, Hydrogen production over titania-based photocatalysts, ChemSusChem 3 (2010) 681-694. https://doi.org/10.1002/cssc.201000014

[9] G. Ma, T. Hisatomi, K. Domen, Semiconductors for photocatalytic and photoelectrochemical solar water splitting, From molecules to materials, Springer2015, pp. 1-56. https://doi.org/10.1007/978-3-319-13800-8_1

[10] J. Wen, J. Xie, X. Chen, X. Li, A review on g-C_3N_4-based photocatalysts, Applied surface science 391 (2017) 72-123. https://doi.org/10.1016/j.apsusc.2016.07.030

[11] A. Kumar, G. Sharma, M. Naushad, H. Ala'a, A. Kumar, I. Hira, T. Ahamad, A.A. Ghfar, F.J. Stadler, Visible photodegradation of ibuprofen and 2, 4-D in simulated waste water using sustainable metal free-hybrids based on carbon nitride and biochar, Journal of environmental management 231 (2019) 1164-1175. https://doi.org/10.1016/j.jenvman.2018.11.015

[12] A. Kumar, S.K. Sharma, G. Sharma, M. Naushad, F.J. Stadler, CeO_2/g-C_3N_4/V_2O_5 ternary nano hetero-structures decorated with CQDs for enhanced photo-reduction capabilities under different light sources: Dual Z-scheme mechanism, Journal of Alloys and Compounds (2020) 155692. https://doi.org/10.1016/j.jallcom.2020.155692

[13] A. Kumar, A. Kumar, G. Sharma, H. Ala'a, M. Naushad, A.A. Ghfar, F.J. Stadler, Quaternary magnetic BiOCl/g-C_3N_4/Cu_2O/Fe_3O_4 nano-junction for visible light and solar powered degradation of sulfamethoxazole from aqueous environment, Chemical Engineering Journal 334 (2018) 462-478. https://doi.org/10.1016/j.cej.2017.10.049

[14] G. Sharma, A. Kumar, S. Sharma, H. Ala'a, M. Naushad, A.A. Ghfar, T. Ahamad, F.J. Stadler, Fabrication and characterization of novel Fe^0@ Guar gum-crosslinked-soya lecithin nanocomposite hydrogel for photocatalytic degradation of methyl violet dye, Separation and Purification Technology 211 (2019) 895-908. https://doi.org/10.1016/j.seppur.2018.10.028

[15] J. Fu, J. Yu, C. Jiang, B. Cheng, g-C_3N_4-Based heterostructured photocatalysts, Advanced Energy Materials 8 (2018) 1701503. https://doi.org/10.1002/aenm.201701503

[16] S. Cao, J. Yu, g-C_3N_4-based photocatalysts for hydrogen generation, The journal of physical chemistry letters 5 (2014) 2101-2107. https://doi.org/10.1021/jz500546b

[17] X. Yang, F. Qian, G. Zou, M. Li, J. Lu, Y. Li, M. Bao, Facile fabrication of acidified g-C_3N_4/g-C_3N_4 hybrids with enhanced photocatalysis performance under visible light irradiation, Applied Catalysis B: Environmental 193 (2016) 22-35. https://doi.org/10.1016/j.apcatb.2016.03.060

[18] W. Liu, M. Wang, C. Xu, S. Chen, Facile synthesis of g-C_3N_4/ZnO composite with enhanced visible light photooxidation and photoreduction properties, Chemical Engineering Journal 209 (2012) 386-393. https://doi.org/10.1016/j.cej.2012.08.033

[19] T. Tyborski, C. Merschjann, S. Orthmann, F. Yang, M.-C. Lux-Steiner, T. Schedel-Niedrig, Crystal structure of polymeric carbon nitride and the determination of its process-temperature-induced modifications, Journal of Physics: Condensed Matter 25 (2013) 395402. https://doi.org/10.1088/0953-8984/25/39/395402

[20] A. Thomas, A. Fischer, F. Goettmann, M. Antonietti, J.-O. Müller, R. Schlögl, J.M. Carlsson, Graphitic carbon nitride materials: variation of structure and morphology and their use as metal-free catalysts, Journal of Materials Chemistry 18 (2008) 4893-4908. https://doi.org/10.1039/b800274f

[21] G. Sharma, B. Thakur, A. Kumar, S. Sharma, M. Naushad, F.J. Stadler, Gum Acacia-cl-poly (acrylamide)@ carbon nitride Nanocomposite Hydrogel for Adsorption of Ciprofloxacin and its Sustained Release in Artificial Ocular Solution, Macromolecular Materials and Engineering 305 (2020) 2000274. https://doi.org/10.1002/mame.202000274

[22] B. Wang, H. Yu, X. Quan, S. Chen, Ultra-thin g-C_3N_4 nanosheets wrapped silicon nanowire array for improved chemical stability and enhanced photoresponse, Materials Research Bulletin 59 (2014) 179-184. https://doi.org/10.1016/j.materresbull.2014.07.011

[23] L. Sun, X. Zhao, C.-J. Jia, Y. Zhou, X. Cheng, P. Li, L. Liu, W. Fan, Enhanced visible-light photocatalytic activity of gC_3N_4–$ZnWO_4$ by fabricating a heterojunction: investigation based on experimental and theoretical studies, Journal of Materials Chemistry 22 (2012) 23428-23438. https://doi.org/10.1039/c2jm34965e

[24] A. Kumari, A. Kumar, G. Sharma, J. Iqbal, M. Naushad, F.J. Stadler, Constructing Z-scheme $LaTiO_2N$/g-C_3N_4@Fe_3O_4 magnetic nano heterojunctions with promoted charge separation for visible and solar removal of indomethacin, Journal of Water Process Engineering 36 (2020) 101391. https://doi.org/10.1016/j.jwpe.2020.101391

[25] G. Dong, Y. Zhang, Q. Pan, J. Qiu, A fantastic graphitic carbon nitride (g-C_3N_4) material: electronic structure, photocatalytic and photoelectronic properties, Journal of Photochemistry and Photobiology C: Photochemistry Reviews 20 (2014) 33-50. https://doi.org/10.1016/j.jphotochemrev.2014.04.002

[26] Q. Dong, N. Mohamad Latiff, V. Mazánek, N.F. Rosli, H.L. Chia, Z.k. Sofer, M. Pumera, Triazine-and Heptazine-Based Carbon Nitrides: Toxicity, ACS Applied Nano Materials 1 (2018) 4442-4449. https://doi.org/10.1021/acsanm.8b00708

[27] X. Wang, K. Maeda, X. Chen, K. Takanabe, K. Domen, Y. Hou, X. Fu, M. Antonietti, Polymer semiconductors for artificial photosynthesis: hydrogen evolution by mesoporous graphitic carbon nitride with visible light, Journal of the American Chemical Society 131 (2009) 1680-1681. https://doi.org/10.1021/ja809307s

Materials Research Forum LLC
https://doi.org/10.21741/9781644901359-5

[28] N. Mansor, A.B. Jorge, F. Corà, C. Gibbs, R. Jervis, P.F. McMillan, X. Wang, D.J. Brett, Graphitic carbon nitride supported catalysts for polymer electrolyte fuel cells, The Journal of Physical Chemistry C 118 (2014) 6831-6838. https://doi.org/10.1021/jp412501j

[29] J. Wang, B. Yang, S. Li, B. Yan, H. Xu, K. Zhang, Y. Shi, C. Zhai, Y. Du, Enhanced photo-electrochemical response of reduced graphene oxide and C_3N_4 nanosheets for rutin detection, Journal of colloid and interface science 506 (2017) 329-337. https://doi.org/10.1016/j.jcis.2017.07.059

[30] Z. Jiang, X. Lv, D. Jiang, J. Xie, D. Mao, Natural leaves-assisted synthesis of nitrogen-doped, carbon-rich nanodots-sensitized, Ag-loaded anatase TiO_2 square nanosheets with dominant {001} facets and their enhanced catalytic applications, Journal of Materials Chemistry A 1 (2013) 14963-14972. https://doi.org/10.1039/c3ta13248j

[31] A. Kumar, A. Kumari, G. Sharma, B. Du, M. Naushad, F.J. Stadler, Carbon quantum dots and reduced graphene oxide modified self-assembled $S@C_3N_4/B@C_3N_4$ metal-free nano-photocatalyst for high performance degradation of chloramphenicol, Journal of Molecular Liquids 300 (2020) 112356. https://doi.org/10.1016/j.molliq.2019.112356

[32] Y.-P. Zhu, T.-Z. Ren, Z.-Y. Yuan, Mesoporous phosphorus-doped g-C_3N_4 nanostructured flowers with superior photocatalytic hydrogen evolution performance, ACS applied materials & interfaces 7 (2015) 16850-16856. https://doi.org/10.1021/acsami.5b04947

[33] X. Chen, X. Zhao, Z. Kong, W.-J. Ong, N. Li, Unravelling the electrochemical mechanisms for nitrogen fixation on single transition metal atoms embedded in defective graphitic carbon nitride, Journal of Materials Chemistry A 6 (2018) 21941-21948. https://doi.org/10.1039/C8TA06497K

[34] D. Zeng, P. Wu, W.-J. Ong, B. Tang, M. Wu, H. Zheng, Y. Chen, D.-L. Peng, Construction of network-like and flower-like 2H-$MoSe_2$ nanostructures coupled with porous g-C_3N_4 for noble-metal-free photocatalytic H_2 evolution under visible light, Applied Catalysis B: Environmental 233 (2018) 26-34. https://doi.org/10.1016/j.apcatb.2018.03.102

[35] G. Sharma, A. Kumar, S. Sharma, M. Naushad, P. Dhiman, D.-V.N. Vo, F.J. Stadler, $Fe_3O_4/ZnO/Si_3N_4$ nanocomposite based photocatalyst for the degradation of dyes from aqueous solution, Materials Letters 278 (2020) 128359. https://doi.org/10.1016/j.matlet.2020.128359

[36] J. Xu, L. Zhang, R. Shi, Y. Zhu, Chemical exfoliation of graphitic carbon nitride for efficient heterogeneous photocatalysis, Journal of Materials Chemistry A 1 (2013) 14766-14772. https://doi.org/10.1039/c3ta13188b

[37] Y. Zeng, C. Liu, L. Wang, S. Zhang, Y. Ding, Y. Xu, Y. Liu, S. Luo, A three-dimensional graphitic carbon nitride belt network for enhanced visible light photocatalytic hydrogen evolution, Journal of Materials Chemistry A 4 (2016) 19003-19010. https://doi.org/10.1039/C6TA07397B

[38] X. Ma, Y. Lv, J. Xu, Y. Liu, R. Zhang, Y. Zhu, A strategy of enhancing the photoactivity of g-C_3N_4 via doping of nonmetal elements: a first-principles study, The Journal of Physical Chemistry C 116 (2012) 23485-23493. https://doi.org/10.1021/jp308334x

[39] A. Mishra, A. Mehta, S. Basu, N.P. Shetti, K.R. Reddy, T.M. Aminabhavi, Graphitic carbon nitride (g–C_3N_4)–based metal-free photocatalysts for water splitting: a review, Carbon 149 (2019) 693-721. https://doi.org/10.1016/j.carbon.2019.04.104

[40] A. Kumar, G. Sharma, M. Naushad, H. Ala'a, A. Garcia-Penas, G.T. Mola, C. Si, F.J. Stadler, Bio-inspired and biomaterials-based hybrid photocatalysts for environmental detoxification: A review, Chemical Engineering Journal 382 (2020) 122937. https://doi.org/10.1016/j.cej.2019.122937

[41] K. Maeda, Z-scheme water splitting using two different semiconductor photocatalysts, Acs Catalysis 3 (2013) 1486-1503. https://doi.org/10.1021/cs4002089

[42] J. Low, C. Jiang, B. Cheng, S. Wageh, A.A. Al-Ghamdi, J. Yu, A review of direct Z-scheme photocatalysts, Small Methods 1 (2017) 1700080. https://doi.org/10.1002/smtd.201700080

[43] Q. Xu, L. Zhang, B. Cheng, J. Fan, J. Yu, S-scheme heterojunction photocatalyst, Chem (2020). https://doi.org/10.1016/j.chempr.2020.06.010

[44] K. Maeda, K. Domen, Photocatalytic water splitting: recent progress and future challenges, The Journal of Physical Chemistry Letters 1 (2010) 2655-2661. https://doi.org/10.1021/jz1007966

[45] Y. Cao, Z. Wu, J. Ni, W.A. Bhutto, J. Li, S. Li, K. Huang, J. Kang, Type-II core/shell nanowire heterostructures and their photovoltaic applications, Nano-Micro Letters 4 (2012) 135-141. https://doi.org/10.1007/BF03353704

[46] Q. Xu, L. Zhang, J. Yu, S. Wageh, A.A. Al-Ghamdi, M. Jaroniec, Direct Z-scheme photocatalysts: Principles, synthesis, and applications, Materials Today 21 (2018) 1042-1063. https://doi.org/10.1016/j.mattod.2018.04.008

[47] R.B. Chandran, S. Breen, Y. Shao, S. Ardo, A.Z. Weber, Evaluating particle-suspension reactor designs for Z-scheme solar water splitting via transport and kinetic

modeling, Energy & Environmental Science 11 (2018) 115-135.
https://doi.org/10.1039/C7EE01360D

[48] H. Tada, T. Mitsui, T. Kiyonaga, T. Akita, K. Tanaka, All-solid-state Z-scheme in CdS–Au–TiO 2 three-component nanojunction system, Nature materials 5 (2006) 782-786. https://doi.org/10.1038/nmat1734

[49] H. Li, H. Yu, X. Quan, S. Chen, Y. Zhang, Uncovering the key role of the fermi level of the electron mediator in a Z-scheme photocatalyst by detecting the charge transfer process of WO_3-metal-gC_3N_4 (Metal= Cu, Ag, Au), ACS Applied Materials & Interfaces 8 (2016) 2111-2119. https://doi.org/10.1021/acsami.5b10613

[50] Y. Sasaki, H. Nemoto, K. Saito, A. Kudo, Solar water splitting using powdered photocatalysts driven by Z-schematic interparticle electron transfer without an electron mediator, The Journal of Physical Chemistry C 113 (2009) 17536-17542. https://doi.org/10.1021/jp907128k

[51] A. Kudo, Y. Miseki, Heterogeneous photocatalyst materials for water splitting, Chemical Society Reviews 38 (2009) 253-278. https://doi.org/10.1039/B800489G

[52] F.E. Osterloh, Inorganic materials as catalysts for photochemical splitting of water, Chemistry of materials 20 (2007) 35-54. https://doi.org/10.1021/cm7024203

[53] X. Chen, S. Shen, L. Guo, S.S. Mao, Semiconductor-based photocatalytic hydrogen generation, Chemical reviews 110 (2010) 6503-6570. https://doi.org/10.1021/cr1001645

[54] J.C. Colmenares, E. Kuna, Photoactive Hybrid Catalysts Based on Natural and Synthetic Polymers: A Comparative Overview, Molecules 22 (2017) 790. https://doi.org/10.3390/molecules22050790

[55] L. Song, C. Guo, T. Li, S. Zhang, C_{60}/graphene/g-C_3N_4 composite photocatalyst and mutually-reinforcing synergy to improve hydrogen production in splitting water under visible light radiation, Ceramics International 43 (2017) 7901-7907. https://doi.org/10.1016/j.ceramint.2017.03.115

[56] L. Ge, C. Han, J. Liu, Novel visible light-induced g-C_3N_4/Bi_2WO_6 composite photocatalysts for efficient degradation of methyl orange, Applied Catalysis B: Environmental 108-109 (2011) 100-107. https://doi.org/10.1016/j.apcatb.2011.08.014

[57] S. Ahmed, M.G. Rasul, W.N. Martens, R. Brown, M.A. Hashib, Heterogeneous photocatalytic degradation of phenols in wastewater: A review on current status and developments, Desalination 261 (2010) 3-18. https://doi.org/10.1016/j.desal.2010.04.062

[58] W. Zou, B. Deng, X. Hu, Y. Zhou, Y. Pu, S. Yu, K. Ma, J. Sun, H. Wan, L. Dong, Crystal-plane-dependent metal oxide-support interaction in CeO_2/g-C_3N_4 for

photocatalytic hydrogen evolution, Applied Catalysis B: Environmental 238 (2018) 111-118. https://doi.org/10.1016/j.apcatb.2018.07.022

[59] W. Chen, Z.-C. He, G.-B. Huang, C.-L. Wu, W.-F. Chen, X.-H. Liu, Direct Z-scheme 2D/2D $MnIn_2S_4$/g-C_3N_4 architectures with highly efficient photocatalytic activities towards treatment of pharmaceutical wastewater and hydrogen evolution, Chemical Engineering Journal 359 (2019) 244-253. https://doi.org/10.1016/j.cej.2018.11.141

[60] D. Xu, L. Li, T. Xia, F. Wan, F. Wang, H. Bai, W. Shi, Heterojunction composites of g-C_3N_4/$KNbO_3$ enhanced photocatalytic properties for water splitting, International Journal of Hydrogen Energy 43 (2018) 16566-16572. https://doi.org/10.1016/j.ijhydene.2018.07.068

[61] H. Li, M. Wang, Y. Wei, F. Long, Noble metal-free NiS_2 with rich active sites loaded g-C_3N_4 for highly efficient photocatalytic H_2 evolution under visible light irradiation, Journal of colloid and interface science 534 (2019) 343-349. https://doi.org/10.1016/j.jcis.2018.09.041

[62] G. Nagaraju, K. Manjunath, S. Sarkar, E. Gunter, S.R. Teixeira, J. Dupont, TiO_2–RGO hybrid nanomaterials for enhanced water splitting reaction, International Journal of Hydrogen Energy 40 (2015) 12209-12216. https://doi.org/10.1016/j.ijhydene.2015.07.094

[63] M.R. Gholipour, F. Béland, T.-O. Do, Graphitic carbon nitride-titanium dioxide nanocomposite for photocatalytic hydrogen production under visible light, International Journal of Chemical Reactor Engineering 14 (2016) 851-858. https://doi.org/10.1515/ijcre-2015-0094

[64] H.Y. Hafeez, S.K. Lakhera, S. Bellamkonda, G.R. Rao, M. Shankar, D. Bahnemann, B. Neppolian, Construction of ternary hybrid layered reduced graphene oxide supported g-C_3N_4-TiO_2 nanocomposite and its photocatalytic hydrogen production activity, international journal of hydrogen energy 43 (2018) 3892-3904. https://doi.org/10.1016/j.ijhydene.2017.09.048

[65] H. Liu, D. Chen, Z. Wang, H. Jing, R. Zhang, Microwave-assisted molten-salt rapid synthesis of isotype triazine-/heptazine based g-C_3N_4 heterojunctions with highly enhanced photocatalytic hydrogen evolution performance, Applied Catalysis B: Environmental 203 (2017) 300-313. https://doi.org/10.1016/j.apcatb.2016.10.014

[66] Y. Tachibana, L. Vayssieres, J.R. Durrant, Artificial photosynthesis for solar water-splitting, Nature Photonics 6 (2012) 511. https://doi.org/10.1038/nphoton.2012.175

[67] S. Sun, W. Wang, D. Li, L. Zhang, D. Jiang, Solar light driven pure water splitting on quantum sized BiVO4 without any cocatalyst, ACS Catalysis 4 (2014) 3498-3503. https://doi.org/10.1021/cs501076a

[68] Y. Zhang, T. Mori, J. Ye, M. Antonietti, Phosphorus-doped carbon nitride solid: enhanced electrical conductivity and photocurrent generation, Journal of the American Chemical Society 132 (2010) 6294-6295. https://doi.org/10.1021/ja101749y

[69] J. Zhang, Y. Wang, J. Jin, J. Zhang, Z. Lin, F. Huang, J. Yu, Efficient visible-light photocatalytic hydrogen evolution and enhanced photostability of core/shell CdS/g-C$_3$N$_4$ nanowires, ACS Applied Materials & Interfaces 5 (2013) 10317-10324. https://doi.org/10.1021/am403327g

[70] H. Liu, Z. Jin, Z. Xu, Z. Zhang, D. Ao, Fabrication of Zn$_2$InS$_4$–gC$_3$N$_4$ sheet-on-sheet nanocomposites for efficient visible-light photocatalytic H 2-evolution and degradation of organic pollutants, RSC Advances 5 (2015) 97951-97961. https://doi.org/10.1039/C5RA17028A

[71] N. Ding, L. Zhang, H. Zhang, J. Shi, H. Wu, Y. Luo, D. Li, Q. Meng, Microwave-assisted synthesis of ZnIn$_2$S$_4$/g-C$_3$N$_4$ heterojunction photocatalysts for efficient visible light photocatalytic hydrogen evolution, Catalysis Communications 100 (2017) 173-177. https://doi.org/10.1016/j.catcom.2017.06.050

[72] Z. Zou, J. Ye, K. Sayama, H. Arakawa, Direct splitting of water under visible light irradiation with an oxide semiconductor photocatalyst, nature 414 (2001) 625. https://doi.org/10.1038/414625a

[73] A. Kumar, G. Sharma, M. Naushad, T. Ahamad, R.C. Veses, F.J. Stadler, Highly visible active Ag$_2$CrO$_4$/Ag/BiFeO$_3$@ RGO nano-junction for photoreduction of CO$_2$ and photocatalytic removal of ciprofloxacin and bromate ions: The triggering effect of Ag and RGO, Chemical Engineering Journal 370 (2019) 148-165. https://doi.org/10.1016/j.cej.2019.03.196

[74] F. Shi, L. Chen, M. Chen, D. Jiang, A gC$_3$N$_4$/nanocarbon/ZnIn$_2$S$_4$ nanocomposite: an artificial Z-scheme visible-light photocatalytic system using nanocarbon as the electron mediator, Chemical Communications 51 (2015) 17144-17147. https://doi.org/10.1039/C5CC05323D

[75] L. Ge, C. Han, X. Xiao, L. Guo, In situ synthesis of cobalt–phosphate (Co–Pi) modified g-C$_3$N$_4$ photocatalysts with enhanced photocatalytic activities, Applied Catalysis B: Environmental 142 (2013) 414-422. https://doi.org/10.1016/j.apcatb.2013.05.051

[76] C.-J. Chang, H.-T. Weng, C.-C. Chang, CuSZnS$_{1-x}$O$_x$/g-C$_3$N$_4$ heterostructured photocatalysts for efficient photocatalytic hydrogen production, International Journal

of Hydrogen Energy 42 (2017) 23568-23577.
https://doi.org/10.1016/j.ijhydene.2017.01.047

[77] S.-W. Cao, Y.-P. Yuan, J. Fang, M.M. Shahjamali, F.Y. Boey, J. Barber, S.C.J. Loo, C. Xue, In-situ growth of CdS quantum dots on g-C_3N_4 nanosheets for highly efficient photocatalytic hydrogen generation under visible light irradiation, International Journal of Hydrogen Energy 38 (2013) 1258-1266.
https://doi.org/10.1016/j.ijhydene.2012.10.116

[78] S.-W. Cao, X.-F. Liu, Y.-P. Yuan, Z.-Y. Zhang, Y.-S. Liao, J. Fang, S.C.J. Loo, T.C. Sum, C. Xue, Solar-to-fuels conversion over In_2O_3/g-C_3N_4 hybrid photocatalysts, Applied Catalysis B: Environmental 147 (2014) 940-946.
https://doi.org/10.1016/j.apcatb.2013.10.029

[79] Q. Li, B. Guo, J. Yu, J. Ran, B. Zhang, H. Yan, J.R. Gong, Highly efficient visible-light-driven photocatalytic hydrogen production of CdS-cluster-decorated graphene nanosheets, Journal of the American Chemical Society 133 (2011) 10878-10884.
https://doi.org/10.1021/ja2025454

[80] J. Hou, C. Yang, H. Cheng, Z. Wang, S. Jiao, H. Zhu, Ternary 3D architectures of CdS QDs/graphene/$ZnIn_2S_4$ heterostructures for efficient photocatalytic H_2 production, Physical Chemistry Chemical Physics 15 (2013) 15660-15668.
https://doi.org/10.1039/c3cp51857d

[81] Q. Han, B. Wang, J. Gao, Z. Cheng, Y. Zhao, Z. Zhang, L. Qu, Atomically thin mesoporous nanomesh of graphitic C_3N_4 for high-efficiency photocatalytic hydrogen evolution, ACS nano 10 (2016) 2745-2751. https://doi.org/10.1021/acsnano.5b07831

[82] J. Ye, Z. Zou, M. Oshikiri, A. Matsushita, M. Shimoda, M. Imai, T. Shishido, A novel hydrogen-evolving photocatalyst $InVO_4$ active under visible light irradiation, Chemical Physics Letters 356 (2002) 221-226. https://doi.org/10.1016/S0009-2614(02)00254-3

[83] L. Liao, Q. Zhang, Z. Su, Z. Zhao, Y. Wang, Y. Li, X. Lu, D. Wei, G. Feng, Q. Yu, Efficient solar water-splitting using a nanocrystalline CoO photocatalyst, Nature nanotechnology 9 (2014) 69. https://doi.org/10.1038/nnano.2013.272

[84] Y. Yan, F. Cai, Y. Song, W. Shi, $InVO_4$ nanocrystal photocatalysts: Microwave-assisted synthesis and size-dependent activities of hydrogen production from water splitting under visible light, Chemical engineering journal 233 (2013) 1-7.
https://doi.org/10.1016/j.cej.2013.06.121

[85] Y. Wang, N. Herron, Nanometer-sized semiconductor clusters: materials synthesis, quantum size effects, and photophysical properties, The Journal of Physical Chemistry 95 (1991) 525-532. https://doi.org/10.1021/j100155a009

[86] B. Hu, F. Cai, T. Chen, M. Fan, C. Song, X. Yan, W. Shi, Hydrothermal synthesis g-C_3N_4/Nano-InVO$_4$ nanocomposites and enhanced photocatalytic activity for hydrogen production under visible light irradiation, ACS applied materials & interfaces 7 (2015) 18247-18256. https://doi.org/10.1021/acsami.5b05715

[87] X. Shi, M. Fujitsuka, S. Kim, T. Majima, Faster Electron Injection and More Active Sites for Efficient Photocatalytic H_2 Evolution in g-C_3N_4/MoS$_2$ Hybrid, Small 14 (2018) 1703277. https://doi.org/10.1002/smll.201703277

[88] H.-J. Li, D.-J. Qian, M. Chen, Templateless infrared heating process for fabricating carbon nitride nanorods with efficient photocatalytic H_2 evolution, ACS Applied Materials & Interfaces 7 (2015) 25162-25170. https://doi.org/10.1021/acsami.5b06627

[89] Y. Tan, Z. Shu, J. Zhou, T. Li, W. Wang, Z. Zhao, One-step synthesis of nanostructured g-C_3N_4/TiO$_2$ composite for highly enhanced visible-light photocatalytic H2 evolution, Applied Catalysis B: Environmental 230 (2018) 260-268. https://doi.org/10.1016/j.apcatb.2018.02.056

[90] Z. Dong, Y. Wu, N. Thirugnanam, G. Li, Double Z-scheme ZnO/ZnS/g-C_3N_4 ternary structure for efficient photocatalytic H_2 production, Applied Surface Science 430 (2018) 293-300. https://doi.org/10.1016/j.apsusc.2017.07.186

[91] Z.-A. Lan, G. Zhang, X. Wang, A facile synthesis of Br-modified g-C_3N_4 semiconductors for photoredox water splitting, Applied Catalysis B: Environmental 192 (2016) 116-125. https://doi.org/10.1016/j.apcatb.2016.03.062

[92] F. Wei, Y. Liu, H. Zhao, X. Ren, J. Liu, T. Hasan, L. Chen, Y. Li, B.-L. Su, Oxygen self-doped gC$_3$N 4 with tunable electronic band structure for unprecedentedly enhanced photocatalytic performance, Nanoscale 10 (2018) 4515-4522. https://doi.org/10.1039/C7NR09660G

[93] J. Fang, H. Fan, M. Li, C. Long, Nitrogen self-doped graphitic carbon nitride as efficient visible light photocatalyst for hydrogen evolution, Journal of Materials Chemistry A 3 (2015) 13819-13826. https://doi.org/10.1039/C5TA02257F

[94] Q. Xiang, J. Yu, M. Jaroniec, Preparation and enhanced visible-light photocatalytic H_2-production activity of graphene/C$_3$N$_4$ composites, The Journal of Physical Chemistry C 115 (2011) 7355-7363. https://doi.org/10.1021/jp200953k

[95] J.-P. Zou, L.-C. Wang, J. Luo, Y.-C. Nie, Q.-J. Xing, X.-B. Luo, H.-M. Du, S.-L. Luo, S.L. Suib, Synthesis and efficient visible light photocatalytic H_2 evolution of a metal-free g-C_3N_4/graphene quantum dots hybrid photocatalyst, Applied Catalysis B: Environmental 193 (2016) 103-109. https://doi.org/10.1016/j.apcatb.2016.04.017

[96] F. He, G. Chen, Y. Zhou, Y. Yu, L. Li, S. Hao, B. Liu, ZIF-8 derived carbon (C-ZIF) as a bifunctional electron acceptor and HER cocatalyst for gC$_3$N$_4$: construction of a

189

metal-free, all carbon-based photocatalytic system for efficient hydrogen evolution, Journal of Materials Chemistry A 4 (2016) 3822-3827. https://doi.org/10.1039/C6TA00497K

[97] F. He, G. Chen, J. Miao, Z. Wang, D. Su, S. Liu, W. Cai, L. Zhang, S. Hao, B. Liu, Sulfur-mediated self-templating synthesis of tapered C-PAN/g-C$_3$N$_4$ composite nanotubes toward efficient photocatalytic H$_2$ evolution, ACS Energy Letters 1 (2016) 969-975. https://doi.org/10.1021/acsenergylett.6b00398

[98] X. Xu, G. Liu, C. Randorn, J.T. Irvine, g-C$_3$N$_4$ coated SrTiO$_3$ as an efficient photocatalyst for H$_2$ production in aqueous solution under visible light irradiation, international journal of hydrogen energy 36 (2011) 13501-13507. https://doi.org/10.1016/j.ijhydene.2011.08.052

[99] N. Xiao, S. Li, S. Liu, B. Xu, Y. Li, Y. Gao, L. Ge, G. Lu, Novel PtPd alloy nanoparticle-decorated g-C$_3$N$_4$ nanosheets with enhanced photocatalytic activity for H2 evolution under visible light irradiation, Chinese Journal of Catalysis 40 (2019) 352-361. https://doi.org/10.1016/S1872-2067(18)63180-8

[100] X. She, L. Liu, H. Ji, Z. Mo, Y. Li, L. Huang, D. Du, H. Xu, H. Li, Template-free synthesis of 2D porous ultrathin nonmetal-doped g-C$_3$N$_4$ nanosheets with highly efficient photocatalytic H$_2$ evolution from water under visible light, Applied Catalysis B: Environmental 187 (2016) 144-153. https://doi.org/10.1016/j.apcatb.2015.12.046

[101] B. Wang, J. Zhang, F. Huang, Enhanced visible light photocatalytic H$_2$ evolution of metal-free g-C$_3$N$_4$/SiC heterostructured photocatalysts, Applied Surface Science 391 (2017) 449-456. https://doi.org/10.1016/j.apsusc.2016.07.056

[102] K. Qi, K. Lv, I. Khan, S.-y. Liu, Photocatalytic H$_2$ generation via CoP quantum-dot-modified g-C$_3$N$_4$ synthesized by electroless plating, Chinese Journal of Catalysis 41 (2020) 114-121. https://doi.org/10.1016/S1872-2067(19)63459-5

[103] Y. Zou, J.-W. Shi, L. Sun, D. Ma, S. Mao, Y. Lv, Y. Cheng, Energy-band-controlled Zn$_x$Cd$_{1-x}$In$_2$S$_4$ solid solution coupled with g-C$_3$N$_4$ nanosheets as 2D/2D heterostructure toward efficient photocatalytic H$_2$ evolution, Chemical Engineering Journal 378 (2019) 122192. https://doi.org/10.1016/j.cej.2019.122192

[104] J. Zhang, F. Huang, Enhanced visible light photocatalytic H2 production activity of g-C$_3$N$_4$ via carbon fiber, Applied Surface Science 358 (2015) 287-295. https://doi.org/10.1016/j.apsusc.2015.08.089

[105] C. Ji, S.-N. Yin, S. Sun, S. Yang, An in situ mediator-free route to fabricate Cu$_2$O/g-C$_3$N$_4$ type-II heterojunctions for enhanced visible-light photocatalytic H2 generation, Applied Surface Science 434 (2018) 1224-1231. https://doi.org/10.1016/j.apsusc.2017.11.233

[106] Y. Zhong, W. Chen, S. Yu, Z. Xie, S. Wei, Y. Zhou, CdSe quantum dots/g-C_3N_4 heterostructure for efficient H_2 production under visible light irradiation, ACS omega 3 (2018) 17762-17769. https://doi.org/10.1021/acsomega.8b02585

[107] J. Meng, Z. Lan, T. Chen, Q. Lin, H. Liu, X. Wei, Y. Lu, J. Li, Z. Zhang, Organic–Organic Hybrid g-C_3N_4/Ethanediamine Nanosheets for Photocatalytic H_2 Evolution, The Journal of Physical Chemistry C 122 (2018) 24725-24731. https://doi.org/10.1021/acs.jpcc.8b07014

[108] Z. Mao, J. Chen, Y. Yang, D. Wang, L. Bie, B.D. Fahlman, Novel g-C_3N_4/CoO nanocomposites with significantly enhanced visible-light photocatalytic activity for H_2 evolution, ACS applied materials & interfaces 9 (2017) 12427-12435. https://doi.org/10.1021/acsami.7b00370

[109] M.S. Akple, J. Low, S. Wageh, A.A. Al-Ghamdi, J. Yu, J. Zhang, Enhanced visible light photocatalytic H_2-production of g-C_3N_4/WS_2 composite heterostructures, Applied Surface Science 358 (2015) 196-203. https://doi.org/10.1016/j.apsusc.2015.08.250

[110] J. Wang, Y. Xia, H. Zhao, G. Wang, L. Xiang, J. Xu, S. Komarneni, Oxygen defects-mediated Z-scheme charge separation in g-C_3N_4/ZnO photocatalysts for enhanced visible-light degradation of 4-chlorophenol and hydrogen evolution, Applied Catalysis B: Environmental 206 (2017) 406-416. https://doi.org/10.1016/j.apcatb.2017.01.067

[111] R. Zhang, L. Bi, D. Wang, Y. Lin, X. Zou, T. Xie, Z. Li, Investigation on various photo-generated carrier transfer processes of SnS_2/g-C_3N_4 heterojunction photocatalysts for hydrogen evolution, Journal of colloid and interface science 578 (2020) 431-440. https://doi.org/10.1016/j.jcis.2020.04.033

[112] Y. Liu, H. Liu, H. Zhou, T. Li, L. Zhang, A Z-scheme mechanism of N-ZnO/g-C_3N_4 for enhanced H_2 evolution and photocatalytic degradation, Applied Surface Science 466 (2019) 133-140. https://doi.org/10.1016/j.apsusc.2018.10.027

[113] W.-K. Jo, S. Moru, S. Tonda, Magnetically responsive $SnFe_2O_4$/g-C_3N_4 hybrid photocatalysts with remarkable visible-light-induced performance for degradation of environmentally hazardous substances and sustainable hydrogen production, Applied Surface Science 506 (2020) 144939. https://doi.org/10.1016/j.apsusc.2019.144939

[114] X. Han, D. Xu, L. An, C. Hou, Y. Li, Q. Zhang, H. Wang, WO_3/g-C_3N_4 two-dimensional composites for visible-light driven photocatalytic hydrogen production, International Journal of Hydrogen Energy 43 (2018) 4845-4855. https://doi.org/10.1016/j.ijhydene.2018.01.117

[115] Y. Hong, Z. Fang, B. Yin, B. Luo, Y. Zhao, W. Shi, C. Li, A visible-light-driven heterojunction for enhanced photocatalytic water splitting over Ta_2O_5 modified g-C_3N_4

photocatalyst, International Journal of Hydrogen Energy 42 (2017) 6738-6745. https://doi.org/10.1016/j.ijhydene.2016.12.055

[116] L. Luo, Z. Gong, J. Ma, K. Wang, H. Zhu, K. Li, L. Xiong, X. Guo, J. Tang, Ultrathin sulfur-doped holey carbon nitride nanosheets with superior photocatalytic hydrogen production from water, Applied Catalysis B: Environmental (2020) 119742. https://doi.org/10.1016/j.apcatb.2020.119742

[117] X. Lin, X. Hou, L. Cui, S. Zhao, H. Bi, H. Du, Y. Yuan, Increasing π-electron availability in benzene ring incorporated graphitic carbon nitride for increased photocatalytic hydrogen generation, Journal of Materials Science & Technology 65 164-170. https://doi.org/10.1016/j.jmst.2020.03.086

[118] T. Song, L. Hou, B. Long, A. Ali, G.-J. Deng, Ultrathin MXene "bridge" to accelerate charge transfer in ultrathin metal-free 0D/2D black phosphorus/g-C_3N_4 heterojunction toward photocatalytic hydrogen production, Journal of colloid and interface science 584 (2020) 474-483. https://doi.org/10.1016/j.jcis.2020.09.103

[119] J. Yi, T. Fei, L. Li, Q. Yu, S. Zhang, Y. Song, J. Lian, X. Zhu, J. Deng, H. Xu, Large-scale production of ultrathin carbon nitride-based photocatalysts for high-yield hydrogen evolution, Applied Catalysis B: Environmental 281 119475. https://doi.org/10.1016/j.apcatb.2020.119475

[120] Z. Xie, S. Yu, X.-B. Fan, S. Wei, L. Yu, Y. Zhong, X.-W. Gao, F. Wu, Y. Zhou, Wavelength-sensitive photocatalytic H_2 evolution from H2S splitting over g-C_3N_4 with S, N-codoped carbon dots as the photosensitizer, Journal of Energy Chemistry (2020). https://doi.org/10.1016/j.jechem.2020.04.051

[121] H.-X. Fang, H. Guo, C.-G. Niu, C. Liang, D.-W. Huang, N. Tang, H.-Y. Liu, Y.-Y. Yang, L. Li, Hollow tubular graphitic carbon nitride catalyst with adjustable nitrogen vacancy: Enhanced optical absorption and carrier separation for improving photocatalytic activity, Chemical Engineering Journal 402 (2020) 126185. https://doi.org/10.1016/j.cej.2020.126185

Photocatalysis
Materials Research Foundations **100** (2021) 193-207

Materials Research Forum LLC
https://doi.org/10.21741/9781644901359-6

Chapter 6

Recent advances in Photocatalytic Nitrogen Fixation

A. Saravanan[1], P. Senthil Kumar[2,3*]

[1]Department of Biotechnology, Rajalakshmi Engineering College, Chennai 602105, India

[2]Department of Chemical Engineering, Sri Sivasubramaniya Nadar College of Engineering, Chennai 603110, India

[3]SSN-Centre for Radiation, Environmental Science and Technology (SSN-CREST), SSN College of Engineering, Chennai 603110, India

*senthilkumarp@ssn.edu.in

Abstract

Nitrogen fixation is a standout amongst the most significant concoction responses in the biological system of our planet. Under the regularly pressure of the petroleum product exhaustion emergency and anthropogenic worldwide environmental change with ceaseless CO_2 emanation in the 21st century, examine focusing on the union of NH_3 under gentle conditions in an economical and condition agreeable way is lively and flourishing. Thusly, the focal point of this survey is the cutting edge designing of effective photocatalysts for dinitrogen (N_2) obsession toward NH_3 amalgamation. Creating green and feasible techniques for NH_3 combination under surrounding conditions, utilizing sustainable power source, is firmly wanted, by both modern and logical scientists. Photosynthesis for ammonia synthesis, which has as of late pulled in noteworthy consideration, straightforwardly creates NH_3 from daylight, and N_2 and H_2O by means of photocatalysis. Photocatalysts containing copious surface oxygen-opportunities and coordinative unsaturated metal locales have been demonstrated to be equipped for actuating N_2 reduction under fitting photoexcitation. A few impetus materials are examined which incorporate metal oxides, metals sulfides, carbon-based impetuses, and metal nitrides which are for the most part right now being sought after for their better use of their synergist property towards nitrogen fixation. This chapter portrays the photocatalytic reduction systems of nitrate towards unwanted items (nitrite, ammonium) and the more alluring item (dinitrogen).

Keywords

Photocatalysis, Nitrogen Fixation, Ammonia Synthesis, Co-Catalyst, Nitrite

Photocatalysis
Materials Research Foundations **100** (2021) 193-207

Materials Research Forum LLC
https://doi.org/10.21741/9781644901359-6

Contents

1.   Introduction

Nitrogen is an essential component to different bio-macromolecules essential to the course of life. The living beings ordinarily get nitrogen component in the types of smelling salts or nitrate rather than straightforwardly utilizing the N_2 atoms, even in spite of the fact that the substance of N_2 is exceptionally high in the environment. The principle wellspring of nitrogen, dinitrogen (N_2), is the biggest single part of the Earth's climate [1-3]. It is of extraordinary essentialness to create green and feasible systems for NH$_3$ synthesis, particularly utilizing sustainable power source under encompassing conditions. Up to this point, different methodologies that can be worked under trifling conditions, including electrocatalysis and photocatalysis, have been investigated with respect to reduction of N_2 for NH$_3$ synthesis, and a few promising advancements has been accomplished [4,5].

Catalysis is a vital procedure for quickening compound responses through giving diverse progress states and bringing down their actuation energies, which has been viewed as compelling for the structure of dominant part of synthetic mixes. With the developing difficulties in vitality and condition, the usage of sun-based vitality through photocatalytic forms has displayed colossal down to earth application esteems. Importantly, planning progressed photocatalysts has pulled in incredible consideration in vitality related fields, for example, water part, CO_2 reduction, nitrogen fixation. In any case, incredible difficulties still stay in catalysis territories, where a noteworthy impediment is the low clear quantum productivity of impetuses, getting from low efficiencies in light reaping and spatial charge partition [6-9].

Counterfeit photosynthesis of NH$_3$ straightforwardly from daylight, N_2 and H_2O by means of photocatalysis, is viewed as a perfect, energy-saving and environmentally- amiable

Photocatalysis
Materials Research Forum LLC
Materials Research Foundations **100** (2021) 193-207
https://doi.org/10.21741/9781644901359-6

procedure for NH_3 creation since it very well may be performed at ordinary temperature and barometrical weight utilizing sustainable sun based vitality [10]. The system of N_2 fixation is as per the following: N_2 assimilates onto the outside of the semiconductor. In the nearness of light sparkling on the semiconductor, the photogenerated electron is exchanged to the N_2 atoms. The solid triple bonds are debilitated so that N_2 can be decreased, with the cooperation of water atoms. Sadly, both the level of N_2 fixation and the change proficiency are very low because of the wasteful electron exchange from the photocatalyst to N_2 particle [11,12].

N_2 photoreduction was considered to occur in nature over plenteous minerals on the outside of the earth; in this manner, prior examinations on the photocatalytic fixation of N_2 fundamentally centred around soil minerals and sand in nature. Figure 1 shows that photochemical nitrogen fixation.

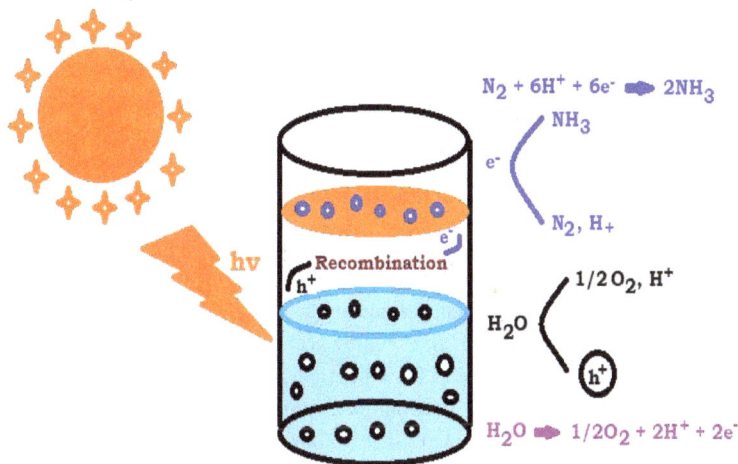

Figure 1. Photochemical nitrogen fixation.

The primary exploratory investigation on as-incorporated TiO_2-based photocatalysts for N_2 fixation under UV light in 1977. A whirlwind of research exercises have been given to the improvement of photocatalytic N2 fixation, particularly in the 21st century, because of the way that the light and water and air of this procedure are moderately perfect and cheap [13,14]. The model photocatalyst TiO_2 as a proof-of-concept to change over N_2 into NH_3 and to oxidize H_2O to oxygen all the while under UV light illumination. From that point

Photocatalysis   Materials Research Forum LLC
Materials Research Foundations **100** (2021) 193-207   https://doi.org/10.21741/9781644901359-6

forward, the examination in this field had pulled in numerous considerations in the last century, however further investigation has turned out to need substantially more testing.

2.   Advances in photocatalytic nitrogen fixation

The photocatalytic nitrogen reduction process under surrounding conditions has been widely considered in trials amid the previous couple of decades. For the most part, there are two kinds of photograph transformation frameworks for Nitrogen Reduction Reaction:

(i) Photochemical cell framework and

(ii) Photo electrochemical cell framework.

The photochemical cell framework uses semiconductor nanoparticle suspensions scattered in N_2-soaked fluid arrangements as photocatalysts. In this framework, both photographs decrease also, photograph oxidation half-responses happen on the various locales of a similar semiconductor nanoparticles, which may result in the invert response or re-oxidization of the alluring items. The photo electrochemical cell framework comprises of photo electrodes with various designs. The photo electrodes reap light to create photograph instigated electrons and openings and advance the charge partition. The initial step of N_2 photo-reduction is photoexcitation, in which the photograph prompted gaps and electrons are created in the valence band and conduction band of semiconductor nanoparticles, separately. The second step incorporates the charge detachment and relocation of photograph actuated bearers. At long last, the electrons diffuse over the mass and the outside of semiconductor photocatalysts furthermore, achieve the dynamic destinations, where the N_2 atoms adsorbed on the locales are diminished to smelling salts by the electrons; in the meantime, H_2O or other conciliatory reagents are oxidized by openings. Various methodologies including heteroatom doping, abandons building, aspect fitting, and surface plasmon adjustment were utilized to improve the photocatalytic efficiencies of the impetuses [15,16].

2.1 TiO_2 and different oxides based photocatalysts

As one of the most punctual considered semiconductor photocatalysts, TiO_2-based photocatalysts has been explored in nitrogen obsession very early, inferable from its ease, high steadiness and non-danger. N_2 on the powder blends of rutile and anatase TiO_2 with soaked H_2O vapour could be diminished to NH_3 under UV illumination alongside follow measures of N_2H_4. TiO_2 assumed a key job in nitrogen photograph obsession. Moreover, iron doping could support the yield of NH_3 and N_2H_4. Aside from filling in as the dopants in TiO_2 network, the progress metal iotas stacked on TiO_2 can go about as co-impetuses. The wasteful charge partition and exchange to the dissolvable, the poor chemisorption of

N_2 on impetus surface and the back reactions are the principle obstructions to the photochemical nitrogen reduction reaction process. To beat these issues, some attainable procedures have been created to build the particular nitrogen reduction reaction process of photocatalysts, for example, by presenting deformities or opportunities, expanding the particular surface territory of impetuses, or building cross breed material interfaces of composite impetuses or heterojunctions [17,18].

2.2   Oxyhalides based photocatalysts

Bismuth oxyhalides have drawn wide consideration as 2D layered semiconductor photocatalysts with tetragonal matlockite structure. The layered structures made up of $[Bi_2O_2]^{2+}$ chunks are isolated by two sections of halogen iotas by the van der Waals collaboration, which is anything but difficult to produce opening helpful to photocatalytic exercises. BiOX with a backhanded bandgap can successfully restrain the recombination of photograph produced electrons and openings. Though, the presented surficial oxygen opportunities on BiOX could be effectively oxidized amid the response, bringing about the quick reduction of photocatalytic nitrogen reduction reaction [19].

2.3   Graphitic nitride carbon based photocatalysts

Graphitic carbon nitride (g-C_3N_4), a novel without metal semiconductor material, has been broadly connected in numerous fields, including photocatalysis. The flexible use of g-C_3N_4 is generally because of its one of a kind physicochemical properties, for example, moderate band gap vitality, vitality stockpiling limit, gas-adsorption limit and uncommon optical properties [20-22]. As a two-dimensional conjugated polymer, graphitic carbon nitride (g-C_3N_4) has gotten extensive consideration as a minimal effort and stable photocatalyst with unmistakable light reaction for sunlight based vitality transformation applications, including photocatalytic water part, CO_2 reduction, natural blend, ecological decontamination.

Classification of photocatalysts for N_2 fixation

Most conventional unmodified semiconductors cannot meet the vitality benchmarks for the decrease possibilities of the halfway responses. Up to this point, many deliberate endeavours have been made concerning decision and alteration of semiconductors, such as doping, presenting opportunities, plasmon acceptance, feature fitting and heterostructure get together, to improve the photocatalytic execution. In the resulting areas, the photocatalysts are grouped dependent on their essential syntheses. Figure 2 shows the classification of photocatalysts for N_2 fixation.

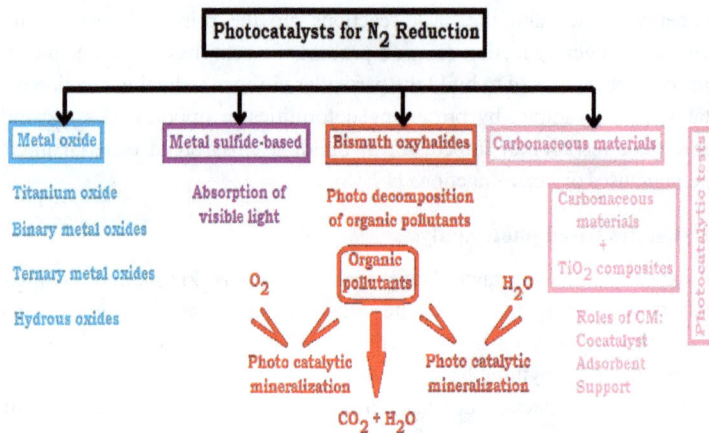

Figure 2. Classification of photocatalysts for N_2 reduction.

Metal oxide

Metal oxides as nano-and micro particles for use as photocatalysts might be gotten by a wide assortment of techniques. The fundamental test in the readiness of nano/micro particles is their security, since these particles tend to total to limit surface strain.

(a) Titanium dioxide

A standout amongst the most significant parts of TiO_2 photocatalysis is that, similar to the photoelectric impact, it relies on the vitality of the occurrence photons, in any case, to a first estimation, not on their power. Hence, regardless of whether these are only a couple of photons of the required vitality, they can actuate photocatalysis. This implies even common room light might be adequate to sanitize the air or to keep the dividers clean in the indoor condition, in light of the fact that the measures of poisons are regularly little. Unmodified TiO_2 displayed N_2 photocatalytic movement after toughening in air in light of the fact that the warm pretreatment created surface imperfections, which presented deformity or debasement states in the band hole of the semiconductor. The shallow Ti^{3+} gave plenteous dynamic locales to N_2 fixation by going about as an electron giver, prompting relative simplicity of separation of the N≡N bond. With a few changes of surficial Ti from Ti^{3+} to Ti^{4+}, the electrons were normally infused into N_2. Meanwhile, Ti^{3+} could be recovered under UV illumination.

Titanium dioxide shows three crystalline structures:

✓ Rutile,

✓ Anatase and

✓ Brookite.

Rutile is the most thermodynamically stable precious stone structure of titanium dioxide, however, anatase is the favoured structure for photograph catalysis since it presents higher photocatalytic movement and it is simpler to get ready. Brookite is the least steady stage and regularly not utilized in photograph catalysis. There are contemplates that show the benefits of mixings distinctive crystalline periods of TiO_2 for acquiring a higher photoactivity. At the point when diverse crystalline stages are coupled, it is for the most part trusted that the development of electrons from the rutile stage to the anatase stage happens, which causes a progressively proficient e-/h + partition and subsequently an expanded photocatalytic movement [23,24].

(b) Metal oxides

Ongoing investigations on photocatalysis centre around colloidal semiconductors, on very little measured particles in which quantization impacts are normal, just as on semiconductors upheld on latent help. Different parallel oxide impetuses likewise appear to be conceivably promising, since it is notable that twofold impetuses regularly display higher synergist action and selectivity than what one can foresee from the properties of their parts.

Metal oxide-based heterogeneous photocatalysts could be connected to a wide scope of applications in natural advancements. Titanium dioxide (TiO_2) is a semiconductor, which has a place with the group of progress metal oxides. It has pulled in much enthusiasm, with a wide scope of uses, for example, photovoltaic cells, gas sensors, colors, and photocatalysis. Other than the four polymorphs of TiO_2 found in nature, two extra high-weight structures have been integrated beginning from rutile: $TiO_2(II)$ which has the PbO_2 structure, and TiO_2 with the hollandite structure. For some utilizations of TiO_2, the molecule estimate, precious stone structure and stage, porosity, and surface region affect the photoactivity. The great properties of nano TiO_2 are because of its low dimensionality and quantum size impact. TiO_2 nanocrystals have a few favourable circumstances over their mass partner as a result of their high surface-to-volume proportion, expanded number of delocalized bearers superficially, improved charge transport furthermore, lifetime managed by their dimensional anisotropy, and the productive detachment of photogenerated openings and electrons. Nitrogen can be effectively brought into the TiO_2 structure, because of its equivalent nuclear estimate with oxygen, little ionization vitality, and high steadiness. In N-doped TiO_2, basic nitrogen saturates to the cross section of TiO_2 and substitutes for a grid oxygen molecule to shape nitride or oxynitride courses of action [25].

(c) Hydrous oxides

Hydrous ferric oxide $Fe_2O_3(H_2O)_n$ is unrivalled to TiO_2 based impetuses in photoreduction of N_2. This material is delicate to unmistakable light and the quantum yields of NH_3 acquired are higher than that of TiO_2 based impetuses. The structure of hydrous ferric oxide is not quite the same as that of the recognizable types of ferric oxide, FeO. Goodness or $Fe(OH)_3$. The synergist action starts from exceedingly negative fiat band potential and solid chemisorption of N_2. In analyses on photocleavage of water, it is notable that composite semiconductor particles, i.e., impetuses stacked with metals and additionally metallic oxides (e.g., RuO_2, NiO) give higher quantum yields, as the electron-opening detachment is increasingly compelling in a composite molecule.

Metal sulfides and different photocatalysts

Like the metal oxide-based photocatalysts, metal sulfides have as of late turned into a hot research subject in the field of photocatalytic NH_3 synthesis because of their solid retention of unmistakable light. Photocatalysts were progressively connected to lessen N_2 to NH_3 under noticeable light. The gaps in the valence band caught the electrons that discharged from RuO_2. To keep the high photocatalytic action of CdS for a more drawn out time, perhaps a few estimates must be taken to stop the corruption of CdS to S and Cd^{2+}. Multicomponent metal sulfides with sulfur opening, for example, $Zn_{0.1}Sn_{0.1}Cd_{0.8}S$ and $Mo_{0.1}Ni_{0.1}Cd_{0.8}S$, could lessen N_2 as photocatalysts under noticeable light, and the centralization of sulfur opportunities catching electrons were straight identified with the NH_3 yields. In addition, G-C_3N_4/ZnSnCdS and g-C_3N_4/ZnMoCdS were, separately, utilized for the obsession of N_2. A tight intersection coupling between g-C_3N_4 and ZnMoCdS was the key for effective charge exchange [26].

The presentation of oxygen opening to copy the reactant focuses of Fe-based impetus for N_2 adsorption could actuate N_2 and advance interfacial electron exchange, hence altogether improving the nitrogen photofixation capacity. Sulfur opening in metal sulfide semiconductors may have a comparative impact on nitrogen photofixation since oxygen and sulfur have the comparative synthetic properties. Compact discs have for quite some time been one of the most alluring unmistakable light dynamic photocatalysts because of its effective light assimilation and appropriate band edge position [27,28]. The Mo and Ni codoped CdS was readied. The doping of Mo and Ni causes the precious stone grid twists, in this manner prompting the development of sulfur opening in as-arranged ternary metal sulfide.

Photocatalysis   Materials Research Forum LLC
Materials Research Foundations **100** (2021) 193-207   https://doi.org/10.21741/9781644901359-6

Bismuth oxyhalides photocatalysts

Bismuth-based photocatalysts that display great photocatalytic action and strength amid the response have pulled in more and more consideration because of their exceptional electronic and auxiliary properties. Right now numerous bismuth-based photocatalysts, for example, $BiVO_4$, $Bi_2W_2O_6$, Bi_2MoO_6, $BiPO_4$, Bi_xTiO_y, Bi_2O_3 and bismuth oxychloride $BiOX$ (X = Cl, Br, I). To date, many fascinating and significant outcomes on $BiOX$ photocatalysts have been gotten. Nonetheless, a few downsides, for example, poor light assimilation, use and adjustment have been found by analysts amid $BiOX$ photocatalysis. The band structure of the material is significant for its vitality usage effectiveness and photocatalytic properties. The positive conduction band (CB) position of $BiOX$ confines its capacity for sub-atomic oxygen actuation, water part to deliver hydrogen, CO_2 decrease, N_2 fixation, natural blend and different applications [29].

Carbonaceous material

The high recombination rate of photoexcited electron-gap sets, low electrical conductivity, restricted light absorbance and little explicit surface zone of this material are primary bottlenecks in the improvement of g-C_3N_4 as a productive photocatalyst. Augmentation of its particular surface region and component doping has been demonstrated to be compelling strategies for improving its photocatalytic movement. Composite or heterojunction photocatalysts, which are gainful for isolating redox locales to improve the photocatalytic action, speak to another technique to illuminating the issue of a high recombination rate. Different g-C_3N_4/semiconductor composite photocatalysts, for example, g-C_3N_4/TiO_2, g-C_3N_4/Ag_3PO_4, g-C_3N_4/$CuFe_2O_4$ and gC_3N_4/WO_3 have been manufactured with higher photocatalytic movement, in which extraordinary components, including the heterojunction type (band-band exchange), p-n intersection charge partition and direct Z-plot component, are included. Notwithstanding the g-C_3N_4/semiconductor composite, the g-C_3N_4/carbonaceous material composite is additionally viewed as an option. These carbonaceous materials give uncommon conductivity, a huge explicit surface region, an expansive absorbance and a quick versatility of charge transporters to advance the photocatalytic action of g-C_3N_4.

Prospects – photoreduction of N_2 to NH_3

The photoreduction of N_2 to NH_3 is viewed as an experimentally testing yet ecologically neighbourly innovation for the supportable development of the human populace.

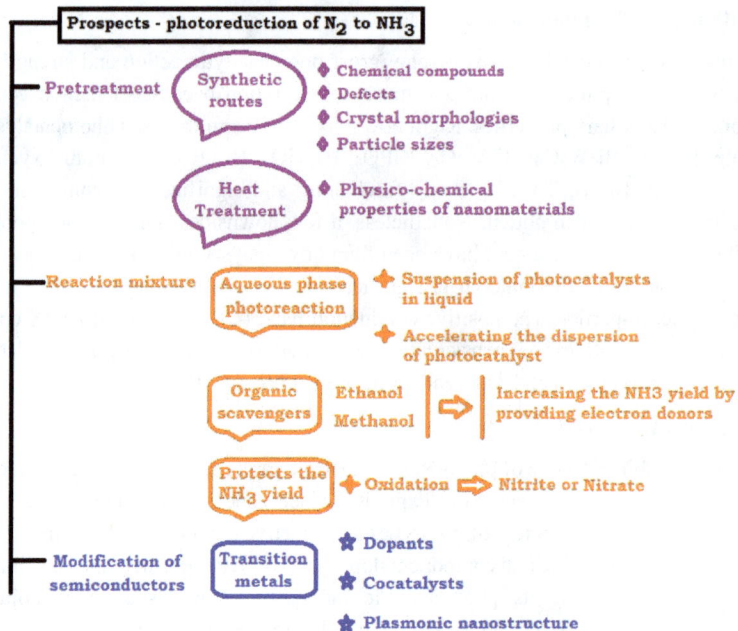

Figure 3. Prospects – photoreduction of N_2 to NH_3.

NH_3 has evoked much research interest and expansive interdisciplinary consideration as a hydrogen bearer notwithstanding its wide use as a modern crude material and compost. Figure 3 shows the prospects of photoreduction of N_2 to NH_3.

Pre-treatment

It is evident that the N2 reduction photocatalytic movement is very subject to the pre-treatment strategy for the as-created impetuses. On the one hand, the engineered courses uniquely affect the photocatalytic action dependent on a plenty of angles, including concoction mixes, absconds, precious stone morphologies and molecule sizes. Also, heat treatment assumes a focal job in the physicochemical properties of the nanomaterials. Remarkably, strengthening at a high temperature presents imperfection states, which upgrade the photoreduction capacity. In another angle, delayed warmth treatment unfavourably influences the stage change and the measure of surface OH gatherings, which is unfavourable to the action of photocatalysts.

Reaction mixture

The gas-stage response, fluid stage photoreaction has additionally been generally utilized by means of the suspension of photocatalysts in fluid. Watery slurry with a suitable convergence of photocatalyst upgrades the photoreaction by quickening the scattering of the photocatalyst for improved mass exchange. Nevertheless, high fixations lead to poor infiltration of photons. The nearness of natural scroungers, for example, ethanol and methanol can clearly expand the NH_3 yield by giving conciliatory electron benefactors to the openings in the semiconductor. Alongside the utilization of photograph created openings in an oxidation procedure, the N_2 decrease rates can be extraordinarily expanded due to the accessibility of inexhaustible photograph created electrons for the decrease responses. In spite of the fact that ethanol is commonly used as a run of the mill natural scrounger, it is vital to extensively think about various alcohols to decide appropriate conciliatory specialists for explicit light-determined synergist frameworks.

Modification of semiconductors

Transition metals are generally utilized as dopants, co-impetuses and plasmonic nanostructures with the point of improving photocatalytic effectiveness. Among these, earth-bounteous iron is the most overwhelming as a proficient metal dopant. Conversely, it has been accounted for those respectable metals as dopants in TiO_2 illustrate hindering impacts for N_2 photoreduction. The prospering examination of oxygen opening actuated in BiOX has propelled numerous specialists because of their remarkable attributes for the proficient partition of charge bearers. Critically, the heterostructure will spatially collect the detachment and exchange of photogenerated charge bearers upon their associations at the reached interface.

Conclusion and future perspective

Photocatalytic nitrogen fixation has drawn rising exploration interests inferable from the accompanying benefits:

- ✓ NH_3 synthesis can be acknowledged by utilizing bottomless N_2 and H_2O as crude materials under gentle conditions, prompting the lower cost as well as the alleviation of vitality emergency;

- ✓ Nitrogen reduction reaction procedure with zero-carbon outflow can proficiently ease ecological issues for example, an Earth-wide temperature boost coming about because of CO_2;

- ✓ It can give a potential course to store spotless and sustainable sun oriented vitality and power as a hydrogen transporter and sans carbon fuel (NH_3).

In any case, the advancement of heterogeneous synergist nitrogen fixation faces extraordinary deterrents, for example, the inherent inactivity of N_2 particles, the frail restricting quality of N_2 to the heterogeneous impetuses, the qualities of multi-electron and multi-proton taken interest pathways. Moreover, the smelling salts creation is ordinarily joined by the synchronous age of hydrogen and hydrazine, bringing about a low selectivity towards NH_3. Even more critically, current comprehension on the essential systems of N2 photograph decrease and electro-decrease is still very constrained because of the extensively muddled response process. To advance the improvement of heterogeneous nitrogen fixation under encompassing conditions, certain viewpoints on the most proficient method to further improve the exhibition of photocatalytic nitrogen reduction reaction frameworks: The plan of progressively productive, cost effective, environment friendly and vigorous impetuses is truly attractive. It is basic to create novel advances for nanomaterial union, in light of the fact that the variation syntheses and morphologies of nanostructured materials play a significant job in improving the reactant nitrogen obsession execution. The interface control, surface building, substance adjustment, etc are promising techniques for the improvement of impetus execution. Particularly, deficient impetus surface has a colossal effect on the reactant execution. Contrasted with the level mass surfaces, the roughened surfaces with rich imperfections and doped heteroatoms have bigger reactant dynamic surface region and significantly progressively low-coordination destinations, which have enormously improved inborn synergist movement. Moreover, the electronic and geometric impacts prompted by compound alteration can offer ascent to the fitting variety and modification of authoritative qualities of response intermediates on impetus surface, giving better synergist properties. The development of heterogeneous single-particle impetuses is another intriguing methodology for advancing the nitrogen reduction reaction execution because of the high exercises of single metal molecules. For precedent, Ru-single iotas were utilized for nitrogen reduction reaction and displayed improved ammonia yield and selectivity. The uplifting news is single-molecule impetuses could be advantageously incorporated from metal-natural structures or metal-natural edifices.

References

[1] V. Rosca, M. Duca, M.T. de Groot, M.T. Koper, Nitrogen cycle electrocatalysis, Chem. Rev. 109 (2009) 2209-2244. https://doi.org/10.1021/cr8003696

[2] D.E. Canfield, A.N. Glazer, P.G. Falkowski, The evolution and future of Earth's nitrogen cycle, Science 330 (2010) 192-196. https://doi.org/10.1126/science.1186120

[3] X. Chen, N. Li, Z. Kong, W-J. Ong, X. Zhao, Photocatalytic fixation of nitrogen to ammonia: state-of-the-art advancements and future prospects, Mater. Horiz. 5 (2017) 9-27. https://doi.org/10.1039/C7MH00557A

[4] C. Guo, J. Ran, A. Vasileff, S-Z. Qiao, Rational design of electrocatalysts and photo(electro)catalysts for nitrogen reduction to ammonia (NH_3) under ambient conditions, Energy Environ. Sci. 11 (2018) 45-56. https://doi.org/10.1039/C7EE02220D

[5] C. Na, G.F. Zheng, Aqueous electrocatalytic N_2 reduction under ambient conditions, Nano Res. 11 (2018) 2992-3008. https://doi.org/10.1007/s12274-018-1987-y

[6] F. Wen, C. Li, Hybrid artificial photosynthetic systems comprising semiconductors as light harvesters and biomimetic complexes as molecular Cocatalysts, Acc. Chem. Res. 46 (2013) 2355-2364. https://doi.org/10.1021/ar300224u

[7] H. Li, J. Shang, Z. Ai, L. Zhang, Efficient visible light nitrogen fixation with BiOBr nanosheets of oxygen vacancies on the exposed {001} facets, J. Am. Chem. Soc. 137 (2015) 6393-6399. https://doi.org/10.1021/jacs.5b03105

[8] Y. Sun, S. Gao, F. Lei, Y. Xie, Atomically-thin two-dimensional sheets for understanding active sites in catalysis, Chem. Soc. Rev. 44 (2015) 623-636. https://doi.org/10.1039/C4CS00236A

[9] H. Wang, X. Zhang, J. Xie, J. Zhang, P. Ma, B. Pan, Y. Xie, Structural distortion in graphitic-C_3N_4 realizing an efficient photoreactivity, Nanoscale, 7 (2015) 5152-5156. https://doi.org/10.1039/C4NR07645A

[10] R. Li, Photocatalytic nitrogen fixation: An attractive approach for artificial photocatalysis, Chinese J. Catal. 39 (2018) 1180-1188. https://doi.org/10.1016/S1872-2067(18)63104-3

[11] D. Zhu, L. Zhang, R.E. Ruther, R.J. Hamers, Photo-illuminated diamond as a solid-state source of solvated electrons in water for nitrogen reduction, Nat. Mater. 12 (2013) 836-841. https://doi.org/10.1038/nmat3696

[12] X. Gao, L. An, D. Qu, W. Jiang, Y. Chai, S. Sun, X. Liu, Z. Sun, Enhanced photocatalytic N_2 fixation by promoting N_2 adsorption with a co-catalyst, Science Bulletin, (2019) In press Accepted. https://doi.org/10.1016/j.scib.2019.05.009

[13] G.N. Schrauzer, T.D. Guth, Photolysis of water and photoreduction of nitrogen on titanium dioxide, J. Am. Chem. Soc. 99 (1977) 7189-7193. https://doi.org/10.1021/ja00464a015

[14] A.J. Medford, M.C. Hatzell, Photon-Driven nitrogen fixation: current progress, thermodynamic considerations, and future outlook, ACS catalysis, 7 (2017) 2624-2643. https://doi.org/10.1021/acscatal.7b00439

[15] T. Hisatomi, J. Kubota, K. Domen, Recent advances in semiconductors for photocatalytic and photoelectrochemical water splitting, Chem. Soc. Rev. 43 (2014) 7520-7535. https://doi.org/10.1039/C3CS60378D

[16] X.X. Chang, T. Wang, J.L. Gong, CO_2 photo-reduction: Insights into CO_2 activation and reaction on surfaces of photocatalysts, Energy Environ. Sci. 9 (2016) 2177-2196. https://doi.org/10.1039/C6EE00383D

[17] W.B. Hou, S.B. Cronin, A review of surface plasmon resonance enhanced photocatalysis, Adv. Funct. Mater. 23 (2013) 1612–1619. https://doi.org/10.1002/adfm.201202148

[18] H. Hirakawa, M. Hashimoto, Y. Shiraishi, T. Hirai, Photocatalytic conversion of nitrogen to ammonia with water on surface oxygen vacancies of titanium dioxide, J. Am. Chem. Soc. 139 (2017) 10929-10936. https://doi.org/10.1021/jacs.7b06634

[19] Y. Bai, L.Q. Ye, T. Chen, L. Wang, X. Shi, X. Zhang, D. Chen, Facet-dependent photocatalytic N2 fixation of bismuth-rich Bi5O7I nanosheets, ACS Appl. Mater. Interfaces, 8 (2016) 27661-27668. https://doi.org/10.1021/acsami.6b08129

[20] T. Xiong, W. Cen, Y. Zhang, F. Dong, Bridging the g-C_3N_4 interlayers for enhanced photocatalysis, ACS Catal. 6 (2016) 2462-2472. https://doi.org/10.1021/acscatal.5b02922

[21] S.Z. Hu, X. Chen, Q. Li, F.Y. Li, Z.P. Fan, H. Wang, Y.J. Wang, B.H. Zheng, G. Wu, Fe^{3+} doping promoted N_2 photofixation ability of honeycombed graphitic carbon nitride: The experimental and density functional theory simulation analysis, Appl. Catal. B-Environ. 201 (2017) 58-69. https://doi.org/10.1016/j.apcatb.2016.08.002

[22] S. Wang, X. Yang, H. Hou, X. Ding, S. Li, F. Deng, Y. Xiang, H. Chen, Highly efficient visible light induced photocatalytic activity of a novel in situ synthesized conjugated microporous poly (benzothiadiazole)–C_3N_4 composite, Catal. Sci. Technol. 7 (2017) 418-426. https://doi.org/10.1039/C6CY02006B

[23] D.C. Hurum, A.G. Agrios, K.A. Gray, T. Rajh, M.C. Thurnauer, Explaining the enhanced photocatalytic activity of Degussa P25 mixed-phase TiO_2 using EPR, J. Phys. Chem. B, 107 (2003) 4545-4549. https://doi.org/10.1021/jp0273934

[24] T.A. Kandiel, L. Robben, A. Alkaim, D. Bahnemann, Brookite versus anatase TiO_2 photocatalysts: Phase transformations and photocatalytic activities, Photochem. Photobiol. Sci. 12 (2013) 602-609. https://doi.org/10.1039/C2PP25217A

[25] A.N. Banerjee, The design, fabrication, and photocatalytic utility of nanostructured semiconductors: focus on TiO_2-based nanostructures, Nanotechnol. Sci. Appl. 4 (2011) 35-65. https://doi.org/10.2147/NSA.S9040

[26] K. Tennakone, S. Punchihewa, R. Tantrigoda, Nitrogen photoreduction with cuprous chloride coated hydrous cuprous oxide, Sol. Energy Mater. 18 (1989) 217-221. https://doi.org/10.1016/0165-1633(89)90055-5

[27] Y. Hu, X. Gao, L. Yu, Y. Wang, J. Ning, S. Xu, X.W. Lou, Carbon-coated CdS petalous nanostructures with enhanced photostability and photocatalytic activity, Angewandte Chemie International Edition in English, 52 (2013) 5636 – 5645. https://doi.org/10.1002/anie.201301709

[28] Y. Cao, S. Hu, F. Li, Z. Fan, J. Bai, G. Lu, Q. Wang, Photofixation of atmospheric nitrogen to ammonia with a novel ternary metal sulfide catalyst under visible light, RSC Adv. 6 (2016) 49862. https://doi.org/10.1039/C6RA08247E

[29] X. Jin, L. Ye, H. Xie, G. Chen, Bismuth-rich bismuth oxyhalides for environmental and energy photocatalysis, Coord. Chem. Rev. 349 (2017) 84-101. https://doi.org/10.1016/j.ccr.2017.08.010

Photocatalysis
Materials Research Foundations **100** (2021) 208-252

Materials Research Forum LLC
https://doi.org/10.21741/9781644901359-7

Chapter 7

Perovskites based Nano Heterojunctions for Photocatalytic Pollutant Removal

Sunil Kumar[1], Amit Kumar[1,2,3*], Gaurav Sharma[1,2,3], Pooja Dhiman[3,4], Genene T. Mola[5]*

[1]School of Advanced Chemical Sciences, Shoolini University, Solan, Himachal Pradesh, India-173229

[2]College of Materials Science and Engineering, Shenzhen Key Laboratory of Polymer Science and Technology, Guangdong Research Center for Interfacial Engineering of Functional Materials, Nanshan District Key Laboratory for Biopolymers and Safety Evaluation, Shenzhen University, Shenzhen, 518055, PR China

[3]International Research Centre of Nanotechnology school of Advanced Chemical Sciences, Shoolini University, Solan, Himachal Pradesh, India-173229

[4]School of Physics and Materials Science, Shoolini Univesity, Solan, Himachal Pradesh, India-173229

[5]School of Chemistry & Physics, University of KwaZulu-Natal, Pietermaritzburg Campus, Private Bag X01, Scottsville 3209, South Africa

* Mola@ukzn.ac.za

Abstract

The development of new generation photocatalytic materials used for the betterment of human as well as environment. Perovskites and perovskites related nano-hetero-junction shows great interest for photocatalytic organic and inorganic pollutant removal. This chapter discusses its crystalline structures varying from cubic (high symmetry) to triclinic (very low symmetry). Various methods have been utilized for synthesis of perovskites such as sol-gel, hydrothermal, vapor deposition methods, solid-state reaction routs from oxide and high pressure technique. The first technique is used for the synthesis of perovskite is ceramic route in which the mixture of oxide was treated at high temperature and processed later by ceramic powder method. Various photocatalyst such as nitrides, sulphides, phosphides, oxide and mixed oxide are employed for photocatalytic water splitting or hydrogen generation. Future perspectives of perovskite-related photocatalyst are included in this chapter.

Photocatalysis
Materials Research Foundations **100** (2021) 208-252

Materials Research Forum LLC
https://doi.org/10.21741/9781644901359-7

Keywords

Perovskites, Nano-Hetero-Junction, Sol-Gel, Hydrothermal, Ceramic Route, Hydrogen Generation

Contents

Materials Research Forum LLC
https://doi.org/10.21741/9781644901359-7

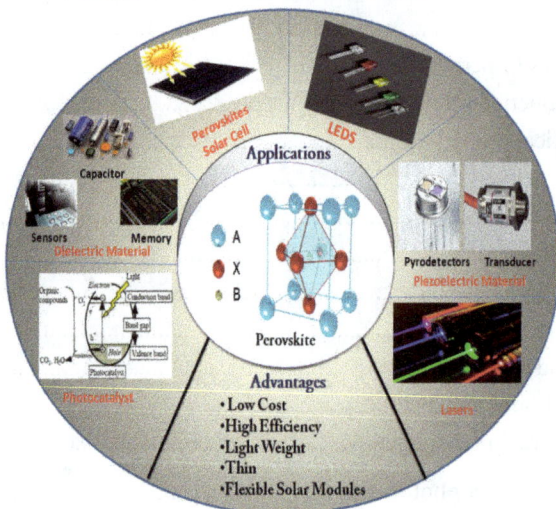

1.   Introduction

Various environmental problems such as water and air contamination and energy shortage are being faced by human being as well as aquatic life. Water pollution has grabbed much attention of environmentalists and researchers worldwide. Large quantities of organic pollutants were discharged into the surrounding soil and water by numerous industries [1]. Perovskite based photocatalysts have found great interest in energy and environmental applications. Perovskites was firstly discovered by German and Russian mineralogist Gustav Rose and Lev Perovski in 1839. Perovskite refers to the crystal structure of calcium

Materials Research Forum LLC
https://doi.org/10.21741/9781644901359-7

titanate and all compounds with the same crystal structure are perovskites. It has a general formula ABX_3, where A is an organic cation, B is a metal cation and X stands for the halide anions [2]. For last few years, various experiments and investigation were done to study the perovskites ABX_3. This type of structure enhance its importance in the field of electrical ceramics, refractories, material science [3], astrophysics [4], particles accelerated [5], heterogeneous catalysis [6] and environment [7] etc. Significant substitution of one or both A and B cation sites accept by the perovskite structured sites, which retained their original crystal structure. This type of process shows partial replacement of the cation sites with foreign metal ions, which modifying their electrical, structural, microstructural, and magnetic properties [8]. Perovskite-type oxides and perovskite-like oxides grasp attention in chemistry as well as in physics.

Perovskites have wide range of crystalline structures varying from cubic to triclinic. Various solid state methods are used for synthesis of perovskites such as sol-gel, hydrothermal, high-pressure technique and vapors deposition methods. Perovskite materials show various fascinating properties for both theoretical and application point of view. These compounds are used in fuel cells as sensors and catalyst electrodes and are candidates for memory devices and spintronics applications. Tanaka and Misono discussed in detail numerous advantages of perovskite photocatalysts [9] as listed follows:

a) They are formed from a large variety of compositional and constituents elements, but their basic essential, basic structure are similar.

b) Their well-defined bulky structures are characterized well and their surface properties are easily explored.

c) Their valency, stoichiometry and vacancy can be varied widely.

d) Physical and solid-state chemical properties have been accumulated [9].

Photocatalytic performance of the catalyst is affected by various factors such as electron and hole effective mass, diffusion length, exciton lifetime, exciton binding energy, electron-hole separation and transport within the lattice. Due to its distinctive crystal structure and electronic properties perovskite oxides show excellent results as a photocatalyst under visible light irradiation. The crystal structure of perovskite shows excellent band gap values and band edge potential which is useful in specific photocatalytic reaction. Separation of photogenerated charge carrier can be influenced by lattice distortions in perovskite. Numerous groups of perovskites such as vanadium-and niobium, titanate, tantalite and ferrites possess high visible-light activity.

The first technique used for perovskite synthesis was ceramic route in which mixture of oxides was treated at high temperature. The first perovskite $CaTiO_3$ was fabricated by sol-gel route. Exceptional chemical tolerance properties were exposed by ion doped calcium titanate, which can be used for LEDs [10]. Also $LaFeO_3$ grabs more attention because of its narrow band gap due to this it is easily excited by visible light and decomposing organic molecule and water [11]. One the other hand $LaFeO_3$ shows good chemisorptions of nitrogen oxide (NO_x) [12].

$BiFeO_3$ shows excellent photocatalytic activity for dye degradation as well as water splitting under visible light irradiation [13]. In addition, it also shows visible-light-driven photo-Fenton degradation activity. Strontium titanate ($SrTiO_3$) is widely used for photocatalysis because of its tremendous structural stability, high photo activity and strong photo-corrosion resistibility [14]. It has wide band gap and low quantum yield which is overcome by formation of heterojunctions via coupling with two or more photocatalyst [15] or by adjusting band gap via doping or co-doping with cations [16, 17]. Rare earth based perovskites with composition $LnFeO_3$ (Ln = rare earth) shows various applications such as gases separators [18], sensors [19], catalyst [20] etc. Various groups investigated their photocatalytic activity because of their excellent photo degradation rate [21] and photo-Fenton like reactions [22].

Hybrid organic and inorganic perovskite (HOIPs) form a subclass of ABX_3 in which the replacement of A-site and X-site are done by the organic amine cations or organic linkers respectively. In HOIPs the organic component in this structure shows additional functionalities and structural flexibility that cannot be achieved by pure inorganic perovskites. Hybrid perovskite have a reported cubic phases of $MAPbX_3$ (MA= methylammonium and X= Cl, Br or I) [23].

The current chapter is focused on the structure, types, properties and synthesis of the perovskites. It also discusses the photocatalytic activity of the perovskites. Different methods are used for the synthesis of perovskites based nano-hetero assemblies. These are useful for organic and inorganic pollutant removal.

2.   Perovksites

2.1   Structure

Generally, the structural formula of perovskite materials is ABX_3 and A_2BX_4, where A and B are cations and X is the anion, by the formation of anionic bond A and B are binds together. A possesses larger ionic radius than the B. The structure of the perovskite is analyzed as follow: A is placed on edge, and at the center of octahedron B is located. The cationic radii are $r_A > 0.09$ nm and $r_B > 0.051$ nm [6]. Anionic X is represented as oxygen or non oxygen material such as fluorine (-F) and methyl (-CH_3).

Photocatalysis   Materials Research Forum LLC
Materials Research Foundations **100** (2021) 208-252   https://doi.org/10.21741/9781644901359-7

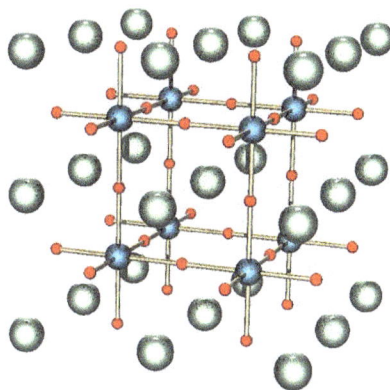

Figure 1: Typical structure of Perovskite.

The structural distortion in perovskite is determined via electronic configuration of metal ions and the ratio of A and B ion size. In perovskites, there are two types of structural distortions, one is off-centering of B ion in BO_6 octahedral and another one is the tilting of BO_6 octahedral. The first type corresponds to displaced phase transition and the second corresponds to order-disorder phase transition [24, 25]. To identify the formability of perovskite structure the tolerance factor (t') was calculated as suggested by Goldschmidt [26].

$$t' = \frac{(rA + r0)}{\sqrt{2}.(rB + r0)} \tag{1}$$

Where r_A and r_B ionic radii of A and B cation and r_0 ionic radius of oxygen anion. When t'<1 gives the tilting mode and t' > 1 show centering of smaller B cation which causes tilting of BO_6 octahedron. Due to contraction of BO_6 octahedron there off-centering occur due to large A and smaller B ion. At room temperature the best ideal perovskite-type structure is cubic [27].

The majority of the perovskite material shows the band gap greater than 3 eV which are further modified its band gaps by the doping of elements, which can be used as high photocatalytic efficiency. Depending on the details of the octahedral rotations there is a transformation of crystal phase in ABX_3 perovskite materials which displayed lattice distortion in following succession such as orthogonal, tetragonal, rhombohedral, monoclinic and triclinic phase, [26]. Some of the perovskite materials examples are given below:

Niobium based perovskites such as $KNbO_3$ having cubic, orthorhombic and tetragonal with band gap of 3.12-3.24 [28]. $CuNbO_3$ is a monoclinic crystalline structure with band gap 2.0 eV [29]. $SrNbO_3$ having cubic structure with band gap of 2.79 eV [30]. Iron based perovskites such as $LaFeO_3$ is a cubic and orthorhombic structure with a band gap of 2.1 eV [31]. Orthorhombic structure was shown by the bismuth based perovskites such as $AgBiO_3$ with 2.5 eV band gap [32]. Table 1 shows various perovskites with their band gaps and crystalline structure.

Table 1. Perovskites with their band gaps and crystalline structure.

Perovskites	Eg (eV)	Crystalline Structure	Reference
$SrTiO_3$	3.1-3.7 eV	Cubic	[33]
$CaTiO_3$	3.6 eV	Orthorhombic	[34]
$PbTiO_3$	2.75 eV	Tetragonal	[35]
$MnTiO_3$	3.1 eV	Rhombohedral	[36]
$KTaO_3$	3.6 eV	Cubic and orthorhombic	[37]
$AgTaO_3$	3.4 eV	Rhombohedral	[38, 39]
$LaTiO_3$	4.7 eV	Cubic and Rhombohedral	[28]
$KNbO_3$	3.12-3.24 eV	Cubic, tetragonal and orthorhombic	[29]
$CuNbO_3$	2.0 eV	Monoclinic	[30]
$SrNbO_3$	2.79 eV	Cubic	[31]
$LaFeO_3$	2.1 eV	Orthorhombic	[32]
$AgBiO_3$	2.5 eV	Orthorhombic	[40]

2.2   Types of perovskites

ABX_3 perovskite crystalline structure shows flexibility and its ability to accommodate a broad range of cation or anion vacancies. And show the origin of large variety of perovskite-based compounds with a wide range of physical properties. The oxide phases have been divided into two types: one is ternary ABO_3 and its solid solutions and the another one is newer complex type compounds $(AB'_xB''_y)O_3$ where B' and B'' are two different element in different oxidation states and $x + y = 1$. On basis of oxidation states the oxides are classified into $A^{1+}B^{5+}O_3$, $A^{2+}B^{4+}O_3$, $A^{3+}B^{3+}O_3$ and oxygen and cation deficient species [41, 42]. The details are sorting in flowchart given below Figure 2.

The complex perovskite type compounds $A(B'_xB''_y)O_3$ divided into different compounds which enclose twice the lower valence state element then the higher valence state element $A(B'_{0.65}B''_{0.33})O_3$, those which have twice the higher state element then the lower state element $A(B'_{0.33}B''_{0.65})O_3$, and at the last on is when both B element in equal amount $(B'_{0.5}B''_{0.5})O_3$ and oxygen deficient phase $A(B'_xB''_y)O_{3-z}$. The valences of the A and B cations are 2^+ and 4^+ respectively. But in some cases if the B^{3+} has six coordinates their valences can be 3^+. The bulking of the $(AO_3)^{4-}$ layers because of variation at the A cation position which causes distortion and displacement of the oxygen anion. Because of bulking, there is a distortion of octahedral when B cation took as a center.

The B cation has an ability of stiffness to endure this effect, and the transition metal oxide, because of its multi valency or the special 3d and 4d electronic configuration, fills the B cation position. Due to this reason, the transition metal oxides show extraordinary properties and perovskite type structure.

2.3   Distinctive properties of perovskites

Perovskite based materials are used for various applications such as sensors, piezoelectric, ultrasonic and under water devices, wastewater treatment, high temperature heating applications, frequency filters for wireless application etc. Depending on the application, perovskites can be synthesized in diverse forms such as bulk, nanocrystalline and thin films. There is a possibility to synthesize to multi component perovskite via partial substitution of cation in position A and B. They acquire several peculiar properties such as pyroelectric, polar properties, ferroelectric, optical properties and many more. Some of the detailed properties are explained below.

2.3.1 Optical properties

Perovskite is a class of materials having excellent optical and photoluminescence properties. Single domain crystal of $BaTiO_3$ shows optical properties studied by Merz at different temperature [43]. At temperature from 20°C to 90°C the crystal has a reflective index at constant value ~2.4 and reached at maximum value 2.46 at 120°C. Thickness of single crystal $BaTiO_3$ is 0.25-1 mm which transmits from 0.5 μ to 6 μ. Higher adsorption was found for wavelength greater than 11 μ and weak adsorption at 8 μ. Noland studied that the single crystal strontium titanate show optical properties produced by a flame fusion [44]. Between 0.20 μ to 17 μ in wavelength the optical coefficient was measured. 70% of better transmission was measured between 0.55 μ to 5 μ. This crystal shows 2.407 of reflective index at 5893 Å.

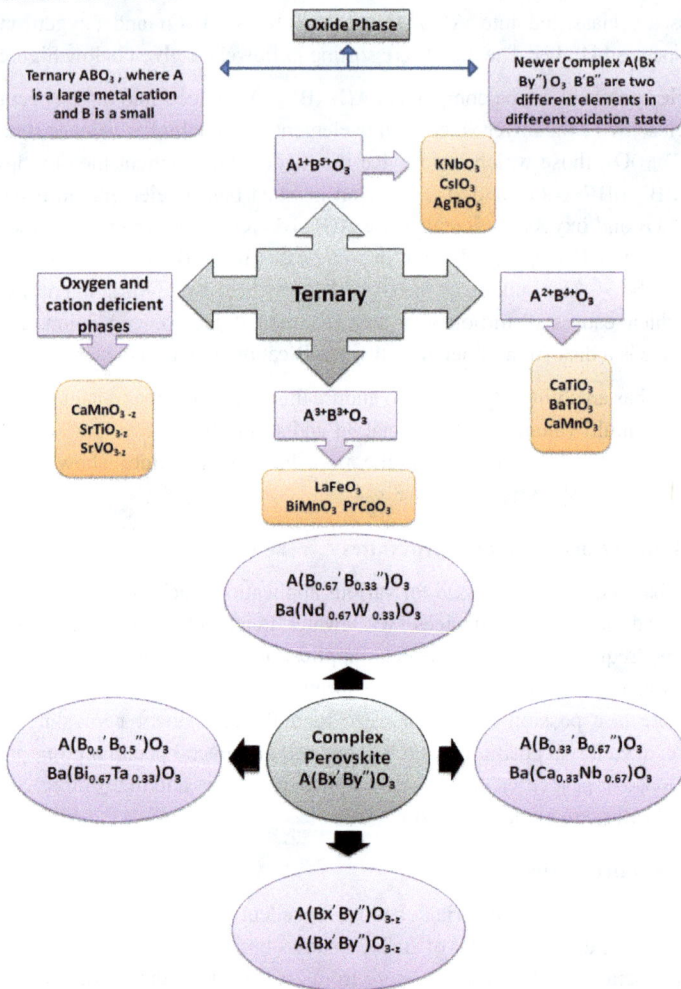

Figure 2: Types of Perovskites.

Materials Research Forum LLC

https://doi.org/10.21741/9781644901359-7

In the paraelectric phase the electro-optic properties of $KTaO_3$, $BaTiO_3$ and $SrTiO_3$ were measured [45]. The electro-optic coefficients of these perovskite are constant with temperature. The term of induced polarization was expressed when the distortion of the optical indicatrix varies from material to material. Last few years researchers show a great interest in materials for laser applications. One of the best deals is using perovskite laser host materials.

Perovskite type oxides phosphors are broadly used in field emission display (FED) and plasma display panel (PDP) devices because on the surface of the phosphors particles a adequately conductive to discharge electric charges [46].

Recently $BaZrO_3$ is well known environment friendly photoluminescence (PL) materials which is used to emit light easily in visible region and prepared easily at low cost [47].

2.3.2 Piezoelectricity

Piezoelectricity is the ability of some materials to generate an electric charge in response to applied mechanical stress. Piezoelectric ceramics is an important group of piezoelectric materials, of which PZT is an example [48]. They show perovskite crystal structure such as tetragonal/rhomohedral close to cubic. The general formula is $A^{2+}B^{4+}O_3^{2-}$ where A is large divalent metal ion i.e. barium along with lead and B denotes a tetravalent metal ion such as titanium or else zirconium. Various examples of piezoelectric materials are topaz, wood, quartz analogic crystal, Gallium orthophosphate ($GaPO_4$), Barium titanate, Lead zirconate titanate etc.

2.3.3 Superconductivity

Superconductivity is a phenomenon of zero electrical resistance and expulsion of magnetic flux fields taking place in certain material cooled below critical temperature [49]. Many important physical properties are demonstrated by the oxide perovskite. Perovskite type structure provides a tremendous support for superconductivity. One of the best examples of superconductivity perovskite is High *Tc* copper oxide [50]. Because of perovskite, various intermetallic compounds used as a superconducting material have been eclipsed.

Some of the best example of superconductor is Sodium, potassium, rubidium and cesium tungsten bronze. Superconducting transition in three samples of reduced strontium titanate was measured by Schooley and co-workers [51].

2.3.4 Multiferroicity

It is class of materials which exhibiting simultaneous ferromagnetic, ferroelectric and ferroelectric ordering. Among other multiferroic, bismuth ferrite ($BiFeO_3$), a rhombohedrally distorted perovskite, is acquiring great attention because of its ferroelectric order and anti-ferromagnetic order above room temperature [52]. In transition metals doped $BaTiO_3$ achieved multiferroic in new era, from experimental as well as theoretical studies [53]. These materials shows both physical and chemical properties which can be enhanced by extrinsic and intrinsic dopants such as transition metal cations doping, excess oxygen vacancies [54].

2.4 Synthesis

2.4.1 Sol-gel method

Various techniques of sol-gel are used for preparing perovskites such as alkoxide, alkoxide-salt and the pechini methods. Due to adaptability in preparing perovskites, pechini method is widely used. One of the greatest advantages of this method is high purity and homogeneity of perovskites structure coupled with an accurate control of the composition of final materials [88].

The flexibility of sol-gel route, principally the pechini method that delivers homogeneous solutions, helps the formation of perovskite structure and makes it very attractive. This method leads formation of pure crystal perovskite structure at 1000°C. For preparing perovskite materials this technique, is one of the most widely used and studied as the chemical routes.

2.4.2 Solid state reactions

Solid state reaction is one of the most conventional methods of synthesis of ceramic compounds. In this method the oxide powder, carbonate or salt are mechanically mixed, heated at high temperature around 1200°C (Figure 3). During crystalline grains this process is carried out for 8 to 24 hours so that the mobility of cation helps to form the perovskite structure [55]. The reaction occurs at the interface of the mixed solids from the bulk to the interface between the particles.

For the preparation of perovskite, the solid-state reaction is one of the oldest methods that we used. In literature, the availability of the experimental data is one of the advantages of this method. The properties of the perovskite are correlated with preparation or sintering conditions. Sintering effect helps perovskites to form crystal structure. For example $BaCo_{0.7}Fe_{0.2}Ta_{0.1}O_{3-\delta}$ sintered above 900°C resulted in pure perovskite structure, where un-reacted materials remain at lower temperature [56].

Solid-state reaction has some limitations such as it related with the particle size and broad particle distribution of final material. This method shows the precursor powder particles undertake widespread milling as well as mixing. After sintering $BaCo_{0.7}Fe_{0.2}Ta_{0.1}O_{3-\delta}$ the grain sizes is approximately 1-2 μm were obtained [57]. To overcome this problem, the fine powder is synthesized by incessant milling at high rotation, which helps to avoid the external heating known as mechanochemical route [58]. Even with these disadvantages, solid state reaction is simple and fast method to synthesized perovskites.

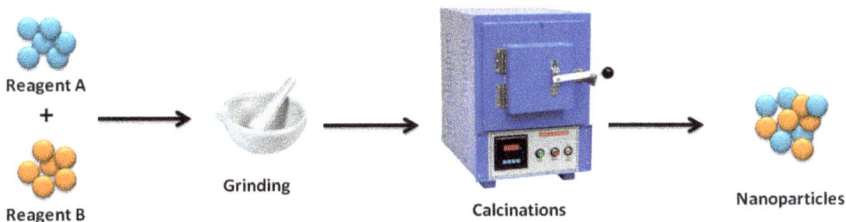

Figure 3: Solid State Reaction.

2.4.3 Hydrothermal method

The hydrothermal method is carried out in autoclave at high-pressure i.e. 15 MPa. At high pressure, it is a combination of both temperature between the boiling point of the water and the material critical temperature. Basically, it is a sol-gel type route which gives a good control of particle size. The basic example of this method is $CaTiO_3$ synthesized at 15 atm and 150°C and further calcined at 1300°C [59].

Through the analysis of XRD pattern it was found that there is no impurities in the sample and the phases has a similar cell parameters. In case of specific material hydrothermal method is not necessary [59].

2.4.4 Co-precipitation method

Supersaturation condition was required by the co-precipitation methods in which solution with soluble cations are mixed with another solution called precipitation agent as shown in Figure 4. In co-precipitation method there is need to control some important parameter such as pH, mixing rate, temperature and concentration to reach desire physical properties such as morphology and particle size.

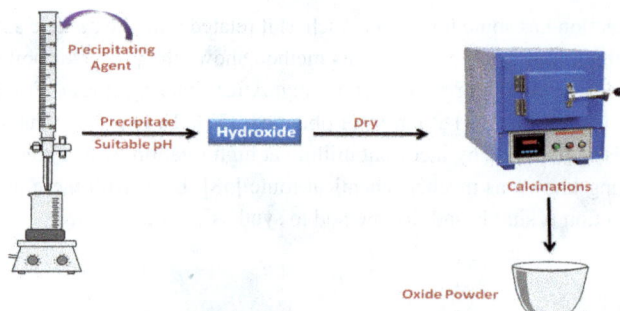

Figure 4: Co-precipitation method.

This method is used in preparing perovskites with good homogeneity and purity. This is used to control on the chemical reaction which is necessary to obtain compounds without deficiency of metal cation.

This method is used while preparing $La_{0.8}Sr_{0.2}Co_{0.6}Fe_{0.4}O_3$(LSCF) by using metal hydrate precursors dissolved in water, where potassium hydroxide used as a precipitation agent [60]. This method is used for synthesis of nanostructure materials such as LSCF [61]. And sphere-like $NdFeO_3$ nanocrystals [61] with modified particle size and morphology.

There are various other methods which are used to prepare advanced ceramics that can also be used for perovskites materials such as spray and freeze drying, spray pyrolysis, microwave assisted synthesis and sintering, combustion synthesis and physical vapor deposition (PVD). These methods are not widely used, and they are under development for the preparation of perovskites materials.

3. Perovskites as a photocatalyst

Photocatalysis is a phenomenon that utilized the energy input from incident radiation and the catalytic properties to be carried or accelerate certain chemical reaction on the surface of the materials [62, 63] (Figure 5). From last many years various chemical reaction have been studied, which are useful for generation of energy and environment cleaning applications. To design and develop the new photocatalytic material we should understand the mechanism of photocatalytic reaction.

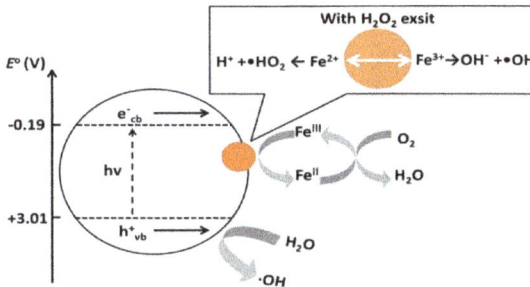

Figure 5: Basic principle of overall photocatalysis [64].

The perovskite crystals structure offers an excellent framework to acquire the band gap values to enable visible light adsorption and band edge potentials helps for precise photocatalytic reactions. The separation of photogenerated charge carrier is influences by the lattice distortion in perovskite compound. Some groups of materials, which show visible light activity, are shown below.

3.1   Titanate based perovskite

To date, titanate perovskites have been mostly studied for photocatalytic applications. Most of the titanate perovskites show band gap (Eg) more than 3.0 eV and have excellent photocatalytic activity under UV radiation [65]. Titanate $MTiO_3$ (M= Sr, Ba, Ca, Mn, Co, Fe, Pb, Cd, Ni) show excellent photostability and corrosion. Calcium Titanate is one of the most widespread perovskites with band gap of 3.6 eV. Cu doped $CaTiO_3$ was widely used as a visible light-driven photocatalytic water deposition [66].

Under visible region, $CoTiO_3$ shows band gap 2.28 eV which has been studied for photocatalytic $64\mu molg^{-1}h^{-1}$ of O_2 evolution reaction was obtained without any co-catalyst [67]. The band gap of $NiTiO_3$ is around 2.16 eV and the light adsorption spectra show peak in visible region which correspond to crystal field splitting [68]. Under visible light irradiation degradation of nitrobenzene is done by the $CaTiO_3$ nanorods [68].

Under visible region $FeTiO_3$ having band gap 2.8 eV. $FeTiO_3/TiO_2$ are used for the photocatalytic degradation of 2-phenol under visible light irradiation The separation of electron hole recombination and hole capturing phase are done by TiO_2 [69]. The junction is shown in Figure 6.

The synthesized $CdTiO_3$ nanofibers having band gap 2.8eV are studied for the degradation for rhodamine [70]. $PbTiO_3$ is visible active photocatalyst having band gap 2.75 eV. Composite of nano- TiO_3 and micro $PbTiO_3$ are studied for photocatalytic performance and effective charge separation [71]. The ferroelectric behaviour of $PbTiO_3$ shows electron-hole separation at interface. Under visible light irradiation $MnTiO_3$ shows improved charge separation and degradation of rhodamine B [36].

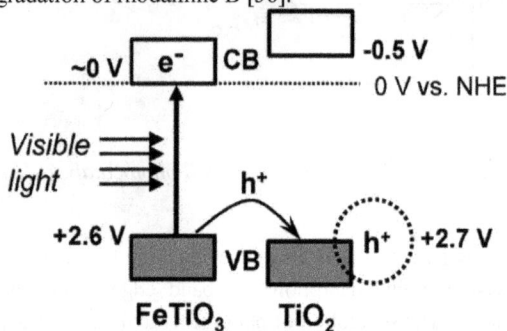

Figure 6: Energy band diagram of Type-B Hetrojunction $TiO_2/FeTiO_3$ [69].

3.2   Ferrite perovskites

Most of the ferrite perovskite have native band gap under visible region. Under visible light irradiation $LaFeO_3$ attain 2.1 eV band gap and it is used for the degradation of pollutants as well as hydrogen evolution. Pt co-catalyst loaded $LaFeO_3$ fabricated by sol-gel method which show exceptional yield of hydrogen evolution under 400W tungsten light source [72] (Figure 7). $LaFeO_3$ shows better photocatalytic activity then Fe_2O_3 because of its comparative electron properties such as electron hole separation, mobility, photoexcited lifetime etc. Under visible light irradiation $GaFeO_3$ has been reported to show overall water splitting having yield 0.10 and 0.04 $\mu mol.h^{-1}$ [73]. $YFeO_3$ having band gap 2.43 eV and showed four times better photocatalytic activity than TiO_2-P25 under visible light irradiation >400 nm for the degradation of RhB [74]. Ferrite based perovskite used for magnetic recovery of particles, which is useful in particle application.

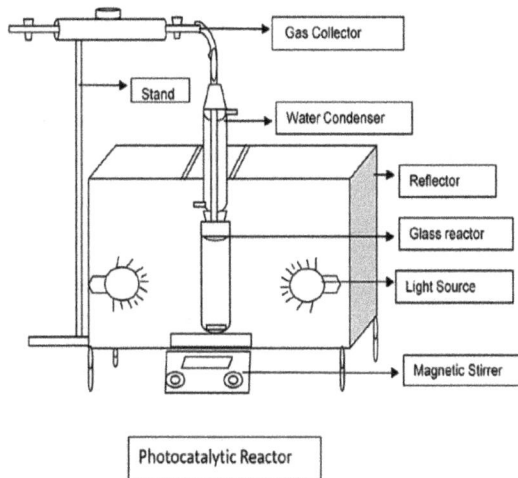

Figure 7: Photocatalytic reactor for hydrogen generation studies [72].

3.3   Vanadium and Niobium based perovskites

Under UV irradiation Vanadium and Niobium based perovskites show excellent photocatalytic activity. Both $KbNO_3$ and $NaNbO_3$ have 3.14 eV and 3.08 eV band gap values respectively. Also the suitable modification of band structure as shown in Figure 8 results in effective visible light photocatalysis [75]. N-doped $NaNbO_3$ used for the degradation of 2-propanol under visible light irradiation [76]. Also, nitrogen doped $KbNO_3$ helps for water splitting and organic pollutant degradation [77].

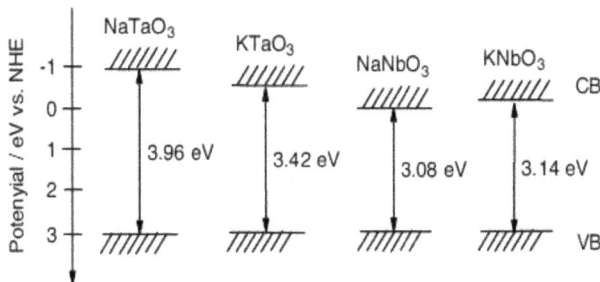

Figure 8: Band structure of $NaTaO_3$, $KTaO_3$, $NaNbO_3$, $KNbO_3$ [75].

AgVO$_3$ show both type of crystalline structure such as α-AgVO$_3$ and β-AgVO$_3$ and having band gaps 2.5 and 2.3 eV respectively [78]. β-AgVO$_3$ shows superior photocatalytic activity then α-phase. Volatile organic compounds (VOCs) were degraded and O$_2$ evolution was done by AgVO$_3$. β-AgVO$_3$ nanowires show tremendous photocatalytic performance in degradation of RhB [79]. Composite of AgBr/AgVO$_3$ show degradation of RhB [80]. The mechanism is shown in Figure 9.

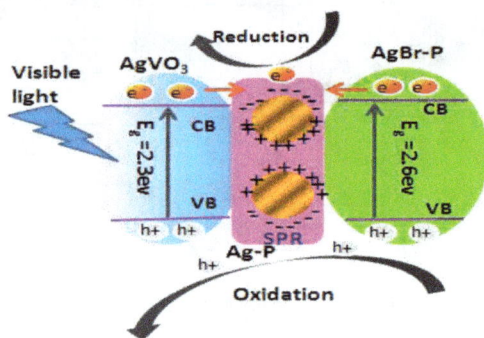

Figure 9: Proposed mechanism of photocatalysis of the AgVO$_3$@AgBr@Ag nanobelt heterostructure [80].

3.4   Oxynitride based perovskites

Perovskite type SrNbO$_2$N nanoparticles were constructed by solvothermal route by post-nitridation treatment. SrNbO$_2$N having narrow band gap of 1.94 eV exhibit high photocatalytic activity for H$_2$ evolution (H$_2$ release date is 148.3 µmol h^{-1}g^{-1}) and high degradation efficiency of methylene blue under visible light irradiation i.e. 100 % in 70 min [81]. Tantalum, niobium and titanium oxynitride perovskites have been reported as visible active photocatalysts for water splitting. Photocatalytic activity of LaTiO$_2$N was first reported by Domen *et al.* [82] for water splitting in the presence of methanol (e$^-$ donor) and Ag$^+$ (e$^-$ acceptor) and has been further investigated for acetone decomposition [83]. BaTaO$_2$N and CaTaO$_2$N are active in water splitting in the presence of methanol or I$^-$ as electron donor [84]. CaNbO$_2$N was found active for H$_2$ and O$_2$ evolution from methanol and aqueous AgNO$_3$ respectively [85]. Table 2 shows various bare perovskites for utilization as photocatalysts.

Table 2: Bare perovskites photocatalysts and their utilization

Sr.No.	Photocatalyst/Dosage	Pollutants/ concentration	% removal	Source of irradiation	Ref.
1.	$SrTiO_3$, sulfur doped $SrTiO_3$, and Ag doped $SrTiO_3$ (0.5 g)	Degradation of 4-chlorophrnol (10 ppm)	99 % within 90 min for Ag doped $SrTiO_3$	Under visible light	[86]
2.	$PbTiO_3$ (50 mg)	Acid red, acid black (5ppm)	69.7 and 96.05 % in 90 min	Ultrasonic probe power of 100-600 W	[87]
3.	Ag-$LaTiO_3$ (1.2 gL^{-1})	Atrazine (50 ppm)	100 % in 40 min	Visible light	[88]
4.	$PbTiO_3$ (500 mg)	Methyl orange (500 mL, 10 mg/L)	91 % in 180 min	Xe lamp 500 W	[89]
5.	$KNbO_3$ (0.5 g)	Rhodamine B (10 mg/L)	89 % in 120 min	Xe lamp 500 W	[90]
6.	$KNbO_3$ (1 g/L)	Acid Red dye (5 mg/L)	69.2 % in 5 h	Ultrasonic irradiation (300 W)	[91]
7.	$LaFeO_3$ 500 mg	Rhodamine B (10 mg/L)	100 % in 180 min	Visible light	[92]
8.	$LaFeO_3$ (10 mg)	Diclofenac (20 mg/L)	98.2 % in 90 min	Visible light	[93]
9.	Erbium doped $CaTiO_3$ (0.1 g)	Methylene Blue (10 mg/L)	Rate: 4.54 $\times 10^{-5} s^{-1}$ in 2.5 h	Hg-Xe lamp 300W	[94]
10.	$AgBiO_3$ (0.1-0.5 g/L)	4-nitrophenol and E.coli (4-ml)	90 % in 5 h and 1 h	Metal halide lamp	[95]
11.	$La(1\text{-}x)$ $AxTiO_3$. $5\text{-}\delta$ (A= Ba, Sr, Ca) (0.1 g)	Congo red dye (100 ppm)	70.92 % in 60 min	Xe lamp 300 W	[96]
12.	Pt-$CaTiO_3$ (0.5 g)	Photo-conversion of nitrobenzene to aniline (8.13×10^{-4} mol/L)	100 % in 60 min	Xe lamp 300 W	[97]
13.	$MnTiO_3$ (0.005 g)	Methylene blue (1×10^{-5} M)	70 % in 240 min	Sunlight irradiation	[98]
14.	F-$MnTiO_3$ (6 mg)	RhB (5×10^{-6} mol/L)	90 % in 240 min	Xe lamp 500 W	[36]
15.	$SrTiO_3$ (20 mg)	CO_2 reduction (10 mL)	26.4 $\mu mol/g$	Xe lamp 300 W	[99]
16.	Cr-$SrTiO_3$ (0.1 g)	Cr(VI) reduction	100 % 3in .5 h	Hg lamp 500 W	[100]
17.	Ag-$SrTiO_3$ (30 mg)	CO_2 reduction (10 mL)	80.24 $\mu mol/g$	Xe lamp 300 W	[101]
18.	$BaTiO_3$ (10 mg)	Tetracycline (20 mg/L)	96 % in 180 min	UV light	[102]

3.5   Other perovskite systems

There are many more perovskites photocatalyst such as compound of bismuth, cobalt, antimony and nickel having band gaps in visible region. Under visible light irradiation pantavalent bismuth perovskites are known to be active photocatalyst. Perovskite such as $LiBiO_3$, $NaBiO_3$, $KBiO_3$ and $AgBiO_3$ having band gaps of 1.63, 2.53, 2.04, 2.5 eV respectively for the degradation of organic pollutant [32]. $NaBiO_3$ shows better photocatalytic performance for phenol and methylene blue [32]. The growth of Microcystis aeruginosa was effectively restricting under solar light source with the help of $AgBiO_3$ photocatalyst [103]. $LaNiO_3$ has been used for degradation of MO (>400) [104]. $AgSbO_3$ has been used for degradation of MB, RhB and 4-chlorophenol [105]. $Gd_2Ti_2O_7/GdCrO_3$ nanoassembly act as a p-n junction photocatalyst [106]. $GdCrO_3$ are responsible for the visible light adsorption with band gap 2.5 eV [107]. Some of the bare perovskite systems are listed in Table 2.

4.   Perovskite based nano-heterojunctions

Heterojunction are analysed as the n-n, p-p and p-n combination [108]. For improving optoelectronic conversion efficiencies, many efforts have been made towards the fabrication of many heterojunctions [109-112]. The heterojunctions are classified into four categories (shown in Figure 10):

I.   Semiconductor-semiconductor (S-S) heterojunction: $SrTiO_3$ based heterojunction such as $SrTiO_3/TiO_2$ [113], $SrTiO_3/SrCO_3$ [114], $SrTiO_3/BiOBr$ [115] and $SrTiO_3/Bi_2O_3$ [116] were synthesized for various photocatalytic applications.

II.   Semiconductor-carbon(S-C) heterojunction, in which carbon can be derivative of graphene oxide (GO) [117], reduced graphene oxides (rGO) [118], Single-walled or multi-walled carbon nanotubes [119] and carbon aerogels [120]. For example, due to its super flexibility, extremely high specific surface and excellent electrical conductivity the Graphene (GR) in heterojunction enhances the charge separation efficiency, reduce the e^- and h^+ recombination and grant a large interface heterogeneous reaction.

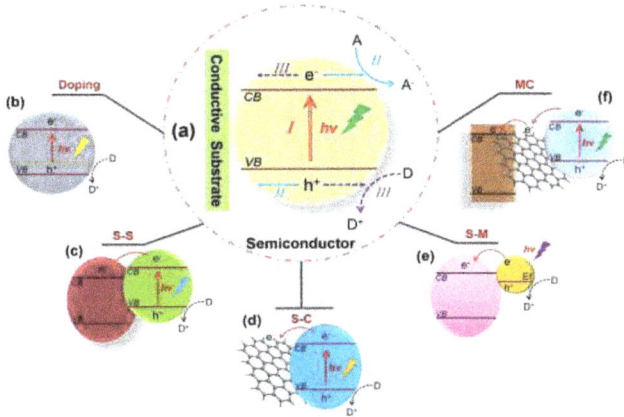

Figure 10: (a) Schematic diagram of semiconductor-based photocurrent generation, (b) Doped semiconductor, (c) S-S heterojunction, (d) S-C heterojunction, (e) S-M heterojunction, (f) M-C heterojunction [121].

III. Semiconductor-metal (S-M) heterojunctions: Ag-LaFeO₃ nanoparticles (ALFO NPs) which match with p-type characteristics exhibited remarkably high selectivity to formaldehyde, by the modification with Ag the LFO enhanced the sensing performance. Ag NPs with localized surface plasmon resonance (LSPR) or SPR properties are well known to increase the visible-light absorption [122, 123]. The surface of the photocatalyst is modified with Ag, Au etc. not only enhances solar absorption but also facilitates surface electron excitation.

IV. Multicomponent (MC) heterojunctions: Various MC Heterojunctions consisting of visible-light active components and an electron transfer system enhance visible light response and transfer of charge carriers. Metallic Ag deposited $BiPO_4/BiOBr/BiFeO_3$ (APBF) MC heterojunction was synthesized. An excellent wide spectrum photo-response gain by the plasmonic junctions is obtained which makes the best use of $BiPO_4$ which is otherwise a poor photocatalyst [124].

4.1 Synthesis of perovskites based heterojunctions

4.1.1 Synthesis of Ag₂S/BiFeO₃

Z-scheme $Ag_2S/BiFeO_3$ heterojunction was successfully synthesized by simple precipitation method. Ag_2S nanoparticle deposited over the $BiFeO_3$ particles inveterate by the morphology and microstructure characterization to form $Ag_2S/BiFeO_3$ heterojunctions [125].

4.1.2 Synthesis of LaFeO$_3$/ g-C$_3$N$_4$

The p-type $LaFeO_3$ and n-type gC_3N_4 was synthesized by facile method to form p-n heterojunctions. The $LaFeO_3$/ g-C$_3$N$_4$ heterojunction acquired higher photocurrent density than pristine g-C$_3$N$_4$ and $LaFeO_3$ [126].

4.1.3 Synthesis of Ag/Fe$_3$O$_4$ bridged SrTiO$_3$/g-C$_3$N$_4$

Z-scheme transfer and surface plasmon resonance effect of Ag augmented by iron oxide was analysed by well-constructed $SrTiO_3$/g-C$_3$N$_4$ junction bridged with Ag/Fe_3O_4 nanoparticles [127]. The synthesis route as shown in Figure 11 follows ultrasonication-photodeposition-volatilization-drying route.

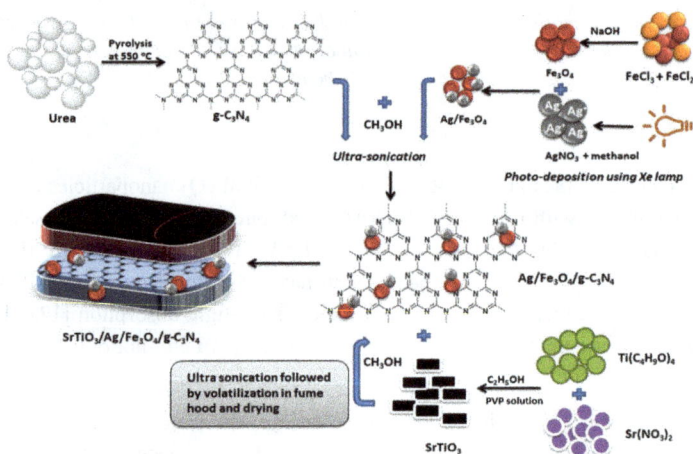

Figure 11: Synthesis Scheme of Ag/Fe$_3$O$_4$ bridged SrTiO$_3$/g-C$_3$N$_4$ [127].

4.1.4 Synthesis of PbTiO$_3$/CQDs

$PbTiO_3$/CQDs were fabricated by hydrothermal route. The Designing of highly-effective $PbTiO_3$/CQDs binary junction possess various properties such as visible light driven photocatalyst, biocompatibility, eco-friendly response and good durability [128].

4.1.5 Synthesis of CoFe₂O₄/BaTiO₃

Novel multiferroic nanostructure of CoFe₂O₄/BaTiO₃ was fabricated via co-precipitation and sol-gel technique. The synthesis of multiferroic nanocomposites, designed to enhance the mechanical coupling between the phases magnetostrictive-piezoelectric core-shell type nanostructure [129].

4.1.6 Synthesis of Ag@BiPO₄/BiOBr/BiFeO₃

Metallic Ag deposited BiPO₄/BiOBr/BiFeO₃ ternary hetero-assembly were constructed by simple precipitation-wet impregnation-photo deposition method. Surface plasmon resonance effect of Ag helps in protection of high conduction band and Valence band in the three semiconductors [124]. The synthesis route is shown in Figure 12.

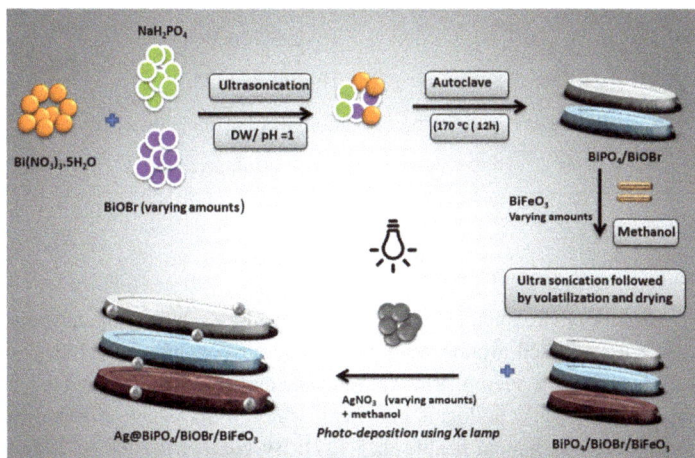

Figure 12: Synthesis scheme of Ag@BiPO₄/BiOBr/BiFeO₃ Kumar et al., 2019 [124].

4.1.6 Synthesis of AgBr/AgTaO₃

AgBr/AgTaO₃ nanocomposites were synthesized by hydrothermal route. It shows strong visible light adsorption, which enhanced photocatalytic performance then their bare counterpart [130]. Table 3 depicts various synthesis methods for perovskites based photocatalysts.

Materials Research Forum LLC
https://doi.org/10.21741/9781644901359-7

Table 3: Synthesis methods for various perovskites based photocatalysts.

Photocatalyst	Synthesis Method	Reference
$Ag_2S/BiFeO_3$	Simple precipitation	[125]
Ag/Fe_3O_4 bridged $SrTiO_3/g$-C_3N_4	Hydrothermal	[127]
$LaFeO_3/ g$-C_3N_4	Facile Method	[126]
$ZnTiO_3@TiO_2$	Hydrothermal	[131]
$PbTiO_3/CQDs$	Hydrothermal	[128]
$CoFe_2O_4/BaTiO_3$	Co-precipitation and sol-gel technique	[129]
$BiVO_4@ \beta$-$AgVO_3$	Hydrothermal	[132]
$AgBr/AgTaO_3$	Hydrothermal	[130]
$Ag@BiPO_4/BiOBr/BiFeO_3$	precipitation-wet impregnation-photo deposition method	[124]

4.2   Photocatalytic organic pollutant removal

Now days various environmental problems are facing by human being correlated to hazardous wastes, polluted ground water, and toxic air contaminants [133, 134]. Many industries such as textiles, dyeing, printing, food and cosmetics use dyes, which discharges a large amount of organic pollutant into land, water which show adversely effect on environment as well as human health [135-137]. By considering the toxicity and wide range of organic contaminants various methods have been developed such as physical separation/transfer [138] and chemical degradation [139]. Under visible light irradiation Cr-doped $SrTiO_3$ and Ag_3PO_4 heterostructured photocatalyst effectively work. The junction was used for the degradation of Isopropanol under visible light irradiation for evaluating their photocatalytic properties. 97% of degradation of isopropanol was observed within 3.8 h of irradiation [140].

The high photocatalytic active $BiFeO_3$ synthesized by the solid-state reaction for the degradation of methylene blue (MB) under visible light irradiation. The catalyst has narrow band gap energy of 2.1 eV, high surface area with hollow structure, high crystallization of perovskite-type $BiFeO_3$. The promotion of separation of photo-induced electron and holes led to near about 86% degradation in 180 min of irradiation [141].

$CaTiO_3$ NPs are used as a highly efficient photocatalyst for the removal of arsenite from aqueous solution. $CaTiO_3$ NPs enhance the photocatalytic activity towards the oxidation of As(III) up to 98.4% within 40 min of irradiation [142].

Under ultraviolet irradiation, photocatalytic decomposition of methylene blue over $CoFe_2O_4/BaTiO_3$ (CFO/BTO) based nano-composite was reported. The synthesized CFO/BTO showed excellent photocatalytic activity for the degradation of methylene blue (MB) which is 99.3% within 90 min in aqueous solution under UV light [143]. The specific surface area of synthesized ZTOCN junction was 1.69 times that of pure g-C_3N_4 with 76% degradation efficiency for methyl orange (MO) under visible light irradiation within 180 min, about 13 times the pure $ZnTiO_3$ and three times pure g-C_3N_4. The complete degradation of MB and RhB under visible light irradiation within 90 min [144].

The photocatalytic activity of $KBiO_3$ showed 100% discoloration of MB within 120 min $KBiO_3$ obtained by the sonochemical method and 67% degradation of CPFX was obtained by $KBiO_3$ synthesized by hydrothermal method [145]. The influence of the enhanced charge separation was displaced in the photocatalytic reaction. Under visible light illumination $AgNbO_3/g$-C_3N_4 photocatalyst showed high photocatalytic H_2-generation [146].

The highly photocatalytic activity for the degradation of Norfloxacin (NFN) was shown by the well-designed $Ag@BiPO_4/BiOBr/BiFeO_3$ under different light sources within 45 min. Under visible light, the photocatalyst show highest 99.1% of degradation of NFN. Under near infrared 30.2% degradation is achieved in 120 min, this is due to presence of Ag which improves the light adsorption [124] (Figure 13).

The highly photoactive $SrTiO_3/(Ag/Fe_3O_4)/g$-C_3N_4 (SFC) nano-hetero-assembly was used for photocatalytic hydrogen production and photodegradation of levofloxacin (LFC) under different light sources (Figure 14). SFC-3 (30wt% g-C_3N_4 and 3% Ag/Fe_3O_4) shows H_2 evolution of 2008 $\mu molg^{-1}h^{-1}$ which is 14 times than the single g-C_3N_4. Under visible light irradiation there is 99.3% of LFC degradation in 90 min [127].

The synthesized photocatalyst show high photocatalytic activity for the degradation of malachite green solution. 4.75% of $NaBiO_3/Bi_2O_3$ shows best photocatalytic activity which is approximately 95.70% of malachite green degradation after 300 min [147].

Water pollution is also caused by the discharge of inorganic cations in the water bodies. Cr^{+6} causes many disease such as epithelial irritant and human cancer inducer [148, 149]. Therefore various approaches were developed for efficient removal of pollutants from

Materials Research Forum LLC

https://doi.org/10.21741/9781644901359-7

water such as adsorption [150] and photocatalysis [151, 152]. The example of binary and ternary perovskite-based photocatalyst listed in Table 4.

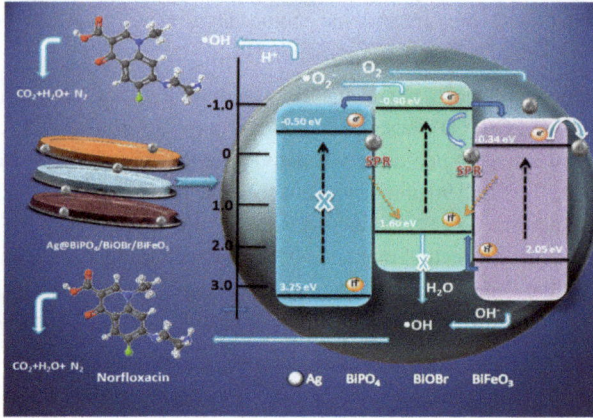

Figure 13: Degradation of Norfloxacin by using Ag@BiPO₄/BiOBr/BiFeO₃ photocatalyst [124].

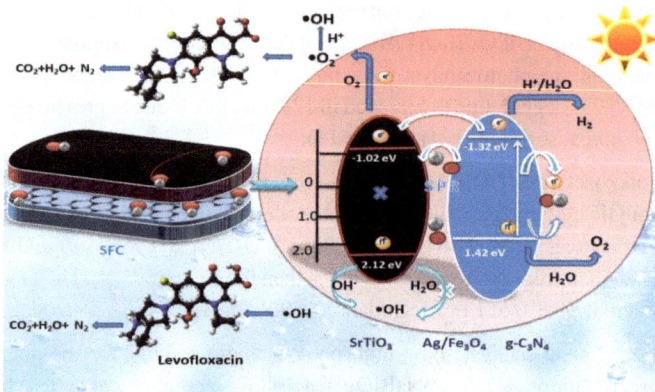

Figure 14: Degradation of levofloxacin by using Ag/Fe₃O₄ bridged SrTiO₃/g-C₃N₄ photocatalyst [127].

Table 4: Binary and ternary perovskites based photocatalyst and their utilization.

Sr.No	Photocatalyst/ Dosage	Pollutants/Concentration	% Removal	Source of irradiation	Reference
1.	$BiVO_4/SrTiO_3$ (0.05 g)	Sulfamethoxazole (10 mg/L)	91 % in 60 min	Xe lamp 500 W	[153]
2.	$Ag_3PO_4/PANI/Cr$: $SrTiO_3$ (50 mg)	Rhodamine B (RhB) (20 mg/L) and Phenol (25 mg/L)	100 % in 10 min and 100 % in 18 min	Xe lamp 300 W	[154]
3.	$SrTiO_3/BiOI$ (50 mg)	Methyl orange (MO), (10 mg/L) Oxytetracycline hydrochloride (20 mg/L)	94.6 % in 40 min 85 % in 90 min	Xe lamp 300 W	[155]
4.	$SrTiO_3/mpg-C_3N_4$ (0.3 g/L)	Basic violet 10 (10 mg/L)	80 % in 120 min	Xe lamp 300 W	[156]
5.	$CoFe_2O_4/SrTiO_3$ (0.1 g)	Azo dye (500 ppm)	98.13 in 120 min	Xe lamp 300 W	[157]
6.	Ag/Fe_3O_4 bridge $SrTiO_3/g-C_3N_4$ (0.2 g/L)	Levofloxacin (10 mg/L)	99.3 % in 90 min	Visible light (500 W Xe) UV (500 W Hg lamp) and NIR	[127]
8.	$BiFeO_3/CuWO_4$ (0.01 g)	Methyl Orange and RhB (50 mg/L)	87 and 85 % in 120 min	LED lamp 5 W	[158]
9.	$BiFeO_3/CuBi_2O_4/BaTiO_3$ (20 mg)	Norfloxacin (10 mg/L)	93.2 % in 60 min	Visible light	[159]
10.	$g-C_3N_4/rGO/BiFeO_3$ (0.5 g)	Chromium (5 mg/L)	100 % in 90 min	Xe lamp 300 W	[160]
11.	$SrTiO_3/(BiFeO_3@ZnS)$ (1 g/L)	2,4-dichlorophenol (10 mg/L) and Cr(VI) (10 mg/L)	94.32 and 97.8 % in 180 min	Visible light and Xe lamp 300 W	[161]
12.	$Ag-AgVO_3/g-C_3N_4$ (0.04 g)	Tetracycline (0.015 g)	83.6 % in 120 min	Xe lamp 300 W	[162]
13.	Graphene quantum dots/ $AgVO_3$ (0.01 g)	Ibuprofen (10 mg/L)	90 % in 120 min	Xe lamp 350 W	[163]
14.	$Ag/AgVO_3/BiOCl$ (50 mg)	Methylene blue (7 mg/L)	93.1 % in 60 min	Xe lamp 300 W	[164]
15.	$\beta-AgVO_3/ZnFe_2O_4$ (24 mg)	Methylene blue (60 mg/L)	87 % in 60 min	Xe lamp 300 W	[165]
16.	$Pt/AgVO_3$ nanowires(0.05 g)	Atrazine (100 ppm)	99 % in 60 min	UV-Visible light	[166]
17.	$ZnO/AgVO_3$ (20 and 50 mg)	Ciprofloxacin and chromium (10 and 20 mg/L)	90 % in 90 and 45 min	Visible light	[167]

18.	Ag/AgVO$_3$/g-C$_3$N$_4$ (0.1 g)	Ciprofloxacin (10 ppm) and hydrogen production	82.6 % and 3.75 mmol/h H$_2$ production	Xe lamp 300 W and 500 W Hg lamp	[168]
19.	MoS$_2$/AgVO$_3$ (50 mg/L)	Rhodamine B (10 mg/L)	83.3 % in 120 min	Xe lamp 300 W	[169]
20.	GO-Ag$_2$O/ Ag$_3$VO$_4$/AgVO$_3$ (0.15 g)	Rhodamine B (15 mg/L) and MO (15 mg/L)	99.2 and 92 % in 45 min	Visible light	[170]
21.	CaTiO$_3$/rGO/ NiFe$_2$O$_4$ (0.1 g)	Methyle orange (5 mg/L)	98.3 % in 90 min	Xe lamp 200 W	[171]
22.	CaTiO$_3$/g-C$_3$N$_4$ (25 mg)	Rhodamine B (1×10^{-5} M)	97 % in 60 min	Xe lamp 300 W	[172]
23.	PbTiO$_3$/CQDs (0.05 g)	Rhodamine B (5×10^{-5} M)	100 % in 120 min	Xe lamp 300 W	[128]
24.	KTaO$_3$/FeVO$_4$/ Bi$_2$O$_3$ (1 g/L)	Ceftriaxone sodium (10 mg/L)	81.30 % in 90 min	Visible light	[173]

Conclusion

Perovskites and perovskites related structures propose a wide range of materials to design novel photocatalyst for energy and environmental applications. The doping and substitution of A- and B- sites as well as the O sites are utilized for reducing band gaps to prepare visible and solar active photocatalysts. The formation of multicomponent heterojunctions is an approach to improve photocatalytic activity for harvesting solar energy and charge carrier separation. The performance of the perovskite and perovskite related photocatalysts gets improved by tailoring of their particle size, crystallinity and morphology. All the physicochemical properties provide perovskites and perovskites-related materials for various applications such as potentials for organic/inorganic contaminants removal, drug delivery, catalyst in modern chemical industry, sensor etc. These represent an alternative class of materials with hybrid character for visible and solar light assisted photodegradation of pollutants and clean energy production.

References

[1] S. Garcia-Segura, E. Brillas, Applied photoelectrocatalysis on the degradation of organic pollutants in wastewaters, Journal of Photochemistry and Photobiology C: Photochemistry Reviews 31 (2017) 1-35. https://doi.org/10.1016/j.jphotochemrev.2017.01.005

[2] Z. Shi, A. Jayatissa, Perovskites-based solar cells: A review of recent progress, materials and processing methods, Materials 11 (2018) 729. https://doi.org/10.3390/ma11050729

[3] A. Navrotsky, D.J. Weidner, Perovskite: a structure of great interest to geophysics and materials science, Washington DC American Geophysical Union Geophysical Monograph Series 45 (1989). https://doi.org/10.1029/GM045

[4] M. Osako, E. Ito, Thermal diffusivity of $MgSiO_3$ perovskite, Geophys. Res. Lett. 18 (1991) 239-242. https://doi.org/10.1029/91GL00212

[5] H. Piel, High Tc superconductors for accelerator cavities, Nuclear Instruments and Methods in Physics Research Section A: Accelerators, Spectrometers, Detectors and Associated Equipment 287 (1990) 294-305. https://doi.org/10.1016/0168-9002(90)91812-P

[6] J. Zhu, H. Li, L. Zhong, P. Xiao, X. Xu, X. Yang, Z. Zhao, J. Li, Perovskite oxides: preparation, characterizations, and applications in heterogeneous catalysis, Acs Catalysis 4 (2014) 2917-2940. https://doi.org/10.1021/cs500606g

[7] B. Seyfi, M. Baghalha, H. Kazemian, Modified $LaCoO_3$ nano-perovskite catalysts for the environmental application of automotive CO oxidation, Chem. Eng. J. 148 (2009) 306-311. https://doi.org/10.1016/j.cej.2008.08.041

[8] B. Xu, K. Yin, J. Lin, Y. Xia, X. Wan, J. Yin, X. Bai, J. Du, Z. Liu, Room-temperature ferromagnetism and ferroelectricity in Fe-doped $BaTiO_3$, Physical Review B 79 (2009) 134109. https://doi.org/10.1103/PhysRevB.79.134109

[9] X. Zhang, J. Zhang, Z. Nie, M. Wang, X. Ren, X.-j. Wang, Enhanced red phosphorescence in nanosized $CaTiO_3$: Pr^{3+} phosphors, Appl. Phys. Lett. 90 (2007) 151911. https://doi.org/10.1063/1.2722205

[10] K. Parida, K. Reddy, S. Martha, D. Das, N. Biswal, Fabrication of nanocrystalline $LaFeO_3$: an efficient sol–gel auto-combustion assisted visible light responsive photocatalyst for water decomposition, Int. J. Hydrogen Energy 35 (2010) 12161-12168. https://doi.org/10.1016/j.ijhydene.2010.08.029

[11] M. Pena, J. Fierro, Chemical structures and performance of perovskite oxides, Chem. Rev. 101 (2001) 1981-2018. https://doi.org/10.1021/cr980129f

[12] S. Bharathkumar, M. Sakar, S. Balakumar, Versatility of electrospinning in the fabrication of fibrous mat and mesh nanostructures of bismuth ferrite ($BiFeO_3$) and their magnetic and photocatalytic activities, PCCP 17 (2015) 17745-17754. https://doi.org/10.1039/C5CP01640A

[13] Y. Jia, C. Wu, D.-H. Kim, B. Lee, S. Rhee, Y.C. Park, C.S. Kim, Q. Wang, C. Liu, Nitrogen doped $BiFeO_3$ with enhanced magnetic properties and photo-Fenton catalytic activity for degradation of bisphenol A under visible light, Chem. Eng. J. 337 (2018) 709-721. https://doi.org/10.1016/j.cej.2017.12.137

[14] H. Bai, J. Juay, Z. Liu, X. Song, S.S. Lee, D.D. Sun, Hierarchical $SrTiO_3/TiO_2$ nanofibers heterostructures with high efficiency in photocatalytic H_2 generation, Applied Catalysis B: Environmental 125 (2012) 367-374. https://doi.org/10.1016/j.apcatb.2012.06.007

[15] Q. Wang, T. Hisatomi, S.S.K. Ma, Y. Li, K. Domen, Core/shell structured La-and Rh-codoped $SrTiO_3$ as a hydrogen evolution photocatalyst in Z-scheme overall water splitting under visible light irradiation, Chem. Mater. 26 (2014) 4144-4150. https://doi.org/10.1021/cm5011983

[16] A. Kumar, M. Naushad, A. Rana, G. Sharma, A.A. Ghfar, F.J. Stadler, M.R. Khan, $ZnSe-WO_3$ nano-hetero-assembly stacked on Gum ghatti for photo-degradative removal of Bisphenol A: Symbiose of adsorption and photocatalysis, Int. J. Biol. Macromol. 104 (2017) 1172-1184. https://doi.org/10.1016/j.ijbiomac.2017.06.116

[17] X. Lü, J. Xie, H. Shu, J. Liu, C. Yin, J. Lin, Microwave-assisted synthesis of nanocrystalline $YFeO_3$ and study of its photoactivity, Materials Science and Engineering: B 138 (2007) 289-292. https://doi.org/10.1016/j.mseb.2007.01.003

[18] M. Asamoto, Y. Iwasaki, S. Yamaguchi, H. Yahiro, Synthesis of perovsite-type oxide catalysts, $Ln(Fe, Co)O_3$ (Ln= La, Pr, Sm, Gd, Dy, Ho, Er, and Yb), from the thermal decomposition of the corresponding cyano complexes, Catal. Today 185 (2012) 230-235. https://doi.org/10.1016/j.cattod.2011.09.023

[19] T. Chen, Z. Zhou, Y. Wang, Surfactant CATB-assisted generation and gas-sensing characteristics of $LnFeO_3$ (Ln= La, Sm, Eu) materials, Sensors and Actuators B: Chemical 143 (2009) 124-131. https://doi.org/10.1016/j.snb.2009.09.031

[20] S. Thirumalairajan, K. Girija, I. Ganesh, D. Mangalaraj, C. Viswanathan, A. Balamurugan, N. Ponpandian, Controlled synthesis of perovskite LaFeO3 microsphere composed of nanoparticles via self-assembly process and their associated photocatalytic activity, Chem. Eng. J. 209 (2012) 420-428. https://doi.org/10.1016/j.cej.2012.08.012

[21] L. Ju, Z. Chen, L. Fang, W. Dong, F. Zheng, M. Shen, Sol–gel synthesis and photo-Fenton-like catalytic activity of $EuFeO_3$ nanoparticles, J. Am. Ceram. Soc. 94 (2011) 3418-3424. https://doi.org/10.1111/j.1551-2916.2011.04522.x

[22] S. Sun, W. Wang, L. Zhang, M. Shang, Visible light-induced photocatalytic oxidation of phenol and aqueous ammonia in flowerlike $Bi_2Fe_4O_9$ suspensions, The Journal of Physical Chemistry C 113 (2009) 12826-12831. https://doi.org/10.1021/jp9029826

[23] D. Weber, CH3NH3PbX3, ein Pb (II)-system mit kubischer perowskitstruktur/$CH_3NH_3PbX_3$, a Pb (II)-system with cubic perovskite structure,

Zeitschrift für Naturforschung B 33 (1978) 1443-1445. https://doi.org/10.1515/znb-1978-1214

[24] N. Ramadass, ABO$_3$-type oxides—Their structure and properties—A bird's eye view, Materials Science and Engineering 36 (1978) 231-239. https://doi.org/10.1016/0025-5416(78)90076-9

[25] E. Fatuzzo, W.J. Merz, Ferroelectricity, North-Holland Pub. Co.1967.

[26] J. Haines, J. Rouquette, V. Bornand, M. Pintard, P. Papet, F. Gorelli, Raman scattering studies at high pressure and low temperature: technique and application to the piezoelectric material PbZr$_{0.52}$Ti$_{0.48}$O$_3$, Journal of Raman Spectroscopy 34 (2003) 519-523. https://doi.org/10.1002/jrs.1009

[27] J. Shi, L. Guo, ABO$_3$-based photocatalysts for water splitting, Progress in Natural Science: Materials International 22 (2012) 592-615. https://doi.org/10.1016/j.pnsc.2012.12.002

[28] T. Zhang, K. Zhao, J. Yu, J. Jin, Y. Qi, H. Li, X. Hou, G. Liu, Photocatalytic water splitting for hydrogen generation on cubic, orthorhombic, and tetragonal KNbO$_3$ microcubes, Nanoscale 5 (2013) 8375-8383. https://doi.org/10.1039/c3nr02356g

[29] P. Kanhere, Z. Chen, A review on visible light active perovskite-based photocatalysts, Molecules 19 (2014) 19995-20022. https://doi.org/10.3390/molecules191219995

[30] I. Shein, V. Kozhevnikov, A. Ivanovskii, First-principles study of cubic perovskites SrMO$_3$ (M= Ti, V, Zr and Nb), arXiv preprint cond-mat/0504286 (2005).

[31] S. Thirumalairajan, K. Girija, N.Y. Hebalkar, D. Mangalaraj, C. Viswanathan, N. Ponpandian, Shape evolution of perovskite LaFeO$_3$ nanostructures: a systematic investigation of growth mechanism, properties and morphology dependent photocatalytic activities, RSC Advances 3 (2013) 7549-7561. https://doi.org/10.1039/c3ra00006k

[32] T. Takei, R. Haramoto, Q. Dong, N. Kumada, Y. Yonesaki, N. Kinomura, T. Mano, S. Nishimoto, Y. Kameshima, M. Miyake, Photocatalytic activities of various pentavalent bismuthates under visible light irradiation, J. Solid State Chem. 184 (2011) 2017-2022. https://doi.org/10.1016/j.jssc.2011.06.004

[33] K. Van Benthem, C. Elsässer, R. French, Bulk electronic structure of SrTiO 3: experiment and theory, Journal of Applied Physics 90 (2001) 6156-6164. https://doi.org/10.1063/1.1415766

[34] H. Zhang, G. Chen, X. He, J. Xu, Electronic structure and photocatalytic properties of Ag–La codoped CaTiO$_3$, J. Alloys Compd. 516 (2012) 91-95. https://doi.org/10.1016/j.jallcom.2011.11.142

[35] J. Joseph, T. Vimala, V. Sivasubramanian, V. Murthy, Structural investigations on Pb (Zr_xT_{1-x}) O_3 solid solutions using the X-ray Rietveld method, Journal of materials science 35 (2000) 1571-1575. https://doi.org/10.1023/A:1004778223721

[36] W. Dong, D. Wang, L. Jiang, H. Zhu, H. Huang, J. Li, H. Zhao, C. Li, B. Chen, G. Deng, Synthesis of F doping $MnTiO_3$ nanodiscs and their photocatalytic property under visible light, Materials Letters 98 (2013) 265-268. https://doi.org/10.1016/j.matlet.2013.02.056

[37] A. Zaleska-Medynska, A. Malankowska, M. Marchelek, B. Bajorowicz, P. Mazierski, T. Klimczuk, $KTaO_3$-based nanocomposites for air treatment, (2014).

[38] L. Ni, M. Tanabe, H. Irie, A visible-light-induced overall water-splitting photocatalyst: conduction-band-controlled silver tantalate, Chemical Communications 49 (2013) 10094-10096. https://doi.org/10.1039/c3cc45222k

[39] H. Wang, F. Wu, H. Jiang, Electronic band structures of $ATaO_3$ (A= Li, Na, and K) from first-principles many-body perturbation theory, The Journal of Physical Chemistry C 115 (2011) 16180-16186. https://doi.org/10.1021/jp2047294

[40] N. Kumada, N. Kinomura, A. Sleight, Neutron powder diffraction refinement of ilmenite-type bismuth oxides: $ABiO_3$ (A= Na, Ag), Mater. Res. Bull. 35 (2000) 2397-2402. https://doi.org/10.1016/S0025-5408(00)00453-0

[41] F.S. Galasso, Structure, properties and preparation of perovskite-type compounds: international series of monographs in solid state physics, Elsevier 2013.

[42] A. Bhalla, R. Guo, R. Roy, The perovskite structure—a review of its role in ceramic science and technology, Mater. Res. Innovations 4 (2000) 3-26. https://doi.org/10.1007/s100190000062

[43] F. Cordero, F. Trequattrini, F. Craciun, H. Langhammer, D. Quiroga, P. Silva Jr, Probing ferroelectricity in highly conducting materials through their elastic response: Persistence of ferroelectricity in metallic $BaTiO_{3-\delta}$, Physical Review B 99 (2019) 064106. https://doi.org/10.1103/PhysRevB.99.064106

[44] P. Chen, W. Xu, Y. Gao, P. Holdway, J.H. Warner, M.R. Castell, Thermal Degradation of Monolayer MoS_2 on $SrTiO_3$ Supports, The Journal of Physical Chemistry C 123 (2019) 3876-3885. https://doi.org/10.1021/acs.jpcc.8b11298

[45] J. Geusic, H. Marcos, L. Van Uitert, Laser oscillations in Nd-doped yttrium aluminum, yttrium gallium and gadolinium garnets, Appl. Phys. Lett. 4 (1964) 182-184. https://doi.org/10.1063/1.1753928

[46] Y. Pan, Q. Su, H. Xu, T. Chen, W. Ge, C. Yang, M. Wu, Synthesis and red luminescence of Pr^{3+}-doped $CaTiO_3$ nanophosphor from polymer precursor, J. Solid State Chem. 174 (2003) 69-73. https://doi.org/10.1016/S0022-4596(03)00175-0

[47] K. Dhahri, M. Bejar, E. Dhahri, M. Soares, M. Graça, M. Sousa, M. Valente, Blue-green photoluminescence in $BaZrO_{3-\delta}$ powders, Chem. Phys. Lett. 610 (2014) 341-344. https://doi.org/10.1016/j.cplett.2014.07.057

[48] F.P. Sun, Z. Chaudhry, C. Liang, C. Rogers, Truss structure integrity identification using PZT sensor-actuator, J. Intell. Mater. Syst. Struct. 6 (1995) 134-139, https://doi.org/10.1177/1045389X9500600117

[49] M. Tinkham, Introduction to superconductivity, Courier Corporation2004.

[50] D. Murphy, S. Sunshine, R. Van Dover, R.J. Cava, B. Batlogg, S. Zahurak, L. Schneemeyer, New superconducting cuprate perovskites, Phys. Rev. Lett. 58 (1987) 1888. https://doi.org/10.1103/PhysRevLett.58.1888

[51] J. Schooley, W. Hosler, M.L. Cohen, Superconductivity in Semiconducting $SrTiO_3$, Phys. Rev. Lett. 12 (1964) 474. https://doi.org/10.1103/PhysRevLett.12.474

[52] C. Ederer, N.A. Spaldin, Weak ferromagnetism and magnetoelectric coupling in bismuth ferrite, Physical Review B 71 (2005) 060401. https://doi.org/10.1103/PhysRevB.71.060401

[53] H. Liu, B. Cao, C. O'Connor, Intrinsic magnetism in $BaTiO_3$ with magnetic transition element dopants (Co, Cr, Fe) synthesized by sol-precipitation method, J. Appl. Phys. 109 (2011) 07B516. https://doi.org/10.1063/1.3556768

[54] N.H. Chan, R. Sharma, D.M. Smyth, Nonstoichiometry in Acceptor-Doped $BaTiO_3$, J. Am. Ceram. Soc. 65 (1982) 167-170. https://doi.org/10.1111/j.1151-2916.1982.tb10388.x

[55] D. Segal, Chemical synthesis of ceramic materials, J. Mater. Chem. 7 (1997) 1297-1305. https://doi.org/10.1039/a700881c

[56] H. Luo, Y. Wei, H. Jiang, W. Yuan, Y. Lv, J. Caro, H. Wang, Performance of a ceramic membrane reactor with high oxygen flux Ta-containing perovskite for the partial oxidation of methane to syngas, Journal of Membrane Science 350 (2010) 154-160. https://doi.org/10.1016/j.memsci.2009.12.023

[57] M. Reichmann, P.-M. Geffroy, J. Fouletier, N. Richet, T. Chartier, Effect of cation substitution in the A site on the oxygen semi-permeation flux in $La_{0.5}A_{0.5}Fe_{0.7}Ga_{0.3}O_{3-\delta}$ and $La_{0.5}A_{0.5}Fe_{0.7}Co_{0.3}O_{3-\delta}$ dense perovskite membranes with A= Ca, Sr and Ba (part I), J. Power Sources 261 (2014) 175-183. https://doi.org/10.1016/j.jpowsour.2014.03.074

[58] S. Ohara, A. Kondo, H. Shimoda, K. Sato, H. Abe, M. Naito, Rapid mechanochemical synthesis of fine barium titanate nanoparticles, Mater. Lett. 62 (2008) 2957-2959. https://doi.org/10.1016/j.matlet.2008.01.083

[59] T. Kutty, R. Vivekanandan, Preparation of $CaTiO_3$ fine powders by the hydrothermal method, Mater. Lett. 5 (1987) 79-83. https://doi.org/10.1016/0167-577X(87)90080-2

[60] Y. Zeng, Y. Lin, S. Swartz, Perovskite-type ceramic membrane: synthesis, oxygen permeation and membrane reactor performance for oxidative coupling of methane, Journal of membrane science 150 (1998) 87-98. https://doi.org/10.1016/S0376-7388(98)00182-3

[61] E. Mostafavi, A. Babaei, A. Ataie, Synthesis of nano-structured $La_{0.6}Sr_{0.4}Co_{0.2}Fe_{0.8}O_3$ perovskite by co-precipitation method, Journal of Ultrafine Grained and Nanostructured Materials 48 (2015) 45-52.

[62] A. Kumar, G. Sharma, M. Naushad, A.a.H. Al-Muhtaseb, A. García-Peñas, G.T. Mola, C. Si, F.J. Stadler, Bio-inspired and biomaterials-based hybrid photocatalysts for environmental detoxification: A review, Chemical Engineering Journal 382 (2020) 122937. https://doi.org/10.1016/j.cej.2019.122937

[63] A. Kumar, A. Rana, G. Sharma, M. Naushad, P. Dhiman, A. Kumari, F.J. Stadler, Recent advances in nano-Fenton catalytic degradation of emerging pharmaceutical contaminants, Journal of Molecular Liquids 290 (2019) 111177. https://doi.org/10.1016/j.molliq.2019.111177

[64] Q. Sun, Y. Hong, Q. Liu, L. Dong, Synergistic operation of photocatalytic degradation and Fenton process by magnetic Fe_3O_4 loaded TiO_2, Applied Surface Science 430 (2018) 399-406. https://doi.org/10.1016/j.apsusc.2017.08.085

[65] K. Maeda, Photocatalytic water splitting using semiconductor particles: history and recent developments, Journal of Photochemistry and Photobiology C: Photochemistry Reviews 12 (2011) 237-268. https://doi.org/10.1016/j.jphotochemrev.2011.07.001

[66] H. Zhang, G. Chen, Y. Li, Y. Teng, Electronic structure and photocatalytic properties of copper-doped $CaTiO_3$, International Journal of Hydrogen Energy 35 (2010) 2713-2716. https://doi.org/10.1016/j.ijhydene.2009.04.050

[67] Y. Qu, W. Zhou, H. Fu, Porous Cobalt Titanate Nanorod: A New Candidate for Visible Light-Driven Photocatalytic Water Oxidation, ChemCatChem 6 (2014) 265-270. https://doi.org/10.1002/cctc.201300718

[68] Y. Qu, W. Zhou, Z. Ren, S. Du, X. Meng, G. Tian, K. Pan, G. Wang, H. Fu, Facile preparation of porous $NiTiO_3$ nanorods with enhanced visible-light-driven photocatalytic performance, Journal of Materials Chemistry 22 (2012) 16471-16476. https://doi.org/10.1039/c2jm32044d

[69] Y.J. Kim, B. Gao, S.Y. Han, M.H. Jung, A.K. Chakraborty, T. Ko, C. Lee, W.I. Lee, Heterojunction of $FeTiO_3$ nanodisc and TiO_2 nanoparticle for a novel visible light

photocatalyst, The Journal of Physical Chemistry C 113 (2009) 19179-19184. https://doi.org/10.1021/jp908874k

[70] M.S. Hassan, T. Amna, M.-S. Khil, Synthesis of high aspect ratio CdTiO₃ nanofibers via electrospinning: characterization and photocatalytic activity, Ceram. Int. 40 (2014) 423-427. https://doi.org/10.1016/j.ceramint.2013.06.018

[71] L. Li, Y. Zhang, A.M. Schultz, X. Liu, P.A. Salvador, G.S. Rohrer, Visible light photochemical activity of heterostructured PbTiO₃–TiO₂ core–shell particles, Catalysis Science & Technology 2 (2012) 1945-1952. https://doi.org/10.1039/c2cy20202f

[72] S.N. Tijare, M.V. Joshi, P.S. Padole, P.A. Mangrulkar, S.S. Rayalu, N.K. Labhsetwar, Photocatalytic hydrogen generation through water splitting on nano-crystalline LaFeO₃ perovskite, Int. J. Hydrogen Energy 37 (2012) 10451-10456. https://doi.org/10.1016/j.ijhydene.2012.01.120

[73] P. Dhanasekaran, N. Gupta, Factors affecting the production of H₂ by water splitting over a novel visible-light-driven photocatalyst GaFeO₃, Int. J. Hydrogen Energy 37 (2012) 4897-4907. https://doi.org/10.1016/j.ijhydene.2011.12.068

[74] P. Tang, H. Chen, F. Cao, G. Pan, Magnetically recoverable and visible-light-driven nanocrystalline YFeO₃ photocatalysts, Catalysis Science & Technology 1 (2011) 1145-1148. https://doi.org/10.1039/c1cy00199j

[75] J. Liu, G. Chen, Z. Li, Z. Zhang, Hydrothermal synthesis and photocatalytic properties of ATaO₃ and ANbO₃ (A= Na and K), Int. J. Hydrogen Energy 32 (2007) 2269-2272. https://doi.org/10.1016/j.ijhydene.2006.10.005

[76] H. Shi, X. Li, H. Iwai, Z. Zou, J. Ye, 2-Propanol photodegradation over nitrogen-doped NaNbO₃ powders under visible-light irradiation, J. Phys. Chem. Solids 70 (2009) 931-935. https://doi.org/10.1016/j.jpcs.2009.05.002

[77] R. Wang, Y. Zhu, Y. Qiu, C.-F. Leung, J. He, G. Liu, T.-C. Lau, Synthesis of nitrogen-doped KNbO₃ nanocubes with high photocatalytic activity for water splitting and degradation of organic pollutants under visible light, Chemical Engineering Journal 226 (2013) 123-130. https://doi.org/10.1016/j.cej.2013.04.049

[78] R. Konta, H. Kato, H. Kobayashi, A. Kudo, Photophysical properties and photocatalytic activities under visible light irradiation of silver vanadates, Physical Chemistry Chemical Physics 5 (2003) 3061-3065. https://doi.org/10.1039/b300179b

[79] W. Jia, H. Dong, J. Zhao, S. Dang, Z. Zhang, T. Li, X. Liu, B. Xu, p-Cu₂O/n-ZnO heterojunction fabricated by hydrothermal method, Applied Physics A 109 (2012) 751-756. https://doi.org/10.1007/s00339-012-7111-3

[80] Y. Sang, L. Kuai, C. Chen, Z. Fang, B. Geng, Fabrication of a visible-light-driven plasmonic photocatalyst of AgVO₃@ AgBr@ Ag nanobelt heterostructures, ACS

applied materials & interfaces 6 (2014) 5061-5068.
https://doi.org/10.1021/am5002019

[81] J. Wang, X. Wang, B. Liu, X. Li, M. Cao, Facile synthesis of $SrNbO_2N$
nanoparticles with excellent visible-light photocatalytic performances, Materials
Letters 152 (2015) 131-134. https://doi.org/10.1016/j.matlet.2015.03.104

[82] A. Kasahara, K. Nukumizu, G. Hitoki, T. Takata, J.N. Kondo, M. Hara, H.
Kobayashi, K. Domen, Photoreactions on $LaTiO_2N$ under visible light irradiation, The
Journal of Physical Chemistry A 106 (2002) 6750-6753.
https://doi.org/10.1021/jp025961+

[83] R. Aguiar, A. Kalytta, A. Reller, A. Weidenkaff, S.G. Ebbinghaus, Photocatalytic
decomposition of acetone using $LaTi(O,N)_3$ nanoparticles under visible light
irradiation, Journal of Materials Chemistry 18 (2008) 4260-4265.
https://doi.org/10.1039/b806794e

[84] M. Higashi, R. Abe, T. Takata, K. Domen, Photocatalytic overall water splitting
under visible light using $ATaO_2N$ (A= Ca, Sr, Ba) and WO_3 in a IO_3^-/I^- shuttle redox
mediated system, Chemistry of Materials 21 (2009) 1543-1549.
https://doi.org/10.1021/cm803145n

[85] B. Siritanaratkul, K. Maeda, T. Hisatomi, K. Domen, Synthesis and photocatalytic
activity of perovskite niobium oxynitrides with wide visible-light absorption bands,
ChemSusChem 4 (2011) 74-78. https://doi.org/10.1002/cssc.201000207

[86] B. Anitha, L.G. Devi, Study of reaction dynamics of photocatalytic degradation of 4-
chlorophenol using $SrTiO_3$, sulfur doped $SrTiO_3$, silver metallized $SrTiO_3$ and silver
metallized sulfur doped $SrTiO_3$ catalysts: Detailed analysis of kinetic results, Surfaces
and Interfaces 16 (2019) 50-58. https://doi.org/10.1016/j.surfin.2019.04.009

[87] O. Amiri, K. Salar, P. Othman, T. Rasul, D. Faiq, M. Saadat, Purification of
wastewater by the piezo-catalyst effect of $PbTiO_3$ nanostructures under ultrasonic
vibration, Journal of Hazardous Materials (2020) 122514.
https://doi.org/10.1016/j.jhazmat.2020.122514

[88] A. Shawky, R. Mohamed, I. Mkhalid, M. Youssef, N. Awwad, Visible light-
responsive Ag/LaTiO3 nanowire photocatalysts for efficient elimination of atrazine
herbicide in water, Journal of Molecular Liquids 299 (2020) 112163.
https://doi.org/10.1016/j.molliq.2019.112163

[89] Y. Li, H. Sun, N. Wang, W. Fang, Z. Li, Effects of pH and temperature on
photocatalytic activity of $PbTiO_3$ synthesized by hydrothermal method, Solid state
sciences 37 (2014) 18-22. https://doi.org/10.1016/j.solidstatesciences.2014.08.003

[90] T.B. Wermuth, S. Arcaro, J. Venturini, T.M.H. Ribeiro, A.d.A.L. Rodriguez, E.L. Machado, T.F. de Oliveira, S.E.F. de Oliveira, M.N. Baibich, C.P. Bergmann, Microwave-synthesized $KNbO_3$ perovskites: photocatalytic pathway on the degradation of rhodamine B, Ceramics International 45 (2019) 24137-24145. https://doi.org/10.1016/j.ceramint.2019.08.122

[91] H. Zhang, C. Wei, Y. Huang, J. Wang, Preparation of cube micrometer potassium niobate ($KNbO_3$) by hydrothermal method and sonocatalytic degradation of organic dye, Ultrasonics sonochemistry 30 (2016) 61-69. https://doi.org/10.1016/j.ultsonch.2015.11.003

[92] M. Ismael, M. Wark, Perovskite-type LaFeO3: photoelectrochemical properties and photocatalytic degradation of organic pollutants under visible light irradiation, Catalysts 9 (2019) 342. https://doi.org/10.3390/catal9040342

[93] W. Meng, Y. Wang, Y. Zhang, C. Liu, Z. Wang, Z. Song, B. Xu, D.C. Tsang, F. Qi, A. Ikhlaq, Degradation Rhodamine B dye wastewater by sulfate radical-based visible light-fenton mediated by $LaFeO_3$: Reaction mechanism and empirical modeling, Journal of the Taiwan Institute of Chemical Engineers (2020). https://doi.org/10.1016/j.jtice.2020.04.005

[94] L. Lozano-Sánchez, S. Obregón, L. Díaz-Torres, S.-W. Lee, V. Rodríguez-González, Visible and near-infrared light-driven photocatalytic activity of erbium-doped $CaTiO_3$ system, Journal of Molecular Catalysis A: Chemical 410 (2015) 19-25. https://doi.org/10.1016/j.molcata.2015.09.005

[95] B. Boruah, R. Gupta, J.M. Modak, G. Madras, Novel insights into the properties of AgBiO3 photocatalyst and its application in immobilized state for 4-nitrophenol degradation and bacteria inactivation, Journal of photochemistry and photobiology A: Chemistry 373 (2019) 105-115. https://doi.org/10.1016/j.jphotochem.2018.11.001

[96] M. Bradha, T. Vijayaraghavan, S. Suriyaraj, R. Selvakumar, A.M. Ashok, Synthesis of photocatalytic $La_{(1-x)}A_xTiO_{3.5-\delta}$ (A= Ba, Sr, Ca) nano perovskites and their application for photocatalytic oxidation of congo red dye in aqueous solution, Journal of Rare Earths 33 (2015) 160-167. https://doi.org/10.1016/S1002-0721(14)60397-5

[97] A. Shawky, M. Alhaddad, K. Al-Namshah, R. Mohamed, N.S. Awwad, Synthesis of Pt-decorated $CaTiO_3$ nanocrystals for efficient photoconversion of nitrobenzene to aniline under visible light, Journal of Molecular Liquids 304 (2020) 112704. https://doi.org/10.1016/j.molliq.2020.112704

[98] S. Alkaykh, A. Mbarek, E.E. Ali-Shattle, Photocatalytic degradation of methylene blue dye in aqueous solution by MnTiO3 nanoparticles under sunlight irradiation, Heliyon 6 (2020) e03663. https://doi.org/10.1016/j.heliyon.2020.e03663

[99] C. Luo, J. Zhao, Y. Li, W. Zhao, Y. Zeng, C. Wang, Photocatalytic CO_2 reduction over SrTiO3: Correlation between surface structure and activity, Applied Surface Science 447 (2018) 627-635. https://doi.org/10.1016/j.apsusc.2018.04.049

[100] J. Jang, J. Lee, P.H. Borse, K. Lim, O. Jung, E. Jeong, M. Won, H. Kim, Platinum nanoparticle co-catalyst-induced improved photoelectrical properties in a chromium-doped $SrTiO_3$ photocatalyst, Journal of the Korean Physical Society 55 (2009) 2470-2475. https://doi.org/10.3938/jkps.55.2470

[101] S. Wan, M. Chen, M. Ou, Q. Zhong, Plasmonic Ag nanoparticles decorated SrTiO3 nanocubes for enhanced photocatalytic CO_2 reduction and H_2 evolution under visible light irradiation, Journal of CO2 Utilization 33 (2019) 357-364. https://doi.org/10.1016/j.jcou.2019.06.024

[102] P. Demircivi, B. Gulen, E.B. Simsek, D. Berek, Enhanced photocatalytic degradation of tetracycline using hydrothermally synthesized carbon fiber decorated $BaTiO_3$, Materials Chemistry and Physics 241 (2020) 122236. https://doi.org/10.1016/j.matchemphys.2019.122236

[103] X. Yu, J. Zhou, Z. Wang, W. Cai, Preparation of visible light-responsive $AgBiO_3$ bactericide and its control effect on the Microcystis aeruginosa, Journal of Photochemistry and Photobiology B: Biology 101 (2010) 265-270. https://doi.org/10.1016/j.jphotobiol.2010.07.011

[104] P.S. Tang, H. Sun, F. Cao, J.T. Yang, S.L. Ni, H.F. Chen, Visible-light driven $LaNiO_3$ nanosized photocatalysts prepared by a sol-gel process, Advanced Materials Research, Trans Tech Publ, 2011, pp. 83-87. https://doi.org/10.4028/www.scientific.net/AMR.279.83

[105] J. Singh, S. Uma, Efficient photocatalytic degradation of organic compounds by ilmenite $AgSbO_3$ under visible and UV light irradiation, The Journal of Physical Chemistry C 113 (2009) 12483-12488. https://doi.org/10.1021/jp901729v

[106] K. Parida, A. Nashim, S.K. Mahanta, Visible-light driven $Gd_2Ti_2O_7/GdCrO_3$ composite for hydrogen evolution, Dalton Transactions 40 (2011) 12839-12845. https://doi.org/10.1039/c1dt11517k

[107] H. Kato, K. Asakura, A. Kudo, Highly efficient water splitting into H_2 and O_2 over lanthanum-doped $NaTaO_3$ photocatalysts with high crystallinity and surface nanostructure, J. Am. Chem. Soc. 125 (2003) 3082-3089. https://doi.org/10.1021/ja027751g

[108] A. Gubanov, Theory of the contact of two semiconductors of the same type of conductivity, Zh. Tekh. Fiz 21 (1951) 304.

[109] S.J. Moniz, S.A. Shevlin, D.J. Martin, Z.-X. Guo, J. Tang, Visible-light driven heterojunction photocatalysts for water splitting–a critical review, Energy & Environmental Science 8 (2015) 731-759. https://doi.org/10.1039/C4EE03271C

[110] A. Kumar, A. Kumar, G. Sharma, A.a.H. Al-Muhtaseb, M. Naushad, A.A. Ghfar, F.J. Stadler, Quaternary magnetic $BiOCl/g-C_3N_4/Cu_2O/Fe_3O_4$ nano-junction for visible light and solar powered degradation of sulfamethoxazole from aqueous environment, Chemical Engineering Journal 334 (2018) 462-478. https://doi.org/10.1016/j.cej.2017.10.049

[111] S.K. Sharma, A. Kumar, G. Sharma, M. Naushad, D.-V.N. Vo, M. Alam, F.J. Stadler, Fe_3O_4 mediated Z-scheme $BiVO_4/Cr_2V_4O_{13}$ strongly coupled nano-heterojunction for rapid degradation of fluoxetine under visible light, Materials Letters 281 (2020) 128650. https://doi.org/10.1016/j.matlet.2020.128650

[112] A. Kumar, G. Sharma, M. Naushad, T. Ahamad, R.C. Veses, F.J. Stadler, Highly visible active $Ag_2CrO_4/Ag/BiFeO_3@RGO$ nano-junction for photoreduction of CO_2 and photocatalytic removal of ciprofloxacin and bromate ions: The triggering effect of Ag and RGO, Chemical Engineering Journal 370 (2019) 148-165. https://doi.org/10.1016/j.cej.2019.03.196

[113] N.X. Huy, D.T.T. Phuong, N. Van Minh, A study on structure, morphology, optical properties, and photocatalytic ability of $SrTiO_3/TiO_2$ granular composites, Physica B: Condensed Matter 532 (2018) 37-41. https://doi.org/10.1016/j.physb.2017.04.028

[114] S. Jin, G. Dong, J. Luo, F. Ma, C. Wang, Improved photocatalytic NO removal activity of $SrTiO_3$ by using $SrCO_3$ as a new co-catalyst, Applied Catalysis B: Environmental 227 (2018) 24-34. https://doi.org/10.1016/j.apcatb.2018.01.020

[115] T. Kanagaraj, S. Thiripuranthagan, Photocatalytic activities of novel $SrTiO_3$– BiOBr heterojunction catalysts towards the degradation of reactive dyes, Applied Catalysis B: Environmental 207 (2017) 218-232. https://doi.org/10.1016/j.apcatb.2017.01.084

[116] H. Che, J. Chen, K. Huang, W. Hu, H. Hu, X. Liu, G. Che, C. Liu, W. Shi, Construction of $SrTiO_3/Bi_2O_3$ heterojunction towards to improved separation efficiency of charge carriers and photocatalytic activity under visible light, Journal of Alloys and Compounds 688 (2016) 882-890. https://doi.org/10.1016/j.jallcom.2016.07.311

[117] J. Liu, H. Bai, Y. Wang, Z. Liu, X. Zhang, D.D. Sun, Self-assembling TiO_2 nanorods on large graphene oxide sheets at a two-phase interface and their anti-recombination in photocatalytic applications, Adv. Funct. Mater. 20 (2010) 4175-4181. https://doi.org/10.1002/adfm.201001391

[118] M.S.A. Sher Shah, A.R. Park, K. Zhang, J.H. Park, P.J. Yoo, Green synthesis of biphasic TiO_2–reduced graphene oxide nanocomposites with highly enhanced photocatalytic activity, ACS applied materials & interfaces 4 (2012) 3893-3901. https://doi.org/10.1021/am301287m

[119] L. Yue, S. Wang, G. Shan, W. Wu, L. Qiang, L. Zhu, Novel MWNTs–Bi_2WO_6 composites with enhanced simulated solar photoactivity toward adsorbed and free tetracycline in water, Applied Catalysis B: Environmental 176 (2015) 11-19. https://doi.org/10.1016/j.apcatb.2015.03.043

[120] Y.-n. Zhang, N. Qin, J. Li, S. Han, P. Li, G. Zhao, Facet exposure-dependent photoelectrocatalytic oxidation kinetics of bisphenol A on nanocrystalline {001} TiO_2/carbon aerogel electrode, Applied Catalysis B: Environmental 216 (2017) 30-40. https://doi.org/10.1016/j.apcatb.2017.05.042

[121] L. Shi, Y. Yin, L.-C. Zhang, S. Wang, M. Sillanpää, H. Sun, Design and engineering heterojunctions for the photoelectrochemical monitoring of environmental pollutants: A review, Applied Catalysis B: Environmental (2019). https://doi.org/10.1016/j.apcatb.2019.02.044

[122] W. Hou, S.B. Cronin, A review of surface plasmon resonance-enhanced photocatalysis, Adv. Funct. Mater. 23 (2013) 1612-1619. https://doi.org/10.1002/adfm.201202148

[123] A. Tanaka, K. Fuku, T. Nishi, K. Hashimoto, H. Kominami, Functionalization of Au/TiO_2 plasmonic photocatalysts with Pd by formation of a core–shell structure for effective dechlorination of chlorobenzene under irradiation of visible light, The Journal of Physical Chemistry C 117 (2013) 16983-16989. https://doi.org/10.1021/jp403855p

[124] A. Kumar, S.K. Sharma, G. Sharma, H. Ala'a, M. Naushad, A.A. Ghfar, F.J. Stadler, Wide spectral degradation of Norfloxacin by Ag@$BiPO_4$/BiOBr/$BiFeO_3$ nano-assembly: Elucidating the photocatalytic mechanism under different light sources, Journal of hazardous materials 364 (2019) 429-440. https://doi.org/10.1016/j.jhazmat.2018.10.060

[125] A. Lagesson, M. Saaristo, T. Brodin, J. Fick, J. Klaminder, J. Martin, B. Wong, Fish on steroids: temperature-dependent effects of 17β-trenbolone on predator escape, boldness, and exploratory behaviors, Environmental pollution 245 (2019) 243-252. https://doi.org/10.1016/j.envpol.2018.10.116

[126] Q. Liang, J. Jin, C. Liu, S. Xu, Z. Li, Constructing a novel pn heterojunction photocatalyst $LaFeO_3$/g-C_3N_4 with enhanced visible-light-driven photocatalytic

activity, J. Alloys Compd. 709 (2017) 542-548.
https://doi.org/10.1016/j.jallcom.2017.03.190

[127] A. Kumar, A. Rana, G. Sharma, M. Naushad, A.a.H. Al-Muhtaseb, C. Guo, A. Iglesias-Juez, F.J. Stadler, High-performance photocatalytic hydrogen production and degradation of levofloxacin by wide spectrum-responsive Ag/Fe_3O_4 bridged $SrTiO_3/g$-C_3N_4 plasmonic nanojunctions: joint effect of Ag and Fe_3O_4, ACS applied materials & interfaces 10 (2018) 40474-40490. https://doi.org/10.1021/acsami.8b12753

[128] H. Kooshki, A. Sobhani-Nasab, M. Eghbali-Arani, F. Ahmadi, V. Ameri, M. Rahimi-Nasrabadi, Eco-friendly synthesis of $PbTiO_3$ nanoparticles and $PbTiO_3$/carbon quantum dots binary nano-hybrids for enhanced photocatalytic performance under visible light, Separation and Purification Technology 211 (2019) 873-881. https://doi.org/10.1016/j.seppur.2018.10.057

[129] V. Corral-Flores, D. Bueno-Baques, R. Ziolo, Synthesis and characterization of novel $CoFe_2O_4$–$BaTiO_3$ multiferroic core–shell-type nanostructures, Acta Mater. 58 (2010) 764-769. https://doi.org/10.1016/j.actamat.2009.09.054

[130] F. Wang, T. Wang, J. Lang, Y. Su, X. Wang, Improved photocatalytic activity and durability of $AgTaO_3$/AgBr heterojunction: the relevance of phase and electronic structure, J. Mol. Catal. A: Chem. 426 (2017) 52-59. https://doi.org/10.1016/j.molcata.2016.11.001

[131] K.S. Ranjith, T. Uyar, Conscientious Design of Zn-S/Ti-N Layer by Transformation of $ZnTiO_3$ on Electrospun $ZnTiO_3@TiO_2$ Nanofibers: Stability and Reusable Photocatalytic Performance under Visible Irradiation, ACS Sustainable Chemistry & Engineering 6 (2018) 12980-12992. https://doi.org/10.1021/acssuschemeng.8b02455

[132] Y. Yang, Y. Liu, B. Huang, R. Zhang, Y. Dai, X. Qin, X. Zhang, Enhanced visible photocatalytic activity of a $BiVO_4@\beta$-$AgVO_3$ composite synthesized by an in situ growth method, RSC Advances 4 (2014) 20058-20061. https://doi.org/10.1039/C4RA02110J

[133] P. Dhiman, T. Mehta, A. Kumar, G. Sharma, M. Naushad, T. Ahamad, G.T. Mola, $Mg_{0.5}Ni_xZn_{0.5-x}Fe_2O_4$ spinel as a sustainable magnetic nano-photocatalyst with dopant driven band shifting and reduced recombination for visible and solar degradation of Reactive Blue-19, Advanced Powder Technology (2020). https://doi.org/10.1016/j.apt.2020.10.010

[134] A. Kumar, G. Sharma, M. Naushad, A.a.H. Al-Muhtaseb, A. Kumar, I. Hira, T. Ahamad, A.A. Ghfar, F.J. Stadler, Visible photodegradation of ibuprofen and 2,4-D in simulated waste water using sustainable metal free-hybrids based on carbon nitride

and biochar, Journal of Environmental Management 231 (2019) 1164-1175.
https://doi.org/10.1016/j.jenvman.2018.11.015

[135] A. Kumar, G. Sharma, A. Kumari, C. Guo, M. Naushad, D.-V.N. Vo, J. Iqbal, F.J.
Stadler, Construction of dual Z-scheme g-C_3N_4/$Bi_4Ti_3O_{12}$/$Bi_4O_5I_2$ heterojunction for
visible and solar powered coupled photocatalytic antibiotic degradation and hydrogen
production: Boosting via I^-/I_3^- and Bi^{3+}/Bi^{5+} redox mediators, Applied Catalysis B:
Environmental 284 (2021) 119808. https://doi.org/10.1016/j.apcatb.2020.119808

[136] A. Kumar, S.K. Sharma, G. Sharma, C. Guo, D.-V.N. Vo, J. Iqbal, M. Naushad,
F.J. Stadler, Silicate glass matrix@Cu_2O/$Cu_2V_2O_7$ p-n heterojunction for enhanced
visible light photo-degradation of sulfamethoxazole: High charge separation and
interfacial transfer, Journal of Hazardous Materials 402 (2021) 123790.
https://doi.org/10.1016/j.jhazmat.2020.123790

[137] S.K. Sharma, A. Kumar, G. Sharma, F.J. Stadler, M. Naushad, A.A. Ghfar, T.
Ahamad, $LaTiO_2N$/Bi_2S_3 Z-scheme nano heterostructures modified by rGO with high
interfacial contact for rapid photocatalytic degradation of tetracycline, Journal of
Molecular Liquids 311 (2020) 113300. https://doi.org/10.1016/j.molliq.2020.113300

[138] B. Huang, C. Lei, C. Wei, G. Zeng, Chlorinated volatile organic compounds (Cl-
VOCs) in environment—sources, potential human health impacts, and current
remediation technologies, Environment international 71 (2014) 118-138.
https://doi.org/10.1016/j.envint.2014.06.013

[139] H. Huang, H. Huang, Q. Feng, G. Liu, Y. Zhan, M. Wu, H. Lu, Y. Shu, D.Y.
Leung, Catalytic oxidation of benzene over Mn modified TiO_2/ZSM-5 under vacuum
UV irradiation, Applied Catalysis B: Environmental 203 (2017) 870-878.
https://doi.org/10.1016/j.apcatb.2016.10.083

[140] J. Guo, S. Ouyang, P. Li, Y. Zhang, T. Kako, J. Ye, A new heterojunction
Ag_3PO_4/Cr-$SrTiO_3$ photocatalyst towards efficient elimination of gaseous organic
pollutants under visible light irradiation, Applied Catalysis B: Environmental 134
(2013) 286-292. https://doi.org/10.1016/j.apcatb.2012.12.038

[141] Y. Huo, Y. Jin, Y. Zhang, Citric acid assisted solvothermal synthesis of $BiFeO_3$
microspheres with high visible-light photocatalytic activity, J. Mol. Catal. A: Chem.
331 (2010) 15-20. https://doi.org/10.1016/j.molcata.2010.08.009

[142] J. Zhuang, Q. Tian, S. Lin, W. Yang, L. Chen, P. Liu, Precursor morphology-
controlled formation of perovskites $CaTiO_3$ and their photo-activity for As (III)
removal, Applied Catalysis B: Environmental 156 (2014) 108-115.
https://doi.org/10.1016/j.apcatb.2014.02.015

[143] H. Dang, D.V. Trinh, N.K. Nguyen, T.T. Le, D.D. Nguyen, H.V. Tran, M.-H. Phan, C.D. Huynh, Enhanced Photocatalytic Activity for Degradation of Organic Dyes Using Magnetite $CoFe_2O_4/BaTiO_3$ Composite, Journal of Nanoscience and Nanotechnology 18 (2018) 7850-7857. https://doi.org/10.1166/jnn.2018.15542

[144] C. Kang, K. Xiao, Z. Yao, Y. Wang, D. Huang, L. Zhu, F. Liu, T. Tian, Hydrothermal synthesis of graphene-$ZnTiO_3$ nanocomposites with enhanced photocatalytic activities, Res. Chem. Intermed. 44 (2018) 6621-6636. https://doi.org/10.1007/s11164-018-3512-z

[145] T. Montalvo-Herrera, D. Sánchez-Martínez, D. Hernandez-Uresti, L.M. Torres-Martínez, The role of the reactive oxygen species and the influence of $KBiO_3$ synthesis method in the photodegradation of methylene blue and ciprofloxacin, Reaction Kinetics, Mechanisms and Catalysis 126 (2019) 561-573. https://doi.org/10.1007/s11144-018-1521-y

[146] P. Chen, P. Xing, Z. Chen, X. Hu, H. Lin, L. Zhao, Y. He, In-situ synthesis of $AgNbO_3/g$-C_3N_4 photocatalyst via microwave heating method for efficiently photocatalytic H2 generation, Journal of colloid and interface science 534 (2019) 163-171. https://doi.org/10.1016/j.jcis.2018.09.025

[147] L. Cheng, Y. Kang, Synthesis of $NaBiO_3/Bi_2O_3$ heterojunction-structured photocatalyst and its photocatalytic mechanism, Mater. Lett. 117 (2014) 94-97. https://doi.org/10.1016/j.matlet.2013.11.124

[148] R.R. Ray, Adverse hematological effects of hexavalent chromium: an overview, Interdisciplinary toxicology 9 (2016) 55-65. https://doi.org/10.1515/intox-2016-0007

[149] P. Dhiman, S. Sharma, A. Kumar, M. Shekh, G. Sharma, M. Naushad, Rapid visible and solar photocatalytic Cr(VI) reduction and electrochemical sensing of dopamine using solution combustion synthesized ZnO–Fe_2O_3 nano heterojunctions: Mechanism Elucidation, Ceramics International (2020). https://doi.org/10.1016/j.ceramint.2020.01.275

[150] G. Sharma, A. Kumar, S. Sharma, A.H. Al-Muhtaseb, M. Naushad, A.A. Ghfar, T. Ahamad, F.J. Stadler, Fabrication and characterization of novel Fe^0@Guar gum-crosslinked-soya lecithin nanocomposite hydrogel for photocatalytic degradation of methyl violet dye, Separation and Purification Technology 211 (2019) 895-908. https://doi.org/10.1016/j.seppur.2018.10.028

[151] Y. Zhang, W. Cui, W. An, L. Liu, Y. Liang, Y. Zhu, Combination of photoelectrocatalysis and adsorption for removal of bisphenol A over TiO_2-graphene hydrogel with 3D network structure, Applied Catalysis B: Environmental 221 (2018) 36-46. https://doi.org/10.1016/j.apcatb.2017.08.076

[152] G. Sharma, Z.A. ALOthman, A. Kumar, S. Sharma, S.K. Ponnusamy, M. Naushad, Fabrication and characterization of a nanocomposite hydrogel for combined photocatalytic degradation of a mixture of malachite green and fast green dye, Nanotechnology for Environmental Engineering 2 (2017) 4. https://doi.org/10.1007/s41204-017-0014-y

[153] J. Li, F. Wang, L. Meng, M. Han, Y. Guo, C. Sun, Controlled synthesis of $BiVO_4/SrTiO_3$ composite with enhanced sunlight-driven photofunctions for sulfamethoxazole removal, Journal of colloid and interface science 485 (2017) 116-122. https://doi.org/10.1016/j.jcis.2016.07.040

[154] X. Yu, Y. Lin, H. Liu, C. Yang, Y. Peng, C. Du, S. Wu, X. Li, Y. Zhong, Photocatalytic performances of heterojunction catalysts of silver phosphate modified by PANI and Cr-doped $SrTiO_3$ for organic pollutant removal from high salinity wastewater, Journal of Colloid and Interface Science 561 (2020) 379-395. https://doi.org/10.1016/j.jcis.2019.10.123

[155] X.-J. Wen, C.-G. Niu, L. Zhang, C. Liang, G.-M. Zeng, An in depth mechanism insight of the degradation of multiple refractory pollutants via a novel $SrTiO_3/BiOI$ heterojunction photocatalysts, Journal of Catalysis 356 (2017) 283-299. https://doi.org/10.1016/j.jcat.2017.10.022

[156] P. Eghbali, A. Hassani, B. Sündü, Ö. Metin, Strontium titanate nanocubes assembled on mesoporous graphitic carbon nitride ($SrTiO_3/mpg-C_3N_4$): preparation, characterization and catalytic performance, Journal of Molecular Liquids 290 (2019) 111208. https://doi.org/10.1016/j.molliq.2019.111208

[157] N. Eskandari, G. Nabiyouni, S. Masoumi, D. Ghanbari, Preparation of a new magnetic and photo-catalyst $CoFe_2O_4–SrTiO_3$ perovskite nanocomposite for photo-degradation of toxic dyes under short time visible irradiation, Composites Part B: Engineering 176 (2019) 107343. https://doi.org/10.1016/j.compositesb.2019.107343

[158] H. Ramezanalizadeh, F. Manteghi, Design and development of a novel $BiFeO_3/CuWO_4$ heterojunction with enhanced photocatalytic performance for the degradation of organic dyes, Journal of Photochemistry and Photobiology A: Chemistry 338 (2017) 60-71. https://doi.org/10.1016/j.jphotochem.2017.02.004

[159] X. Zhang, X. Wang, J. Chai, S. Xue, R. Wang, L. Jiang, J. Wang, Z. Zhang, D.D. Dionysiou, Construction of novel symmetric double Z-scheme $BiFeO_3/CuBi_2O_4/BaTiO_3$ photocatalyst with enhanced solar-light-driven photocatalytic performance for degradation of norfloxacin, Applied Catalysis B: Environmental (2020) 119017. https://doi.org/10.1016/j.apcatb.2020.119017

[160] X. Hu, W. Wang, G. Xie, H. Wang, X. Tan, Q. Jin, D. Zhou, Y. Zhao, Ternary assembly of g-C$_3$N$_4$/graphene oxide sheets/BiFeO$_3$ heterojunction with enhanced photoreduction of Cr (VI) under visible-light irradiation, Chemosphere 216 (2019) 733-741. https://doi.org/10.1016/j.chemosphere.2018.10.181

[161] Z. Qu, Z. Liu, A. Wu, C. Piao, S. Li, J. Wang, Y. Song, Preparation of a coated Z-scheme and H-type SrTiO$_3$/(BiFeO$_3$@ ZnS) composite photocatalyst and application in degradation of 2, 4-dichlorophenol with simultaneous conversion of Cr (VI), Separation and Purification Technology 240 (2020) 116653. https://doi.org/10.1016/j.seppur.2020.116653

[162] D. Chen, B. Li, Q. Pu, X. Chen, G. Wen, Z. Li, Preparation of Ag-AgVO$_3$/g-C$_3$N$_4$ composite photo-catalyst and degradation characteristics of antibiotics, Journal of Hazardous Materials 373 (2019) 303-312. https://doi.org/10.1016/j.jhazmat.2019.03.090

[163] Z.-d. Lei, J.-j. Wang, L. Wang, X.-y. Yang, G. Xu, L. Tang, Efficient photocatalytic degradation of ibuprofen in aqueous solution using novel visible-light responsive graphene quantum dot/AgVO$_3$ nanoribbons, Journal of Hazardous Materials 312 (2016) 298-306. https://doi.org/10.1016/j.jhazmat.2016.03.044

[164] L. Zhang, X. Yuan, H. Wang, X. Chen, Z. Wu, Y. Liu, S. Gu, Q. Jiang, G. Zeng, Facile preparation of an Ag/AgVO$_3$/BiOCl composite and its enhanced photocatalytic behavior for methylene blue degradation, RSC advances 5 (2015) 98184-98193. https://doi.org/10.1039/C5RA21453J

[165] R. Abazari, A.R. Mahjoub, Potential applications of magnetic β-AgVO$_3$/ZnFe$_2$O$_4$ nanocomposites in dyes, photocatalytic degradation, and catalytic thermal decomposition of ammonium perchlorate, Industrial & engineering chemistry research 56 (2017) 623-634. https://doi.org/10.1021/acs.iecr.6b03727

[166] A.Y. Malkhasian, Synthesis and characterization of Pt/AgVO$_3$ nanowires for degradation of atrazine using visible light irradiation, Journal of Alloys and Compounds 649 (2015) 394-399. https://doi.org/10.1016/j.jallcom.2015.07.167

[167] S. Song, K. Wu, H. Wu, J. Guo, L. Zhang, Synthesis of Z-scheme multi-shelled ZnO/AgVO$_3$ spheres as photocatalysts for the degradation of ciprofloxacin and reduction of chromium (VI), Journal of Materials Science 55 (2020) 4987-5007. https://doi.org/10.1007/s10853-019-04316-8

[168] M.F.R. Samsudin, C. Frebillot, Y. Kaddoury, S. Sufian, W.-J. Ong, Bifunctional Z-Scheme Ag/AgVO$_3$/g-C$_3$N$_4$ photocatalysts for expired ciprofloxacin degradation and hydrogen production from natural rainwater without using scavengers, Journal of

Environmental Management 270 (2020) 110803.
https://doi.org/10.1016/j.jenvman.2020.110803

[169] L. Cao, Novel MoS_2-modified $AgVO_3$ composites with remarkably enhanced photocatalytic activity under visible-light irradiation, Materials Letters 188 (2017) 252-256. https://doi.org/10.1016/j.matlet.2016.10.120

[170] R. Ran, X. Meng, Z. Zhang, Facile preparation of novel graphene oxide-modified $Ag_2O/Ag_3VO_4/AgVO_3$ composites with high photocatalytic activities under visible light irradiation, Applied Catalysis B: Environmental 196 (2016) 1-15. https://doi.org/10.1016/j.apcatb.2016.05.012

[171] X. Chen, L. Di, H. Yang, T. Xian, A magnetically recoverable $CaTiO_3$/reduced graphene oxide/$NiFe_2O_4$ nanocomposite for the dye degradation under simulated sunlight irradiation, Journal of the Ceramic Society of Japan 127 (2019) 221-231. https://doi.org/10.2109/jcersj2.18168

[172] A. Kumar, C. Schuerings, S. Kumar, A. Kumar, V. Krishnan, Perovskite-structured $CaTiO_3$ coupled with g-C_3N_4 as a heterojunction photocatalyst for organic pollutant degradation, Beilstein journal of nanotechnology 9 (2018) 671-685. https://doi.org/10.3762/bjnano.9.62

[173] J. Qiao, M. Lv, Z. Qu, M. Zhang, X. Cui, D. Wang, C. Piao, Z. Liu, J. Wang, Y. Song, Preparation of a novel Z-scheme $KTaO_3$/$FeVO_4$/Bi_2O_3 nanocomposite for efficient sonocatalytic degradation of ceftriaxone sodium, Science of The Total Environment 689 (2019) 178-192. https://doi.org/10.1016/j.scitotenv.2019.06.416

Photocatalysis
Materials Research Foundations **100** (2021) 253- 272

Materials Research Forum LLC
https://doi.org/10.21741/9781644901359-8

Chapter 8

Photocatalytic Membranes

P. Senthil Kumar[1,2*], P.R. Yaashikaa[1]
[1]Department of Chemical Engineering, Sri Sivasubramaniya Nadar College of Engineering,
Chennai 603110, India
[2]SSN-Centre for Radiation, Environmental Science and Technology (SSN-CREST), SSN
College of Engineering, Chennai 603110, India
* senthilkumarp@ssn.edu.in

Abstract

A photocatalytic membrane can be characterized as a blend between a photocatalyst and membrane; it is promising for taking care of the issues experienced in detachment and photocatalysis. The photocatalyst can deliver, by retention of bright, infrared, or obvious light, compound changes of response accomplices, continually accompanying them into different compound communications without the event of a perpetual change of its synthetic synthesis. There has been significant advancement in the improvement of photocatalytic membrane through joining of metal-oxide photocatalysts to upgrade the presentation of the membranes. An ideal measure of the photocatalyst ought to be consolidated into the membrane to acknowledge sensible photocatalytic action with insignificant outcomes. New improvements in structure and assembling of photocatalytic membranes have made an incredible commitment to the photocatalytic application. Hybridizing photocatalysis with membrane offers photocatalytic response and products partition in a solitary advance and well control of the product maintenance. This section features a portion of the ongoing advances in photocatalytic membrane - kinds of photocatalysts hybridized with the membrane frameworks, reactor design, and average strategies for the creation of photocatalytic membranes, manufacture and membrane application in cleansing and pollutant expulsion from wastewater.

Keywords

Photocatalysis, Membrane, Metal-Oxide Photocatalysts, Wastewater, Purification

Materials Research Forum LLC
https://doi.org/10.21741/9781644901359-8

Contents

1.   Introduction

A photocatalytic membrane can be characterized as a blend between a photocatalyst and a membrane. The photocatalyst can deliver, by retention of bright, infrared, or unmistakable light, synthetic changes of response accomplices, over and again accompanying them into different concoction associations without the event of a lasting adjustment of its synthetic arrangement [1,2]. The membrane has the assignment to immobilize the photocatalyst and to go about as an atomic partition obstruction for the reagents or potentially response items. This synergic mix can have a few points of interest in examination with conventional reactors relying upon the particular capacities performed by the membrane. Membrane

Photocatalysis   Materials Research Forum LLC
Materials Research Foundations **100** (2021) 253- 272   https://doi.org/10.21741/9781644901359-8

partitions are in certainty regularly portrayed by lower working temperature, in correlation with warm partition forms, for example, refining, and they might offer an answer because of impetuses or items with a constrained warm strength [3,4].

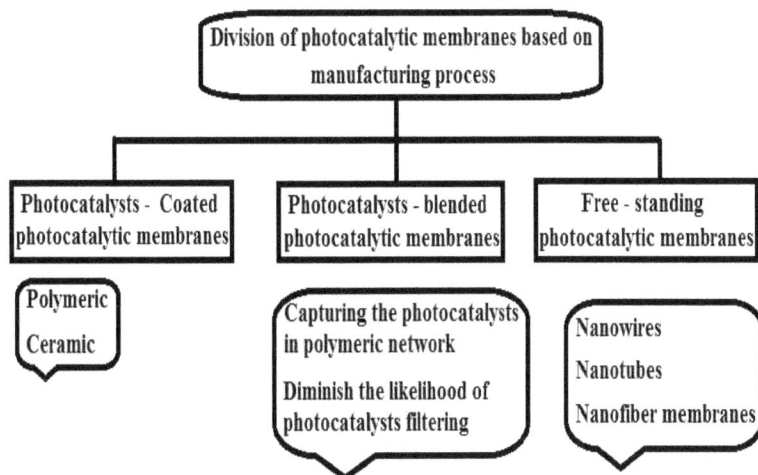

Figure 1. Division of photocatalytic membrane.

Furthermore, membrane division forms can isolate non-volatile parts by a distinction in measurements, charge, or instability. Figure 1 shows that division of photocatalytic membranes based on manufacturing process. The membrane can likewise characterize the response volume giving a reaching zone to two immiscible stages (e.g., in stage exchange catalysis) keeping away from the utilization of contaminating helpers, similar to solvents, in concurrence with the green science standards. The plan of the membrane reactor requires a multidisciplinary approach in which extraordinary disciplines, similar to: science, compound building, membrane designing, and procedure building, give their commitment to accomplish a synergic blend of the partition and response forms that permits ideal exhibitions regarding profitability and maintainability [5].

Consolidating membrane and photocatalytic innovation has been proposed as of late, in particular photocatalytic membrane. New advancements in structure and assembling of photocatalytic membranes have made an extraordinary commitment to the photocatalytic application [6]. Hybridizing photocatalysis with membrane offers photocatalytic response and products division in a solitary advance and well control of the products and results maintenance [7,8]. Photocatalytic membranes utilized in response frameworks spare extensive vitality and time, making critical commitments to upgrade photocatalytic forms. Photocatalytic membranes have extraordinary potential for water refinement, wastewater treatment, and self-cleaning applications in light of their antifouling, antimicrobial, and super hydrophilic properties, just as simultaneous photocatalytic oxidation and division [9,10].

The present key issues for the utilization of photocatalytic half and half membrane are the means by which to create frameworks that manage reasonable streams while guaranteeing the majority of the photocatalyst particles are dynamic and how to ensure the membrane protection from fouling [11]. Fouling is a procedure of adsorption or affidavit of particles, colloids, salt, and macromolecules on the surface or inside pores of membranes, causing an abatement of saturation transition, selectivity, and detachability of the membrane and notwithstanding shortening the membrane life. It has been exhibited that the characteristic hydrophobicity of membrane materials is one of the fundamental purposes behind fouling because natural foulants can be effectively adsorbed on the hydrophobic membrane. The basic cleaning or discharging cannot expel these natural foulants. To take care of this issue, numerous endeavours have been given to improve the surface properties of the membrane in the creation procedure [12,13]. Photocatalytic membrane reactor with solubilized/suspended photocatalyst can be additionally characterized in:

✓ Integrative-type photocatalytic membrane and

✓ Split sort photocatalytic membrane

In the first, the photocatalytic response and membrane division forms occur in one mechanical assembly, i.e., an inorganic or polymeric membrane is submerged in the slurry photocatalytic reactor. In split-type photocatalytic, membrane the photocatalytic response and membrane partition occur into two separate mechanical assemblies, i.e., the photocatalysis module and the membrane module, which are suitably coupled.

The designs examined were

(i) Light of the cell containing the membrane, with three sub-cases:

✓ impetus kept on the membrane;

✓ impetus in suspension, restricted by methods for the membrane;

✓ entanglement of the photocatalyst in a polysulfone membrane; and

(ii) Illumination of the distribution tank and impetus in suspension bound by methods for the membrane. Besides, the setup where the distribution tank was lighted and the impetus was utilized in suspension seemed, by all accounts, to be the most fascinating for modern applications. For instance, in reactor improvement, high light productivity, high membrane saturate stream rate, and selectivity can be gotten by measuring independently the "photocatalytic framework" and the "Membrane framework" and exploiting all the best research results for every framework.

2.   Photocatalytic membrane reactor

2.1   Conversion of organic compounds

Photocatalytic blend in membrane reactors is still at a starter look into stage, regardless of the extraordinary probability of photocatalytic forms, particularly when they are combined with a membrane partition framework. The likelihood of utilizing a membrane photoreactor for natural union, building up a crossover framework in which the photocatalytic response and the division of the ideal item happens in one stage is of high intrigue. TiO_2 has been utilized as impetus, benzene as both reactant and extraction dissolvable, and a polypropylene membrane to isolate the natural stage from the watery receptive condition. To maintain a strategic distance from auxiliary items and to acquire a proficient phenol generation, the utilization of a membrane framework, with high phenol penetrability and complete dismissal of the impetus, combined with the photocatalytic procedure, appears a helpful arrangement. The membrane photoreactor worked in this investigation comprises of an outside light set on a group reactor containing the watery arrangement with the impetus in suspension; by methods for a peristaltic siphon, the arrangement is pulled back from the photocatalytic reactor to a membrane contactor module in which a benzene arrangement is available as strip stage.

2.2   Corruption of organic compounds

Debasement of natural mixes in photocatalytic membrane reactors has been one of the principal applications in the blend of photocatalysis and membranes. The likelihood of utilizing a membrane photoreactor for debasement of organics, building up a mixture framework in which the photocatalytic response and the division of the ideal item happen in just a single gadget, is of high intrigue.

The impact of characteristic natural issue (i.e., humic acids) and cross-stream speeds on UF transitions and natural expulsion was investigated with and without UV light in the photocatalytic reactor. The communication between the two solutes in the framework, humic acids and TiO_2 photocatalysts, assumed a noteworthy job in the development of thick cake membranes at the membrane surface, prompting a more prominent motion decrease amid ultrafiltration of TiO_2 particles. As indicated by visual perceptions of the utilized membranes and the estimation of back-transport speeds of the solutes, a

Materials Research Forum LLC
https://doi.org/10.21741/9781644901359-8

considerable measure of TiO_2 saved on the membrane initiated a higher measure of humic acids to aggregate at the membrane through the adsorption onto TiO_2 particles. The humic-corrosive loaded TiO_2 particles offered multiple occasion's higher explicit cake obstruction with a considerably expanded compressibility coefficient than TiO_2 particles alone. The higher the cross-stream speeds, the more prominent the UV_{254} evacuation accomplished. This was because the ascent of cross-stream speeds added to the decrease of focus polarization at the membrane surface, accordingly bringing about a diminishing of the main thrust for humic acids to go through the membrane. At the point when photocatalytic responses occurred with UV light, UV_{254} expulsion efficiencies of the pervade were improved extraordinarily, and furthermore the saturate transition was kept at a steady dimension with no indication of fouling. Albeit humic acids were not totally mineralized by photocatalysis, the corruption of the humic acids upgraded the UF motion, as they were changed to less absorbable mixes.

2.3   Obstinate pollutants degradation

The expanding emanation of unmanageable natural contaminations has tested the regular water treatment. As one of the propelled treatment frameworks, photocatalytic oxidation has attracted critical consideration water treatment inquire about in view of its natural similarity and amazing oxidation capacity. Photocatalysis is an increasing speed of photo-induced response by the nearness of a photocatalyst. Among the photocatalysts, titanium dioxide (TiO_2) has been demonstrated to be an alluring and promising semiconductor impetus in heterogeneous photocatalysis and in innovative oxidation forms claim to its extraordinary photocatalytic effectiveness. Notwithstanding, a primary confinement remains: the partition of the impetus from treated. Photocatalytic membrane reactor with incorporated impetus is along these lines an extremely encouraging technique to defeat photocatalysis disadvantages. In such design, the membrane assumes the job of a hindrance for the photocatalyst and a particular boundary for the particle to be corrupted. The fundamental preferred position is to keep the photocatalyst in the response condition by methods for the transmembrane motion, the control of the living arrangement time of the atom in the reactor, and the accomplishment of a ceaseless procedure with synchronous impetus and item division from the response condition [12]. It likewise permits some extra tasks for impetus reusing to be dodged, for example, coagulation, flocculation, sedimentation, or filtration, and in this way prompts vitality sparing and empowers the span of the establishment to be limited. Finally, this design empowers the principle downside of membrane procedures to be survived: the way that it just exchanges toxin from a stage to another without treating it.

PMR can be separated into two primary gatherings: photocatalyst in suspension or immobilized in/on the membrane. The real downside of the main setup is the membrane fouling (prompting a reduction of the penetrate motion) and the crumbling of the membrane surface when natural membranes are utilized. The second arrangement empowers this downside to be stayed away from as oxidation by hydroxyl radicals happens on the outside surface and inside the pores of the membrane while reactants are saturating in a one-pass stream.

2.4   Photocatalytic nanomaterials: Preparation and properties

Photocatalysis, in which sun oriented photons are utilized to drive redox responses to create synthetic substances on the outside of an illuminated semiconductor, is one of the principle elective answers for sustainable power source age and natural remediation.

In the class of photocatalytic materials, semiconductor materials have pulled in extensive consideration in the fields of sun oriented vitality change and ecological insurance since it speaks to a simple method to use the vitality of either normal daylight or counterfeit indoor enlightenment. It has been the most broadly examined and utilized in numerous applications due to its solid oxidizing capacities for the decay of natural contaminations, concoction dependability, long solidness, nontoxicity, minimal effort, and straightforwardness to obvious light. Doping of TiO_2 is one of the answers for improve/broaden its photocatalytic properties. Using Ag related to TiO_2 could conceivably enable a few distinctive inactivation components to work in show. Along these lines, a synergism happens between Ag and TiO_2 when Ag/TiO_2 is utilized for inactivating microorganisms under UV radiation. Nanostructuring makes conceivable to plan nanomaterials with improved properties or even age of new functionalities, which have not been acquired in traditional materials. This is the second answer for propose superior photocatalytic nanomaterials. Little material size is commonly advantageous for surface-subordinate catalysis since it improves explicit surface region and furthermore expanded receptive locales. Nonetheless, littler material size is not constantly higher synergist proficient in light of the fact that it can cause solid quantum constrainment impact that builds the recombination likelihood of photogenerated electron–opening sets.

Along these lines, photogenerated electron–gap pair relocation requires a decreased recombination of appropriate focus inclination or potential slope from the centre to surface of materials that is corresponded with morphology, structure, and surface properties of nanostructured materials. These days, on account of their improved or even novel properties, photocatalytic nanomaterials have made more open doors in expanding the uses of photocatalytic materials in different fields, for example, water and air sanitizations, hydrogen age, CO_2 decrease, colour sharpened sun based cells, antifogging surfaces, heat exchange and warmth scattering, anticorrosion, lithography, photochromism, sun based synthetic concoctions generation, and numerous others. They can be readied utilizing synthetic courses, for example, sol–gel procedure and alkali treatment of sub-atomic antecedents to exist in numerous pieces, specifically in oxide, oxynitride, and nitride frameworks.

2.5 Photocatalytic process

A photocatalytic procedure is many decrease and oxidation photocatalytic responses happening within the sight of a photocatalyst that offers ascend to at least one items beginning from natural or inorganic substrate(s). The heterogeneous photocatalytic technique has all the earmarks of being productive just in those situations where low convergences of poisons are available in light of the fact that the nearness of high fixations hinders the response rates. Besides, it ought to be checked, with incredible consideration, both the nonappearance of mass exchange wonders controlling the response energy and the conceivable photocatalyst deactivation, because of conceivable amassing of side-effects and additionally intermediates onto the photocatalyst surface which could obstruct the (photo)active locales (especially for durable keeps running in gas-strong frameworks).

These issues are not emotional in the watery frameworks. A moderate blending of the responding suspension is adequate to ensure the nonattendance of a mass control on the response energy. The photocatalyst surface is cleaned by generation of a high number of oxidant radicals that prompt back-to-back assaults to the intermediates offering ascend to their total mineralization by methods for a few adsorption-desorption steps. Moreover, the mineralization can continue through intermediates that do not desorb into the main part of the arrangement, yet are exposed to oxidant assaults onto the surface until the development of CO_2. Backhanded proof of this conduct as a rule is the recognition of CO_2 not long after the beginning of light.

2.6 Photocatalytic processes by membrane operations

A photocatalytic procedure joined with membrane activities is many decrease and oxidation photocatalytic responses happening within the sight of a photocatalyst. Photocatalysis and, specifically, heterogeneous photocatalysis is a procedure of incredible probability for toxins reduction. To improve the general execution of the photo process, heterogeneous photocatalysis has been frequently joined with physical or substance tasks, which influence the concoction energy as well as the general proficiency. Different conceivable outcomes to couple heterogeneous photocatalysis with different advances to photodegrade natural and inorganic poisons broke up in genuine or engineered fluid effluents. These mixes increment the photo process proficiency by diminishing the response time as for the isolated tasks or they decline the expense as for heterogeneous photocatalysis alone, for the most part as far as light vitality. Contingent upon the activity combined with heterogeneous photocatalysis, two classifications of mixes exist. At the point when the coupling is with ultrasonic light, photograph Fenton response, ozonation, or electrochemical treatment, the blend influences the photocatalytic instruments in this way improving the productivity of the photocatalytic procedure. At the point when the coupling is with natural treatment, membrane reactor, membrane photoreactor, or physical adsorption, the blend does not influence the photocatalytic systems, yet it improves the effectiveness of the general procedure.

3.   Working parameters and limits of photocatalytic membranes

Working under appropriate working conditions is critical to acquire great Photocatalytic Membrane Reactor (PMR) execution with both solubilized/suspended and immobilized photocatalyst [14,15]. Figure 2 shows the schematic representation of parameters and limits of photocatalytic membranes.

The most significant components influencing the photocatalytic membrane execution, influencing the photocatalytic procedure as well as the membrane procedure can be depicted as pursues

3.1   Working mode

Photocatalytic membrane with immobilized photocatalyst can be worked in both impasse and cross flow modes. At the point when the photocatalytic membrane is worked in impasse mode, the substrates are held by the membrane and collect on its surface along these lines shaping a cake membrane. Accordingly, the membrane penetrability and the photocatalytic execution are diminished [16]. Moderately low photocatalytic efficiency is generally seen by working under impasse mode. This disadvantage can be clarified, as pursues since the feed contaminated water was kept into the reactor, a not satisfactory contact between the poisons to be corrupted, the photocatalyst and the light source was accomplished. By working along these lines, the feed stage flows digressively to the membrane.

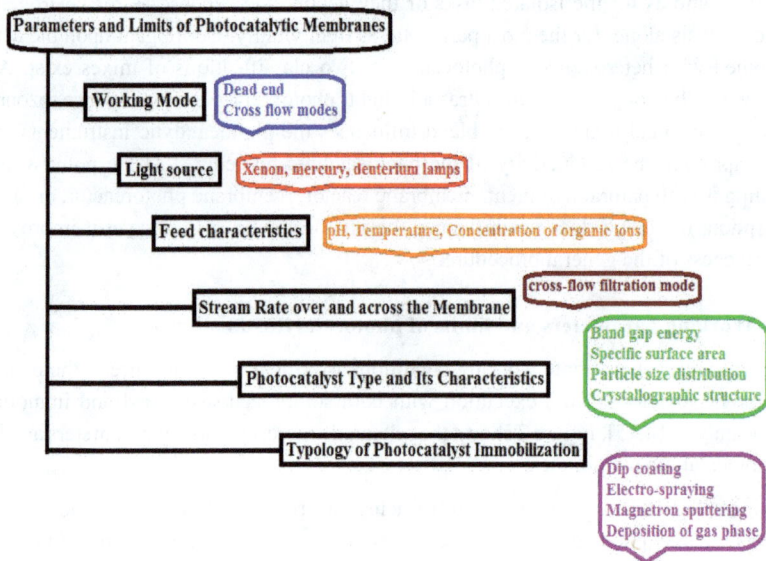

Fig. 2. Parameters and limits of photocatalytic membranes.

3.2   Light source

Because of advancements in solid-state semiconductor innovation, there is currently an incredible potential for the utilization of bright and noticeable LEDs as a light hotspot for photocatalytic applications to natural remediation methods. The beginning advance in a photocatalytic procedure is the illumination of the photocatalyst surface with photons having vitality equivalent to or higher than the band hole. The photocatalytic execution depends emphatically on the estimation of the light power [17,18]. Largely, the response rate increments with the light force at the point when the photocatalytic procedure is worked at low light power. This pattern can be clarified while thinking about that, by working with this light power, the response giving the arrangement of electron-opening couples is overwhelming with regard to the electron-gap recombination. By working with a light source producing at medium light power, the response rate is capacity of the square foundation of light power. This pattern can be attributed to the challenge impact between the electron-gap development and recombination responses. The response rate is not influenced by this parameter when the photocatalytic process is performed with a high light force. Figure 3 shows that control strategies of fouling.

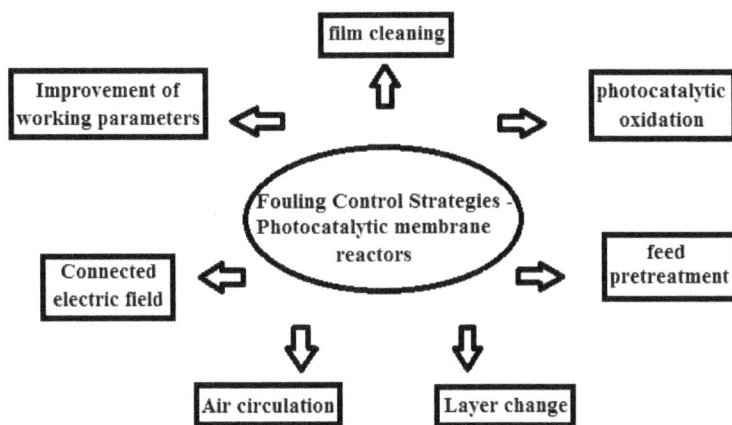

Figure 3. Control Strategies of fouling.

The impact of light force on poisons evacuation and fouling control can be recorded as pursues:

- ✓ At low light forces, electron– gap development responses are prevailing and electron–gap recombination is unimportant; therefore contaminations debasement rate increments directly with light power, and foulants can be totally disposed of;

- ✓ During moderate light forces, the toxins debasement rate is identified with the light power and the foulants corruption rate progresses toward becoming lower, since electron– gap pair partition rivals recombination;

- ✓ At high light forces, contaminations and debasement rate are free of light force.

3.3   Feed characteristics

The materials for the creation of photocatalytic membranes essentially rely upon their warm and synthetic solidness and on the comparing application field. The majority of the semiconductor photocatalysts are inorganic materials such as Bi_2O_3, ZnO_2, TiO2, etc. It is known that the most ordinarily considered membranes could be delivered from various materials counting polymer, metal, clay, or even zeolite. Polymeric membranes have great warm dependability, high vitality proficiency and low value, which are appealing for mechanical applications [19,20]. In any case, some polymeric membranes could not avoid against the UV light and oxidation of dynamic radicals in the photocatalytic procedure.

Looking at polymeric membranes, fired membranes show higher sturdiness in reasonable activity because of their great compound and warm security and mechanical quality. In any case, the earthenware membranes are substantially more costly than the regular polymer membranes in light of the overwhelming expenses of crude materials and membrane creation.

3.4   Stream rate over and across the membrane

On account of PMRs that worked under cross-stream filtration mode, where the feed arrangement is consistently recycled extraneously to the membrane, the distribution stream rate is a significant working parameter influencing the photocatalytic execution of the framework. More often than not, the photocatalytic productivity increments by expanding the distribution stream rate. This pattern can be clarified by considering the bigger choppiness in the arrangement, which advances the mass exchange from the greater part of the feed answer for the outside of the photocatalytic membrane, while decreasing the membrane fouling. Stream rate over the membrane rely upon the connected main thrust (e.g., transmembrane weight in a weight driven membrane partition), membrane structure, and creation. It is a key parameter for the PMR because it decides the contact times between the photocatalyst and the reagents/items. Mass exchange of the reagents to the reactant destinations, and of the item away from them, ought to be quick enough to keep away from response constraint in the meantime the contact time impetus/reagent ought to be additionally suitable to control response selectivity.

3.5   Photocatalyst type

Photocatalyst is the key part of a photocatalytic procedure. The properties of photocatalyst and its fixation in the responding condition speak to key parameters, assuming a significant job on photocatalytic execution. A photocatalyst is a semiconductor material ready to change over the vitality of illuminated photons in the substance important to energize electrons from its valence band to its conduction band. Key properties of photocatalyst, effects affecting photocatalytic proficiency, are the band hole vitality, the particular surface territory and molecule measure conveyance, the crystallographic structure and arrangement, and so forth.

3.5.1 TiO$_2$ photocatalyst

TiO2 remains the most prominent material and can be considered as a benchmark in the field of photocatalysis. Both spatial organizing of TiO$_2$ to build the viable way length of episode light and band structure designing to upgrade the optical reaction in the UV to obvious light range can incite TiO$_2$ to ingest more photons. After excitation by light, the photograph produced electrons and openings move to the surface. For TiO_2 terminals, the current electric field in the exhaustion district encourages charge partition; subsequently this consumption locale contributes the majority of the electrons and gaps that are accessible for the photocatalytic response [21,22]. In any case, doping can legitimately expand the transporter thickness and result in a smaller consumption locale, which enormously upsets the electron– opening division process. The steady lower-vitality mid-hole expresses that outcome from the surface issue can confine both photograph energized electrons and openings and along these lines maintain a strategic distance from quick recombination. Hence, suitable adjustment is important to additionally calibrate the neighbourhood electronic structure of doped TiO_2 and accomplish better charge division [23].

Until this point, astounding advancement has been made to improve the assimilation of light in TiO_2 materials, not just in the characteristic assimilation area of TiO_2, yet in addition to broaden the ghastly range into the obvious/close infrared area. Light ingestion can be improved by applying physical ideas to spatially nanostructured TiO_2 to expand the successful optical way length through the material. After photoexcitation, the created electrons and gaps relocate to the TiO_2 surface; superior charge partition productivity is thus ideal for photocatalytic responses. Doping can slender the bandgap of TiO_2 to permit obvious light gathering. In any case, the unwanted charge recombination that happens as often as possible at the dopant destinations seriously debases the photocatalytic execution [24]. For TiO_2 photoelectrodes, doping can increment the electronic conductivity and anticipate it to improve the photo electrochemical properties. On the other hand, a huge number of studies have uncovered that the monochromatic episode photon-to-electron change productivity (IPCE) diminishes in the natural light ingestion area of TiO_2 after doping, in light of the fact that noteworthy charge recombination happens in the exhaustion district of doped TiO_2 [25,26].

3.5.2 ZnO photocatalyst

Zinc oxide, with a high surface reactivity inferable from an enormous number of local imperfection destinations emerging from oxygen nonstoichiometry, has developed to be an effective photocatalyst material contrasted with other metal oxides. ZnO displays nearly higher response and mineralization rates and can create hydroxyl particles more proficiently than titanium oxide (TiO2). ZnO has been frequently viewed as a substantial option to TiO2 because of its great optoelectronic, reactant and photochemical properties alongside its minimal effort [27]. Zinc oxide has gotten much consideration in light of the fact that of its extraordinary properties and potential application in different zones. The blast hole vitality of ZnO is near that of TiO_2, the most utilized and traditional photocatalytic material, so it hypothetically has the equivalent photocatalytic capacity as TiO_2. Strong based strategies are promising options for delivering nanostructured materials because of their effortlessness and high yield [28]. Contrasted and arrangement stage forms, strong based procedures have no constraint of reagent focus, in this manner permitting the arrangement of materials in high return. Nevertheless, the domain of ZnO photocatalysts has been encountering an unparalleled quick advancement of arrangement stage procedures and less improvement of strong stage strategies.

3.5.3 WO₃ photocatalyst

WO_3 has pulled in huge logical intrigue because of its novel synthetic, useful, physical properties, ease, little band hole vitality, stable physicochemical properties, more profound valence band and solid photo corrosion solidness in watery arrangement just as steady recyclability execution [29,30]. Nevertheless, WO_3 is still a long way from turning into a handy semiconductor for photocatalysis applications because of its quick recombination of photograph prompted electron– opening sets. Since the procedure of electrons exchange to the conduction band of the oxide is a surface marvel, it is foreseen that the bigger the particular surface territory of the semiconductor is, the more the photograph incited electron– opening sets can be exchanged to the oxide's surface for actuating the photocatalysis responses. Hence, change of WO_3 membrane to improve the recombination misfortunes of charge transporters is important and essential. Various examinations have explored furthermore, announced that planning and altering WO_3 structure into nanodimension or nanostructure may defeat the recombination misfortunes of charge bearers and improve the photocatalysis oxidation execution essentially [31]. Hypothetically, work of an oxide semiconductor material at nuclear, atomic and macromolecular scales will change its physical, substance and useful properties radically, which is diverse to those at bigger scale. As a result, it could be unsurprising that band hole narrowing impacts could develop the scope of excitation light to the unmistakable district and give destinations that back off the recombination of transporters that can be foreseen. Consequently, it is restrictive that high explicit surface territory of WO_3 nanostructure could upgrade the photocatalytic action.

3.6   Typology of photocatalyst immobilization

Photocatalyst immobilization, to get a photocatalytic membrane, swapping from the ordinary slurry-type frameworks, could improve the presentation of decontamination innovation based on coupling photocatalysis and membrane filtration in perspective on genuine applications. In such a system, the membrane has the concurrent assignment of supporting the photocatalyst just as going about as a particular obstruction for the species to be debased [32,33].

Photocatalysis   Materials Research Forum LLC
Materials Research Foundations **100** (2021) 253- 272   https://doi.org/10.21741/9781644901359-8

Based on the distinctive strategy of photocatalyst immobilization, photocatalytic membrane with immobilized photocatalyst can be further classified in three sub-classifications.

✓The first sub-classification includes the photocatalytic membrane wherein the photocatalyst is covered on the photocatalytic membrane. The technique of covering can be performed by utilizing various strategies, for example, plunge covering, electro-showering of photocatalyst particles, magnetron sputtering of the photocatalyst, and testimony of gas stage photocatalyst nanoparticles. The fundamental downsides of utilizing this methodology are related to the reduction of membrane penetrability after the covering techniques and to photocatalyst filtering amid the photocatalytic runs.

✓The second sub-classification incorporates the photocatalytic membrane wherein the photocatalyst is mixed into the membrane framework. For all intents and purposes in this sort of photocatalytic membrane, amid the membrane planning the photocatalyst is captured in the polymeric framework. By working with this kind of photocatalytic membrane, the likelihood of photocatalyst filtering was diminished as for the photocatalytic membrane that was set up by photocatalyst covering.

✓The third sub-class of photocatalytic membrane with immobilized photocatalyst depends on the utilization of free-standing photocatalytic membrane. For this situation, the photocatalytic membrane is fabricated with an unadulterated photocatalyst, so the immobilization of the photocatalyst in/on the membrane support is superfluous. Accordingly, a further decrease of the probability of photocatalyst draining was acquired.

In any case, a few disadvantages of photocatalytic membrane are:

✓Moderate loss of photoactivity likewise identified with the low photocatalyst accessibility to light;

✓Need to illuminate the outside of the membrane, bringing about specialized troubles and in conceivable membrane photodegradation;

✓Limited preparing limits inferable from mass exchange restrictions and

✓Unacceptable framework lifetime owing to the conceivable impetus deactivation and wash out.

At that point, it is principal to produce frameworks with advantageous porosity and successful scattering of the impetus particles. The likelihood to capture the photocatalyst in a polymeric membrane, finding a lower productivity with respect to the suspended setup,

Photocatalysis   Materials Research Forum LLC
Materials Research Foundations **100** (2021) 253- 272   https://doi.org/10.21741/9781644901359-8

because the nearness of the polymer around the particles of impetus diminishes the successful surface territory. Another disservice of the framework with photocatalyst captured in polymeric membranes is the danger of a conceivable membrane oxidation by OH radicals assault. Inorganic membranes are best over the customary polymeric materials, attributable to their brilliant warm, concoction, furthermore, mechanical security. The need to create unmistakable light-dynamic photocatalytic frameworks speaks to another key point in perspective on extensive scale application of photocatalytic membrane frameworks, allowing the utilization of a greener light source. Since genuine modern effluents ordinarily contain salts and broke up natural issue together dyes, the impacts of these substances on photocatalytic membrane exhibitions have additionally to be considered. Potential influences of salts and broke down natural issue nearness on framework execution can be:

- Abatement of photocatalytic execution;

- Increment of membrane fouling.

4. Future tendencies

TiO_2 photocatalytic membranes show much improved filtration execution compared with the conventional membranes. TiO_2 can be covered on the permeable backings or manufactured into unsupported membranes. Both polymer and artistic membranes have been generally contemplated as backings of photo catalysts; however, the strength of polymer membrane under the UV brightening is a worry and requires further examination. Attributable to the constraint of the wide bandgap, TiO_2 can be doped with non-metal components or altered with different metals also, semiconductors to improve its photocatalytic execution and the filtration proficiency of photocatalytic membranes. The photocatalytic membranes likewise have exceptional properties including super hydrophilicity, antifouling execution, photocatalytic oxidation and decrease, partition, and antimicrobial capacity. The photocatalytic membranes can be created by means of different strategies: plunge covering, turn covering, electro spraying and electrospinning, submersion precipitation, vacuum filtration. The creation technique is the way to photocatalytic membrane properties and exhibitions. Despite the fact that photocatalytic membranes have appeared potential in water treatment, antifouling, cleansing, and vitality transformation, numerous specialized issues still should be tended to before this strategy can be connected in the business. More endeavours are required to improve the photocatalytic productivity of photocatalytic membranes, for example, immobilized photo catalysts. Photocatalytic membranes are confronting the test of the lower contact zone among photo catalyst, target poison, and light than in the traditional slurry photo catalysis frameworks. A large portion of the photocatalytic membranes in little scale applications still work under UV light, despite the fact that countless light absorption photo catalysts have been accounted for. Accordingly, sun oriented light induced photocatalytic membranes are yet to be researched later on. The union of novel photo catalysts will be constantly helpful for creating photocatalytic membrane with high photocatalytic movement, high saturate transition, and attractive antifouling properties.

References

[1] R. Molinari, L. Palmisano, E. Drioli, M. Schiavello, Studies on various reactor configurations for coupling photocatalysis and membrane processes in water purification, J. Membr. Sci. 206 (2002) 399 – 415. https://doi.org/10.1016/S0376-7388(01)00785-2

[2] S.A. Braslavsky, A.M. Braun, A.E. Cassano, A.V. Emeline, M. Litter, L. Palmisano, V.N. Parmon, N. Serpone, Glossary of terms used in photocatalysis and radiation catalysis (IUPAC recommendations 2011). Pure Appl. Chem. 83 (2011) 931-1014. https://doi.org/10.1351/PAC-REC-09-09-36

[3] A. Stankiewicz, Reactive separations for process intensification: An industrial perspective, Chem. Eng. Process 42 (2003) 137-144. https://doi.org/10.1016/S0255-2701(02)00084-3

[4] J-C. Charpentier, Modern Chemical Engineering in the Framework of Globalization, Sustainability, and Technical Innovation, Ind. Eng. Chem. Res. 46 (2007) 3465–3485. https://doi.org/10.1021/ie061290g

[5] I.F.J. Vankelecom, Polymeric Membranes in Catalytic Reactors, Chem. Rev. 102 (2002) 3779-3810. https://doi.org/10.1021/cr0103468

[6] M. Roso, A. Lorenzetti, S. Besco, M. Monti, G. Berti, M. Modesti, Application of empirical modelling in multi-membranes membrane manufacturing, Comput. Chem. Eng. 35 (2011) 2248-2256. https://doi.org/10.1016/j.compchemeng.2011.03.007

[7] C. Zamfirescu, I. Dincer, G.F. Naterer, Analysis of a photochemical water-splitting reactor with supramolecular catalysts and a proton exchange membrane, Int. J. Hydrog. Energy 36 (2011) 11273-11281. https://doi.org/10.1016/j.ijhydene.2010.12.126

[8] J.H. Jhaveri, Z.V.P. Murthy, A comprehensive review on anti-fouling nanocomposite membranes for pressure driven membrane separation processes, Desalination 379 (2016) 137-154. https://doi.org/10.1016/j.desal.2015.11.009

[9] J.W. Lee, T.O. Kwon, R. Thiruvenkatachari, I.S. Moon, Adsorption and photocatalytic degradation of bisphenol A using TiO_2 and its separation by submerged hollow fiber ultrafiltration membrane, J. Environ. Sci. 18 (2006) 193-200.

[10] M. Ziegmann, F. Saravia, P.A. Torres, F.H. Frimmel, The hybrid process TiO_2/PAC: performance of membrane filtration, Water Sci. Technol. 62 (2010) 1205-1212. https://doi.org/10.2166/wst.2010.420

[11] Y. Huo, Z. Xie, X. Wang, H. Li, M. Hoang, R.A. Caruso, Methyl orange removal by combined visible-light photocatalysis and membrane distillation, Dyes Pigm. 98 (2013) 106-112. https://doi.org/10.1016/j.dyepig.2013.02.009

[12] S. Mozia, Photocatalytic membrane reactors (PMRs) in water and wastewater treatment. A review, Sep. Purif. Technol. 73 (2010) 71–91. https://doi.org/10.1016/j.seppur.2010.03.021

[13] R. Molinari, C. Lavorato, P. Argurio, Photocatalytic reduction of acetophenone in membrane reactors under UV and visible light using TiO2 and Pd/TiO_2 catalysts, Chem. Eng. J. , 274 (2015) 307-316. https://doi.org/10.1016/j.cej.2015.03.120

[14] S.O. Ganiyu, E.D. van Hullebusch, M. Cretin, G. Esposito, M.A. Oturan, Coupling of membrane filtration and advanced oxidation processes for removal of pharmaceutical residues: a critical review, Sep. Purif. Technol. 156 (2015) 891 - 914. https://doi.org/10.1016/j.seppur.2015.09.059

[15] W. Zhang, L. Ding, J. Luo, M.Y. Jaffrin, B. Tang, Membrane fouling in photocatalytic membrane reactors (PMRs) for water and wastewater treatment: A critical review, Chem. Eng. J. 302 (2016) 446-458. https://doi.org/10.1016/j.cej.2016.05.071

[16] W.Y. Wang, A. Irawan, Y. Ku, Photocatalytic degradation of Acid Red 4 using a titanium dioxide membrane supported on a porous ceramic tube, Water Res. 42 (2008) 4725-4732. https://doi.org/10.1016/j.watres.2008.08.021

[17] S. Zhou, A.K. Ray, Kinetic studies for photocatalytic degradation of eosin B on a thin membrane of titanium oxide, Ind. Eng. Chem. Res. 42 (2003) 6020 – 6033. https://doi.org/10.1021/ie030366v

[18] P.A. Pekakis, N.P. Xekoukoulotakis, D. Mantzavinos, Treatment of textile dyehouse wastewater by TiO_2 photocatalysis, Water Res. 40 (2006) 1276-1286. https://doi.org/10.1016/j.watres.2006.01.019

[19] J.M. Herrmann, Heterogeneous Photocatalysis: State of the Art and Present Applications, Top. Catal. 34 (2005) 49-65. https://doi.org/10.1007/s11244-005-3788-2

[20] L. Aoudjit, P.M. Martins, F. Madjene, D.Y. Petrovykh, S. Lanceros-Mendez, Photocatalytic reusable membranes for the effective degradation of tartrazine with a solar photoreactor, J. Hazard. Mater. 344 (2018) 408 – 416. https://doi.org/10.1016/j.jhazmat.2017.10.053

[21] A. Fujishima, X.T. Zhang, D.A. Tryk, TiO_2 Photocatalysis and Related Surface Phenomena, Surf. Sci. Rep. 63 (2008) 515-582. https://doi.org/10.1016/j.surfrep.2008.10.001

[22] H. Xu, S. Ouyang, L. Liu, P. Reunchan, N. Umezawa, J. Ye, Recent advances in TiO_2 - based photocatalysis, J. Mater. Chem. A 2 (2014) 12642. https://doi.org/10.1039/C4TA00941J

[23] X. Chen, S.S. Mao, Titanium Dioxide Nanomaterials: Synthesis, Properties, Modifications, and Applications, Chem. Rev. 107 (2007) 2891-2959. https://doi.org/10.1021/cr0500535

[24] X. Chen, C. Burda, The Electronic Origin of the Visible-Light Absorption Properties of C-, N- and S-Doped TiO_2 Nanomaterials, J. Am. Chem. Soc. 130 (2008) 5018-5019. https://doi.org/10.1021/ja711023z

[25] C.G. Silva, R. Juarez, T. Marino, R. Molinari, H. Garcia, Influence of excitation wavelength (UV or visible light) on the photocatalytic activity of titania containing gold nanoparticles for the generation of hydrogen or oxygen from water, J. Am. Chem. Soc. 133 (2011) 595-602. https://doi.org/10.1021/ja1086358

[26] T. Kaur, A. Sraw, R.K. Wanchoo, A.P. Toor, Visible –Light Induced Photocatalytic Degradation of Fungicide with Fe and Si Doped TiO2 Nanoparticles, Mater. Today: Proceedings, 3 (2016) 354-361. https://doi.org/10.1016/j.matpr.2016.01.020

[27] M. Qamar, M. Muneer, A comparative photocatalytic activity of titanium dioxide and zinc oxide by investigating the degradation of vanillin, Desalination, 249 (2009) 535-540. https://doi.org/10.1016/j.desal.2009.01.022

[28] R. Kitture, S.J. Koppikar, R. Kaul-Ghanekar, S.N. Kale, Catalyst efficiency, photostability and reusability study of ZnO nanoparticles in visible light for dye degradation, J. Phys. Chem. Solids 72 (2011) 60-66. https://doi.org/10.1016/j.jpcs.2010.10.090

[29] P. Dong, G. Hou, X. Xi, R. Shao, F. Dong, WO_3 - based photocatalysts: morphology control, activity enhancement and multifunctional applications, Environmental Science, 4 (2017) 539 – 557. https://doi.org/10.1039/C6EN00478D

Materials Research Forum LLC
https://doi.org/10.21741/9781644901359-8

[30] M.A. Gondal, M.S. Sadullah, T.F. Qahtan, M.A. Dastageer, U. Baig, G.H. McKinley, Fabrication and Wettability Study of WO_3 Coated Photocatalytic Membrane for Oil-Water Separation: A Comparative Study with ZnO Coated Membrane, Sci. Rep. 7 (2017) 1686. https://doi.org/10.1038/s41598-017-01959-y

[31] N. Shafaei, M. Peyravi, M. Jahanshahi, Improving surface structure of photocatalytic self-cleaning membrane by WO_3/PANI nanoparticles, Polym. Adv. Technol. 27 (2016) 1325-1337. https://doi.org/10.1002/pat.3800

[32] I. Horovitz, D. Avisar, M.A. Baker, R. Grilli, L. Lozzi, D. Di Camillo, H. Mamane, Carbamazepine degradation using a N-doped TiO_2 coated photocatalytic membrane reactor: Influence of physical parameters, J. Hazard. Mater. 310 (2016) 98-107. https://doi.org/10.1016/j.jhazmat.2016.02.008

[33] O. Iglesias, M.J. Rivero, A.M. Urtiaga, I. Ortiz, Membrane-based photocatalytic systems for process intensification, Chem. Eng. J. 305 (2016) 136-148. https://doi.org/10.1016/j.cej.2016.01.047

Photocatalysis
Materials Research Foundations **100** (2021) 273-298

Materials Research Forum LLC
https://doi.org/10.21741/9781644901359-9

Chapter 9

Composite Ion Exchangers as New Age Photocatalyst

Manita Thakur [1*], Ajay Sharma [1], Manisha Chandel [1], Suresh Kumar [2], Ajay Kumar [3]

[1] Department of Chemistry, IEC University Baddi

[2] Department of Chemistry, MMU Solan

[3] Department of Chemistry, SILB Solan

* manitathakur1989@gmail.com

Abstract

The innovation of different technologies and emphasis on development of new techniques is indispensable to improve the quality of water globally. Photocatalysis is one of the major techniques explored now a days for the exclusion of water impurities using solar light. Different types of photocatalysts have been employed for the removal of dyes, heavy metals, pesticides from aqueous system. During the last few years, nanocomposite ion exchangers were used as a photocatalyst for the removal of organic pollutants. This chapter includes detailed information about introduction of pollutants into the water system, nanocomposite ion exchangers and photocatalysis removal. Nanocomposite ion exchangers effectively degrade various pollutants present in the marine system. These nanocomposites have also been used in different areas such as fuel cell, sensor, nuclear separation and heavy metal removal etc. Therefore, nanocomposite ion exchangers are a new age photocatalyst with unique and effective properties.

Keywords

Nanocomposite, Photocatalysis, Ion Exchanger, Dyes, Heavy Metals, Photocatalyst

Contents

Materials Research Forum LLC
https://doi.org/10.21741/9781644901359-9

1.   Introduction

The incessant human activities consistently degrade the water resources all over the world. The contamination of water sources due to various industrial activities is a global issue, especially in the developing countries. The scarcity of clean water in terms of quantity and quality becomes a substantial threat to the living organisms. Some sources of industrial pollutants include mining, metal plating, fertilizer industries, tanneries, batteries, pesticides, power regeneration and electronic industries etc. A variety of toxic organic and inorganic pollutants such as heavy metals, azo dyes, anthraquinone, formazan, fertilizers and pesticides, organic pollutants and other biological impurities have been discharged into water system [1-5]. The consumption of these toxins through drinking water, food and air causes serious effects in living beings. These pollutants are highly toxic, non-biodegradable, carcinogenic and persist in the environment. Therefore, the removal of these hazardous materials from industrial waste is necessary to protect the plants, animals and human beings from their adverse effects [6-8]. Different industrial processes discharge waste products in water sources and reach humans through food, water and air as shown in Figure 1.

Photocatalysis
Materials Research Foundations **100** (2021) 273-298

Materials Research Forum LLC
https://doi.org/10.21741/9781644901359-9

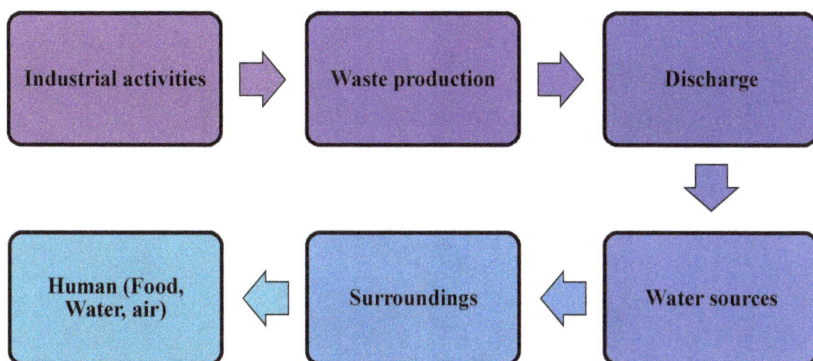

Figure 1: *Industrial activities and their exposure to surroundings.*

Among various contaminants, inorganic and organic toxins are major class of pollutants. Inorganic chemicals are one of the most important categories of water pollutants. These inorganic pollutants include heavy metals, salt and minerals etc. Among these, heavy metals are of major concern because of high toxicity even at low concentrations. These heavy metals enter in the food chain through a number of ways and cause severe problems in human beings. On the other hand, persistent organic compounds are used in agriculture, disease control, manufacturing and industrial processes [9-12]. These are bio-accumulative and resistant to degradation via chemical, biological and photolytic processes. These organic pollutants are very toxic and cause adverse health effects in humans and animals. Some common organic pollutants include dyes, phenols, pesticides, antibiotics and polychlorinated biphenyls etc. Figure 2 shows different pollutants which continuously degrade the quality of drinking water [14-20].

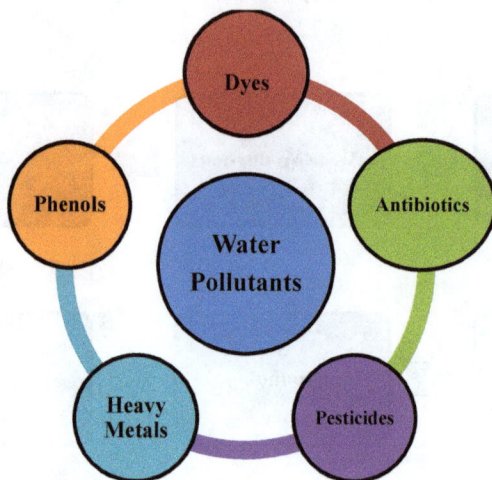

Figure 2: Various types of water pollutants.

1.1 Heavy metals

Industrial processes generate wastewater which contains a large amount of toxic heavy metal ions which must be removed before getting discharged [21-25]. Numerous industries including battery, paints, tanning, ceramics, electroplating, mining, glassware and photographic industries generated polluted products in wastewater. The presence of heavy metal ions such as Pb, Ni, Cd, Cr, Mo, Cu, Fe, Mn, Hg, Co and Bi in wastewater are directly associated with industrial processes. For example, metal ions Cu, Zn, Cr and Cd are generated from metal plating and presence of Cr metal ions in wastewater from tanneries waste. Also, Hg is generated by fossil fuel combustion and smelting processes. Industrial and mining sources are responsible for the production of Pb metal. The concentration of toxic heavy metal ions in wastewater is up to maximum level which is much higher than the safe limits. Therefore, these must be removed before getting mixed into the environment [26-28].

1.2 Dyes

Currently, dyes are quite important materials in either industrial or domestic purposes. In 1856, synthetic dyes were invented. But now days several forms more than 8000 have been manufactured for specific purposes [29]. Dyeing, textile, paper and pulp, tannery and paint industries are big consumers of dyes. Therefore, waste products of industries have been considered as unpleasant pollutant because they impart color in water and due to the toxic nature which affects life unfavorably [30-35].

1.3   Phenols

Like other pollutants such as dyes and metals, phenols are also measured as major pollutants due to their bad taste and odour in water [36-38]. These are very toxic in nature. Therefore, determination and deduction of these pollutants from water is very important. Phenols are produced from chemical, paint, pesticides, resin, paper and pulp, gas and coke manufacture and dyeing industries.

1.4   Pesticides

Rural activities are mainly responsible for the generation of pesticides. According to their functional properties these are classified into some categories such as molluscides, nematicides, rodenticides, pesticides, insecticides, fungicides, bactericides and herbicides [39]. Among all categories of pesticides, fungicides and insecticides are very important to human exposure in food because their time period is very short during harvesting. Significant increase in the production of herbicide is due to the gradually use of chemicals to control weeds in cultivation process. Although DDT is banned, some substitutes like malathion, heptachlor, endrin, lindane and parathion causing environmental pollution [40-42]. Different techniques have been used for the remediation of organic and inorganic pollutants including adsorption, ion exchange, membrane filtration, photocatalysis etc. Ion exchange method has been gained considerable attention due to its low cost, simplicity, and versatility. It is widely used for the removal of organic and inorganic pollutants from aqueous solutions.

The ion exchange process is a reversible process that takes place between charges. The exchanger must have an open network structure which carries the ions for the exchange. The ion exchange process takes place in both aqueous and non-aqueous solution, molten salts and with vapours. The exchange of ions between the solution and ion exchangers is a physicochemical process occuring on the basis of equivalency and principle of electro neutrality. Figure 3 shows the process of ion exchange [43-48].

Materials Research Forum LLC

https://doi.org/10.21741/9781644901359-9

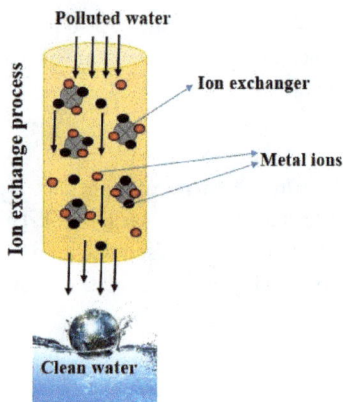

Figure 3: Mechanism of ion exchange process.

2.   Ion exchanger

Ion exchangers are insoluble and charged species containing loosely held ions. These ions are exchanged with other ions in the solution when come in contact with them. The ion exchangers contain either positively or negatively charged ions so called cation and anion exchangers. Ion exchange resins are water insoluble solid substances which absorb both cations and anions from an electrolyte solution and release ions of same charges into the solution. The positively charged ions such as hydrogen and sodium ions are exchanged with positively charged ions such as nickel, copper and zinc ions present in the solution. Similarly, the negative ions such as hydroxyl and chloride ions are exchanged with the negatively charged ions such as sulphate, chromate, nitrate, cyanide etc. of the solution [49-51].

Ion exchangers have been widely used due to their exclusive properties such as higher resistance and selectivity at higher temperatures. Ion exchangers have been used for selective membrane, adsorption, catalysis, drug delivery, antimicrobial action and chemical sensors. The most important application of ion exchangers was purification and demineralization of water. The characteristics of newly synthesised material are ion exchange capacity, effect of eluent concentration, elution behaviour, effect of temperature, pH titration, chemical stability, composition, structural properties and distribution studies [52-55].

Photocatalysis
Materials Research Foundations **100** (2021) 273-298

Materials Research Forum LLC
https://doi.org/10.21741/9781644901359-9

2.1 Classification of ion exchangers

The ion exchanges have drawn considerable attention due to their wide range of applications. Ion exchangers are classified into three categories on the basis of chemical nature as shown in Figure 4.

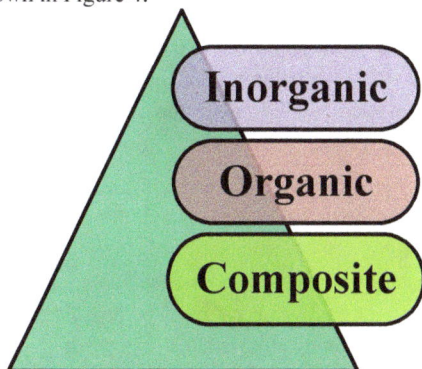

Figure 4: *Classification of ion exchanger on the basis of chemical nature.*

2.2 Composite ion exchangers

Recently, hybrid or composite ion exchangers have been prepared due to certain limitations existed in inorganic and organic ion exchangers. The inorganic ion exchangers are non-reproducible and not suitable for column chromatography. However, the organic ion exchangers are highly responsive to immense radiation doses which lead to major changes in their chain scission, crosslinking, capacity and selectivity. The combinations of polymers and inorganic materials form hybrid ion exchange materials. The organic polymeric part of the composite material provides mechanical and chemical stability. While the inorganic part increases the ion exchange behaviour, thermal stability and electrical conductivity [56, 57].

The hybrid ion exchangers with controlled functionality and hydrophobicity have opened new possibilities in different areas like catalysis, organometallic chemistry, organic host guest chemistry, hydrometallurgy, water treatment, pollution control, antibiotic purification, separation of radioactive isotopes and analytical chemistry [58-60]. The hybrid ion exchangers have been utilised in the field of chemistry, biochemistry, engineering and material science.

Photocatalysis   Materials Research Forum LLC
Materials Research Foundations **100** (2021) 273-298   https://doi.org/10.21741/9781644901359-9

The composite ion exchange materials possess superior properties such as mechanical, chemical and radiation stability. The multifunctional nature of composite materials makes them potentially useful in different fields such as fabrication of selective membranes, chemical sensors, potentiometric sensors, electro-deionization, storage batteries, chromatography, electrodialysis, fuel cells and ion exchange [61-63]. The composite or hybrid materials have been prepared by sol-gel method universally. These materials are generally granular in nature and suitable for column operations. The composite materials with nano-scale dimensions have been widely used due to their multifunctionality, specificity and selectivity.

Nanocomposite materials are more effective due to greater surface area and volume ratio of nanoparticles. The nano-ranged materials find vast applicability in different fields such as drug delivery, chromatography, photocatalysis, ion selective electrode and environmental remediation. In recent years, bio-polymer based composite materials have been studied due to their low cost and renewable nature. The organic polymers such as polyaniline, polyprrole, starch, polystyrene, chitosan, cellulose, alginate, gelatin, and polyacrylonitrile have been used for the fabrication of composite materials.

2.3   Preparative methods for nanocomposites

Polymeric nanocomposites have been usually synthesized through solution route and because of this a large amount of organic solvents has been required and this may pollute the environment. Hence green synthetic methods have attracted much attention. Polymer or inorganic filler nanocomposites have been synthesized in three different ways such as

1. Direct mixing or blending of the polymer

2. In-situ polymerization of monomers in the presence of fillers

3. Sol-gel process

2.3.1 Solution and melt processing

In this method, the layered silicates have been dispersed in a solvent in which the polymer is soluble. On evaporation of the solvent, the silicate sheets have been introduced into the polymer to make a multi-layered and ordered structure as shown in Figure 5. This process is also known as ethe xfoliation or adsorption process. While in melt processing, layered silicate has been mixed to the polymer in the molten state. If the layer's surfaces are compatible with the polymer it results in the formation of nanocomposites by the insertion of selected polymer into interlayer space. There is no requirement of solvent in this method [64].

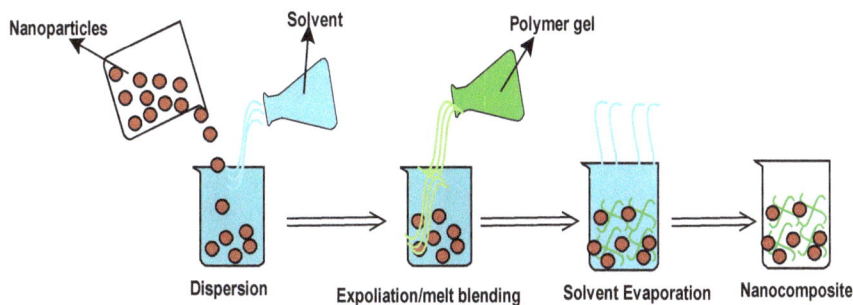

Figure 5: *Preparation of nanocomposite by solution/melt processing.*

2.3.2 In-situ polymerization

In this case monomer units with initiators have been taken and allowed to polymerize in presence of clays. The growth of polymer units separate clays and enter into the interlayer space [65]. The formation of polymer clay nanocomposite has been shown in Figure 6.

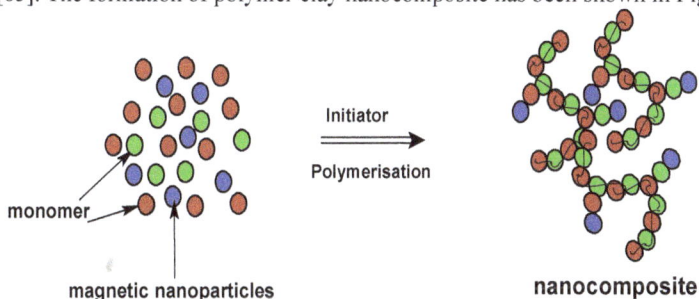

Figure 6: *In-situ polymerization for the synthesis of nanocomposite.*

2.3.3 Sol-gel method

Sol-gel method has been tremendously explored for the fabrication of nanostructure materials like nano-powders, nanocomposites etc. Composites have been synthesized by combining different types of materials (organic and inorganic) using the sol-gel method as shown in Figure 7. In this method, two methodologies can be used such as metal alkoxides hydrolysis and polycondensation of hydrolyzed intermediates. Transparent films of organic-inorganic composite materials have been made by co-hydrolysis and polycondensation of alkyltrimethoxy silane - tetramethoxysilane mixtures [66].

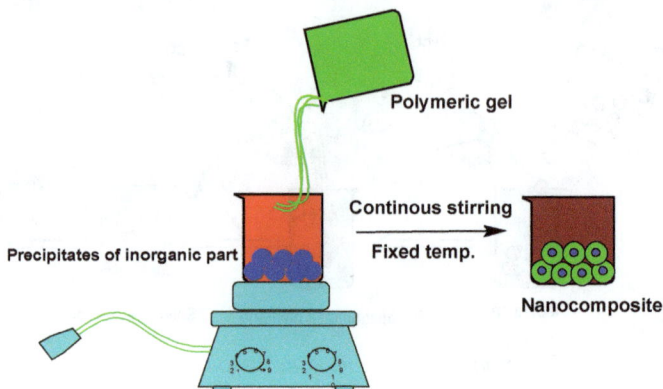

Figure 7: Synthesis of nanocomposites using sol-gel method.

3.   Methods for the removal of pollutants

Different methods including ion exchange, adsorption, membrane filtration and photocatalysis have been developed for the removal of organic and inorganic impurities from wastewater. Photocatalysis is a relatively simple and very effective technique for the exclusion of toxic contaminants using solar radiations. Photocatalytic process is attaining more attention in the field of wastewater treatment to obtain complete mineralization of the pollutant achieved under mild conditions of temperature and pressure [67-70].

3.1   Photocatalysis

Heterogeneous photocatalysis has been used for the removal and recovery of heavy metals and organic impurities. The semiconductor particles are illuminated by ultraviolet light with energy greater than the band gap energy of the photocatalyst, electron-hole pairs are generated. In this method, light energy is used to stimulate the semiconductor material to produce electron-hole pairs which results in detoxification of pollutants. The degradation of pollutants is due to reactive oxidising species (O_2^*, OH^*) generated by two simultaneous reactions.

When sunlight in the visible region falls on a semiconductor, electrons get promoted from valence band to conduction band with the formation of e^- - h^+ pairs. These photogenerated holes oxidize the dissociated adsorbed water. The excited electrons move towards the surface of semiconductor and transferred to surface of adsorbed oxygen where it forms O_2^*

radicals. These free radicals react with water and produce OH* [71-75]. The proposed mechanism of photocatalysis is shown in Figure 8.

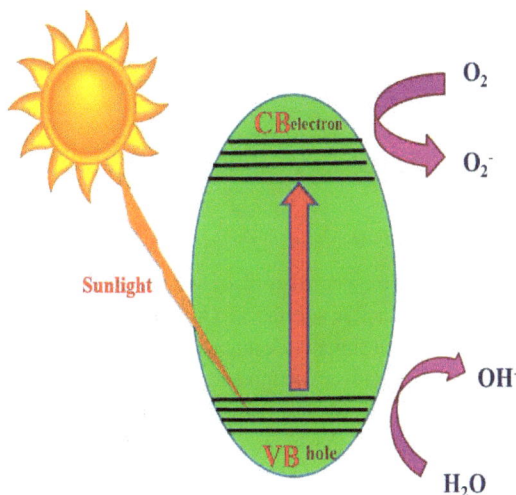

Figure 8: *Mechanism of photoctalaysis.*

3.2   Oxidation and reduction mechanism

The absorbed water is oxidized by positive holes produced in the valence band due to the electrons shifting to the conduction band. Due to irradiation of light, there was formation of hydroxyl (OH) and these hydroxyl radicals react with organic matter present in the dyes. If oxygen is present, the intermediate radicals in the organic compounds along with the oxygen molecules and experience radical chain reactions [76-78]. The organic matter decomposes at the end of the day becoming carbon dioxide and water. Organic compounds can react directly with positive holes resulting in oxidative decomposition as shown in Figure 9.

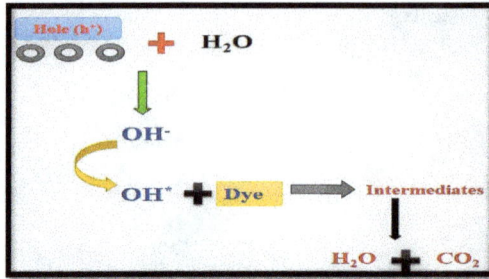

Figure 9: *Mechanism for oxidation process.*

On the other hand, reduction of oxygen takes place as an alternative to hydrogen generation because oxygen is an easily reducible substance. The conduction band electrons react with dissolved oxygen species to form superoxide anions. These superoxide anions attach to the intermediate products in the oxidative reaction and form peroxide or changing to H_2O_2 and then into water as shown in Figure 10. The reduction occurs more easily in organic matter as compared to water. Therefore, the higher concentration of organic matter tends to increase the number of positive holes and reduces the carrier recombination and improves photocatalytic activity [79, 80].

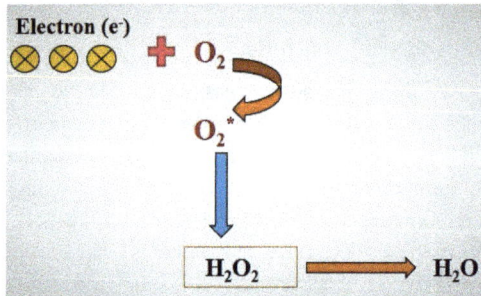

Figure 10: *Mechanism for reduction process.*

Different types of nanocomposites have been synthesised for the removal of dyes, pesticides, heavy metals, phenols and other pollutants using photocatalaysis. In the last few decades, nanocomposites were also explored for the photocatalytic degradation of different dyes in water systems. It has been found that these nanocomposite ion exchangers

Photocatalysis   Materials Research Forum LLC
Materials Research Foundations **100** (2021) 273-298   https://doi.org/10.21741/9781644901359-9

effectively degraded dyes in lesser time. Different nanocomposites have been employed for the degradation of different dyes including methylene blue, malachite green, methyl orange, fast green, methyl violet etc. as shown in Table 1.

Table 1: Different nanocomposite ion exchangers explored for photocatalytic activity

S.No.	Nanocomposite	Target Pollutant	References
1.	Polyaniline Zr(IV) selenotungstophosphate	Methylene blue and malachite green	[81]
2.	Polyacrylamide@Zr(IV) vanadophosphate nanocomposite	Congo red	[82]
3.	Pectin @ zirconium (IV) silicophosphate nanocomposite	Methylene blue	[83]
4.	Polyaniline zirconium (IV) silicophosphate nanocomposite	Methylene blue	[84]
5.	Pectin–thorium (IV) tungstomolybdate nanocomposite	Methylene blue	[85]
6.	Polyaniline zirconium(IV) silicophosphate	Methylene blue	[86]
7.	Pectin Zr(IV) selenotungstophosphate	Methylene blue and malachite green	[87]
8.	Polyacrylamide Ce(IV) silicophosphate	Methylene blue	[88]
9.	Cellulose acetate–zirconium (IV) phosphate nano-composite	Congo red	[89]
10.	Gelatin-Zr(IV) phosphate nanocomposite	Fast green and methylene blue	[90]
11.	Cellulose acetate–tin (IV) phosphate nanocomposite	Methylene blue	[91]
12.	Pectin thorium(IV) tungstomolybdate	Malachite green	[92]
13.	Styrene–tin (IV) phosphate nanocomposite	Methylene blue	[93]
14.	Alginate-Zr (IV) phosphate nanocomposite	Fast green and methylene blue	[94]
15.	lactic acid–Zr(IV) phosphate nanocomposite	Methylene blue	[95]

16.	Tin (IV) phosphate/poly(gelatin-cl-alginate) nanocomposite	Methyl orange	[96]
17.	Gelatin-tin (IV) phosphate nanocomposite	Methylene blue	[97]
18.	Gelatin–zirconium(IV) tungstophosphate nanocomposite	Methyl voilet	[98]
19.	Zirconium (IV) phosphate/poly(gelatin-cl-alginate) nanocomposite	Methylene blue and methylene green	[99]
20.	Gelatin-zirconium Dioxide Nanocomposite	Methylene blue	[100]
21.	Zirconium phosphoborate nanocomposite	Methylene blue	[101]

4.   Other applications of nanocomposite ion exchangers

The ion exchangers possess wide range of applications in various areas such as pollution control, separation of radioisotopes, antibiotic purification, fuel cell etc. Therefore, composite ion exchangers received special attention due to stability in column operations especially for separation, pre-concentration and filtration of ionic species. The composite ion exchangers have good ion exchange capacity, reproducibility, high thermal stability and selectivity for the removal of heavy metal ions. The applications of composite ion exchangers in different fields are shown in Figure 11.

Conclusion

In recent years, various processes including population growth, unrestrained water usage, climate change and infrequent rainfall, etc. continuously affect our surroundings. Due to these activities, demand for water has been increased but there is only small percentage of water which we can use. Water pollution is one of the major causes for water dearth. Therefore, we need to develop different techniques for the removal of toxic impurities present in water. In this, we have discussed photocatalytic processes for their exclusion using nanocomposite ion exchangers. It has been observed that nanocomposite ion exchangers act as photocatalyst and effectively degraded different dyes present in the marine system. These nanocomposites open new prospects for researchers to improve the quality of water.

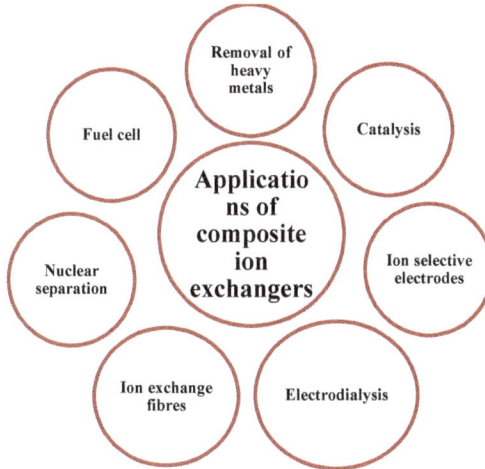

Figure 11: *Applications of composite or hybrid ion exchangers in different fields.*

References

[1]   Q. Wang, Z. Yang, Industrial water pollution, water environment treatment, and health risks in China, Environmental pollution, 218 (2016) 358-365. https://doi.org 10.1016/j.envpol.2016.07.011

[2]   A.K. Awasthi, X. Zeng, J. Li, Environmental pollution of electronic waste recycling in India: a critical review, Environmental pollution, 211 (2016) 259-270. https://doi.org/10.1016/j.envpol.2015.11.027

[3]   L. Zhang. J. Gao, Exploring the effects of international tourism on China's economic growth, energy consumption and environmental pollution: Evidence from a regional panel analysis, Renewable and Sustainable Energy Reviews, 53 (2016) 225-234. https://doi.org/10.1016/j.rser.2015.08.040

[4]   R. Michalski, A. Ficek, Environmental pollution by chemical substances used in the shale gas extraction—a review, Desalination and water treatment, 57 (2015) 1336-43. https://doi.org/10.1080/19443994.2015.1017331

Materials Research Forum LLC

https://doi.org/10.21741/9781644901359-9

[5]   W.A. Suk, H. Ahanchian, K.A. Asante, D.O. Carpenter, F. Diaz-Barriga, E.H. Ha, X. Huo, M. King, M. Ruchirawat, E.R. da Silva, L. Sly, Environmental pollution: an under-recognized threat to children's health, especially in low-and middle-income countries, Environmental health perspectives, 124 (2016) 41-50. https://doi.org/10.1289/ehp.1510517

[6]   T.K. Parmar, D. Rawtani, Y.K. Agrawal. Bioindicators: the natural indicator of environmental pollution, Frontiers in life science, 9 (2016) 110-118. https://doi.org/10.1080/21553769.2016.1162753

[7]   L. M. Chiesa, G. F. Labella, A. Giorgi, S. Panseri, R. Pavlovic, S. Bonacci, F. Arioli, The occurrence of pesticides and persistent organic pollutants in Italian organic honeys from different productive areas in relation to potential environmental pollution, Chemosphere, 154 (2016) 482-490. https://doi.org/10.1016/j.chemosphere.2016.04.004

[8]   S. Dudka, D. C. Adriano, Environmental impacts of metal ore mining and processing: a review, Journal of environmental quality, 26 (1997) 590-602. https://doi.org 10.2134/jeq1997.00472425002600030003x

[9]   Y. M. Hsueh, C. Y. Lee, S. N. Chien, W. J. Chen, H. S. Shiue, S. R. Huang, M. I. Lin, S. C. Mu, R. L. Hsieh, Association of blood heavy metals with developmental delays and health status in children, Scientific reports, 7 (2017) 43608. https://doi.org /10.1038/srep43608 (2017)

[10]   K. Kumar, S. C. Gupta, S. K. Baidoo, Y. Chander, C. J. Rosen, Antibiotic uptake by plants from soil fertilized with animal manure, Journal of environmental quality, 34 (2005) 2082-5. doi:10.2134/jeq2005.0026

[11]   I. Zadnipryany, O. Tretiakova, T. P. Sataieva, W. Zukow, Experimental review of cobalt induced cardiomyopathy, Russian Open Medical Journal, 6 (2017) 103. https://doi.org/10.15275/rusomj.2017.0103

[12]   L. Makarichi, V. Jutidamrongphan, K. A. Techato, The evolution of waste-to-energy incineration: A review, Renewable and Sustainable Energy Reviews, 91 (2018) 812-21. https://doi.org/10.1016/j.rser.2018.04.088

[13]   R. Verma, K. S. Vinoda, M. Papireddy, A. N. Gowda, Toxic pollutants from plastic waste-a review, Procedia Environmental Sciences, 35 (2016) 701-8. https://doi.org/10.1016/j.proenv.2016.07.069

Photocatalysis   Materials Research Forum LLC
Materials Research Foundations **100** (2021) 273-298   https://doi.org/10.21741/9781644901359-9

[14] X. Wang, D. Sun, T. Yao, Climate change and global cycling of persistent organic pollutants: a critical review, Science China Earth Sciences, 59 (2016) 1899-911. https://doi.org/10.1007/s11430-016-5073-0

[15] J. P. Bonde, E. M. Flachs, S. Rimborg, C. H. Glazer, A. Giwercman, C. H. Ramlau-Hansen, K. S. Hougaard, B. B. Høyer, K. K. Hærvig, S. B. Petersen, L. Rylander, The epidemiologic evidence linking prenatal and postnatal exposure to endocrine disrupting chemicals with male reproductive disorders: a systematic review and meta-analysis, Human Reproduction Update, 23 (2016) 104-25. https://doi.org/10.1093/humupd/dmw036

[16] M. Brits, J. De Vos, J. M. Weiss, E. R. Rohwer, J. De Boer, Critical review of the analysis of brominated flame retardants and their environmental levels in Africa, Chemosphere, 164 (2016) 174-89. https://doi.org/10.1016/j.chemosphere.2016.08.097

[17] D. Megson, E. J. Reiner, K. J. Jobst, F. L. Dorman, M. Robson, J. F. Focant. A review of the determination of persistent organic pollutants for environmental forensics investigations, Analytica Chimica Acta, 941 (2016) 10-25. https://doi.org/10.1016/j.aca.2016.08.027

[18] V. Filimonova, F. Gonçalves, J. C. Marques, M. De Troch, A. M. Goncalves, Fatty acid profiling as bioindicator of chemical stress in marine organisms: a review, Ecological indicators, 67 (2016) 657-72. https://doi.org/10.1016/j.ecolind.2016.03.044

[19] G. Cagnetta, H. Liu, K. Zhang, J. Huang, B. Wang, S. Deng, Y. Wang, G. Yu, Mechano-chemical conversion of brominated POPs into useful oxybromides: a greener approach, Scientific reports, 6 (2016) 28394. https://doi.org/10.1038/srep28394

[20] J. Ma, H. Hung, R. W. Macdonald, The influence of global climate change on the environmental fate of persistent organic pollutants: A review with emphasis on the Northern Hemisphere and the Arctic as a receptor, Global and Planetary Change, 146 (2016) 89-108. https://doi.org/10.1016/j.gloplacha.2016.09.011

[21] M. K. Uddin, A review on the adsorption of heavy metals by clay minerals, with special focus on the past decade, Chemical Engineering Journal, 308 (2017) 438-62. https://doi.org/10.1016/j.cej.2016.09.029

[22] Y. Zhao, G. Mao, H. Hongxia, L. Gao, Effects of EDTA and EDDS on heavy metal activation and accumulation of metals by soybean in alkaline soil, Soil and Sediment Contamination: An International Journal, 24(4) (2015) 353-67. https://doi.org/10.1080/15320383.2015.962125

[23]   C. F. Carolin, P. S. Kumar, A. Saravanan, G. J. Joshiba, M. Naushad, Efficient techniques for the removal of toxic heavy metals from aquatic environment: A review, Journal of Environmental Chemical Engineering, 5 (2017) 2782-99. https://doi.org/10.1016/j.jece.2017.05.029

[24]   D. L. Lake, P. W. Kirk, J. N. Lester, Fractionation, characterization and speciation of heavy metals in sewage sludge and sludge-amended soils: a review, Journal of Environmental Quality, 2 (1984) 175-83. https://doi.org/10.2134/jeq1984.00472425001300020001x

[25]   A. C. Bosch, B. O'Neill, G. O. Sigge, S. E. Kerwath, L. C. Hoffman, Heavy metals in marine fish meat and consumer health: a review, Journal of the Science of Food and Agriculture. 96 (2016) 32-48. https://doi.org/10.1002/jsfa.7360

[26]   S. H. Liu, G. M. Zeng, Q. Y. Niu, Y. Liu, L. Zhou, L. H. Jiang, X. F. Tan, P. Xu, C. Zhang, M. Cheng, Bioremediation mechanisms of combined pollution of PAHs and heavy metals by bacteria and fungi: A mini review, Bioresource technology. 224 (2017) 25-33. https://doi.org/10.1016/j.biortech.2016.11.095

[27]   E. Da'na, Adsorption of heavy metals on functionalized-mesoporous silica: A review, Microporous and Mesoporous Materials, 247 (2017) 145-57. https://doi.org/10.1016/j.micromeso.2017.03.050

[28]   M. Abtahi, Y. Fakhri, G. Oliveri Conti, H. Keramati, Y. Zandsalimi, Z. Bahmani, R. HosseiniPouya, M. Sarkhosh, B. Moradi, N. Amanidaz, S. M. Ghasemi, Heavy metals (As, Cr, Pb, Cd and Ni) concentrations in rice (Oryzasativa) from Iran and associated risk assessment: a systematic review, Toxin reviews, 36 (2017) 331-41. https://doi.org/10.1080/15569543.2017.1354307

[29]   G. Richhariya, A. Kumar, P. Tekasakul, B. Gupta, Natural dyes for dye sensitized solar cell: A review, Renewable and Sustainable Energy Reviews, 69 (2017) 705-18. https://doi.org/10.1016/j.rser.2016.11.198

[30]   H. A. Shindy. Fundamentals in the chemistry of cyanine dyes: A review, Dyes and Pigments, 145 (2017) 505-13. https://doi.org/10.1016/j.dyepig.2017.06.029

[31]   A. M. Ghaedi, A. Vafaei, Applications of artificial neural networks for adsorption removal of dyes from aqueous solution: a review, Advances in colloid and interface science, 245 (2017) 20-39. https://doi.org/10.1016/j.cis.2017.04.015

[32]   C. A. Martínez-Huitle, E. Brillas, Decontamination of wastewaters containing synthetic organic dyes by electrochemical methods: a general review, Applied Catalysis B: Environmental, 87 (2009) 105-45. https://doi.org/10.5772/58333

Materials Research Forum LLC

https://doi.org/10.21741/9781644901359-9

[33]   G. Annadurai, R.S. Juang, D.J. Lee, Use of cellulose-based wastes for adsorption of dyes from aqueous solutions, Journal of hazardous materials, 92(3) (2002) 263-74. https://doi.org/10.1016/S0304-3894(02)00017-1

[34]   I. Khurana, A. Saxena, J. M. Khurana, P. K. Rai, Removal of dyes using graphene-based composites: a review, Water, Air & Soil Pollution, 228 (2017) 180. https://doi.org/10.1007/s11270-017-3361-1

[35]   T. Robinson, G. McMullan, R. Marchant, P. Nigam, Remediation of dyes in textile effluent: a critical review on current treatment technologies with a proposed alternative, Bioresource technology, 77(3) (2001) 247-55. https://doi.org/10.1016/S0960-8524(00)00080-8

[36]   Z. Cao, J. Engelhardt, M. Dierks, M. T. Clough, G. H. Wang, E. Heracleous, A. Lappas, R. Rinaldi, F. Schüth, Catalysis meets nonthermal separation for the production of (alkyl) phenols and hydrocarbons from pyrolysis oil, Angewandte Chemie International Edition, 56 (2017) 2334-9. https://doi.org/10.1002/anie.201610405

[37]   S. D. Schimler, M. A. Cismesia, P. S. Hanley, R. D. Froese, M. J. Jansma, D. C. Bland, M. S. Sanford, Nucleophilic deoxy fluorination of phenols via aryl Fluorosulfonate intermediates, Journal of the American Chemical Society, 139 (2017) 1452-5. DOI: 10.1021/jacs.6b12911

[38]   J. Zhou, M. L. Wang, X. Gao, G. F. Jiang, Y. G. Zhou, Bifunctional squaramide-catalyzed synthesis of chiral dihydrocoumarins via ortho-quinonemethides generated from 2-(1-tosylalkyl) phenols, Chemical Communications, 53 (2017) 3531-4.

[39]   R. D. Wauchope, The pesticide content of surface water draining from agricultural fields—a review, Journal of environmental quality, 7 (1978) 459-72. https://doi.org/10.2134/jeq1978.00472425000700040001x

[40]   R. J. Hillocks, Farming with fewer pesticides: EU pesticide review and resulting challenges for UK agriculture, Crop Protection, 31 (2012) 85-93. https://doi.org/10.1016/j.cropro.2011.08.008

[41]   M. Cycoń, A. Mrozik, Z. Piotrowska-Seget, Bio augmentation as a strategy for the remediation of pesticide-polluted soil: a review, Chemosphere, 172 (2017) 52-71. https://doi.org/10.1016/j.chemosphere.2016.12.129

[42]   M. L. Xu, Y. Gao, X. X. Han, B. Zhao, Detection of pesticide residues in food using surface-enhanced Raman spectroscopy: A review, Journal of agricultural and food chemistry, 65 (2017) 6719-26. https://doi.org/10.1021/acs.jafc.7b02504

[43]   K. Khoiruddin, A. N. Hakim, I. G. Wenten, Advances in electrode ionization technology for ionic separation-A review, Membrane Water Treatment, 5 (2014) 87-108. https://doi.org/10.12989/mwt.2014.5.2.087

[44]   O. Oleksiienko, C. Wolkersdorfer, M. Sillanpää, Titanosilicates in cation adsorption and cation exchange–a review, Chemical Engineering Journal, 317 (2017) 570-85. https://doi.org/10.1016/j.cej.2017.02.079

[45]   J. Ran, L. Wu, Y. He, Z. Yang, Y. Wang, C. Jiang, L. Ge, E. Bakangura, T. Xu, Ion exchange membranes: New developments and applications, Journal of Membrane Science, 522 (2017) 267-91. https://doi.org/10.1016/j.memsci.2016.09.033

[46]   M. Endo, E. Yoshikawa, N. Muramatsu, N. Takizawa, T. Kawai, H. Unuma, A. Sasaki, A. Masano, Y. Takeyama, T. Kahara, The removal of cesium ion with natural Itaya zeolite and the ion exchange characteristics, Journal of Chemical Technology & Biotechnology, 88 (2013) 1597-602. https://doi.org/10.1002/jctb.4140

[47]   M. A. Hickner, B. S. Pivovar, The chemical and structural nature of proton exchange membrane fuel cell properties, Fuel cells, 5 (2005) 213-29. https://doi.org/10.1002/fuce.200400064

[48]   T. Xu, Ion exchange membranes: state of their development and perspective, Journal of membrane science, 263(1-2) (2005) 1-29. https://doi.org/10.1016/j.memsci.2005.05.002

[49]   G. Zhao, X. Huang, Z. Tang, Q. Huang, F. Niu, X. Wang, Polymer-based nanocomposites for heavy metal ions removal from aqueous solution: a review, Polymer Chemistry, 9 (2018) 3562-82. https://doi.org/10.1039/C8PY00484F

[50]   W. Jia, B. Tang, P. Wu, Novel composite proton exchange membrane with connected long-range ionic nanochannels constructed via exfoliated nafion–boron nitride nanocomposite, ACS applied materials & interfaces, 9 (2017) 14791-800. https://doi.org/10.1021/acsami.7b00858

[51]   M. Naushad, G. Sharma, A. Kumar, S. Sharma, A. A. Ghfar, A. Bhatnagar, F. J. Stadler, M. R. Khan, Efficient removal of toxic phosphate anions from aqueous environment using pectin based quaternary amino anion exchanger, International Journal of Biological Macromolecules, 106 (2018) 1-0. https://doi.org/10.1016/j.ijbiomac.2017.07.169

[52]   Y. Tanaka, S. H. Moon, V. V. Nikonenko, T. Xu, Ion-exchange membranes, International Journal of Chemical Engineering, (2012). https://doi.org/10.1155/2012/528290

[53]   L. Curkovic, S. Cerjan-Stefanovic, T. Filipan, Metal ion exchange by natural and modified zeolites, Water research, 31(6) (1997) 1379-82. https://doi.org/10.1016/S0043-1354(96)00411-3

[54]   A. Dąbrowski, Z. Hubicki, P. Podkościelny, E. Robens, Selective removal of the heavy metal ions from waters and industrial wastewaters by ion-exchange method, Chemosphere, 56 (2004) 91-106. https://doi.org/10.1016/j.chemosphere.2004.03.006

[55]   Y. F. Yao, J. T. Kummer, Ion exchange properties of and rates of ionic diffusion in beta-alumina, Journal of Inorganic and Nuclear Chemistry, 29 (1967) 2453-75. https://doi.org/10.1016/0022-1902(67)80301-4

[56]   A. Nilchi, R. Saberi, M. Moradi, H. Azizpour, R. Zarghami, Adsorption of cesium on copper hexacyanoferrate–PAN composite ion exchanger from aqueous solution, Chemical Engineering Journal, 172 (2011) 572-80. https://doi.org/10.1016/j.cej.2011.06.011

[57]   F. Šebesta, V. Štefula, Composite ion exchanger with ammonium molybdophosphate and its properties, Journal of Radioanalytical and Nuclear Chemistry, 140 (1990) 15-21. https://doi.org/10.1007/BF02037360

[58]   A. Mardan, R. Ajaz, A. Mehmood, S. M. Raza, A. Ghaffar, Preparation of silica potassium cobalt hexacyanoferrate composite ion exchanger and its uptake behavior for cesium, Separation and purification technology, 16 (1999) 147-58. https://doi.org/10.1016/S1383-5866(98)00121-X

[59]   S. A. Nabi, M. Shahadat, R. Bushra, A. H. Shalla, A. Azam, Synthesis and characterization of nano-composite ion-exchanger; its adsorption behaviour, Colloids and Surfaces B: Biointerfaces, 87 (2011) 122-8. https://doi.org/10.1016/j.colsurfb.2011.05.011

[60]   A. A. Khan, M. M. Alam, Synthesis, characterization and analytical applications of a new and novel 'organic–inorganic' composite material as a cation exchanger and Cd (II) ion-selective membrane electrode: polyaniline Sn (IV) tungstoarsenate, Reactive and Functional Polymers, 55 (2003) 277-90. https://doi.org/10.1016/S1381-5148(03)00018-X

[61]   J. K. Moon, K. W. Kim, C. H. Jung, Y. G. Shul, E. H. Lee, Preparation of organic-inorganic composite adsorbent beads for removal of radionuclides and heavy metal ions, Journal of Radioanalytical and Nuclear Chemistry, 246 (2000) 299-307. https://doi.org/10.1023/A:1006714322455

[62]   S. A. Nabi, M. Shahadat, R. Bushra, A. H. Shalla, F. Ahmed, Development of composite ion-exchange adsorbent for pollutants removal from environmental wastes, Chemical Engineering Journal, 165 (2010) 405-12. https://doi.org/10.1016/j.cej.2010.08.068

[63]   W. W. Ngah, L.C. Teong, M. A. Hanafiah, Adsorption of dyes and heavy metal ions by chitosan composites: A review, Carbohydrate polymers, 83 (2011) 1446-56. https://doi.org/10.1016/j.carbpol.2010.11.004

[64]   Z. Shen, G. P. Simon, Y. B. Cheng, Comparison of solution intercalation and melt intercalation of polymer–clay nanocomposites, Polymer, 43 (2002) 4251-60. https://doi.org/10.1016/S0032-3861(02)00230-6

[65]   S. S. Ray, M. Okamoto, Polymer/layered silicate nanocomposites: a review from preparation to processing, Progress in polymer science, 28 (2003) 1539-641. https://doi.org/10.1016/j.progpolymsci.2003.08.002

[66]   I. A. Rahman, V. Padavettan, Synthesis of silica nanoparticles by sol-gel: size-dependent properties, surface modification, and applications in silica-polymer nanocomposites-a review, Journal of Nanomaterials, (2012) 8. https://doi.org/10.1155/2012/132424

[67]   W. Hou, S. B. Cronin, A review of surface plasmon resonance-enhanced photocatalysis, Advanced Functional Materials, 23 (2013) 1612-9. https://doi.org/10.1002/adfm.201202148

[68]   K. Kabra, R. Chaudhary, R. L. Sawhney, Treatment of hazardous organic and inorganic compounds through aqueous-phase photocatalysis: a review, Industrial & engineering chemistry research, 43 (2004) 7683-96. https://doi.org/10.1021/ie0498551

[69]   Y. Wang, Q. Wang, X. Zhan, F. Wang, M. Safdar, J. He, Visible light driven type II heterostructures and their enhanced photocatalysis properties: a review, Nanoscale, 5 (2013) 8326-39. https://doi.org/10.1039/C3NR01577G

[70]   X. Li, J. Yu, S. Wageh, A. A. Al-Ghamdi, J. Xie, Graphene in photocatalysis: a review, Small, 12 (2016) 6640-96. https://doi.org/10.1002/smll.201600382

[71]   K. Wenderich, G. Mul, Methods, mechanism, and applications of photodeposition in photocatalysis: a review, Chemical reviews, 116 (2016) 14587-619. https://doi.org/10.1021/acs.chemrev.6b00327

[72]   L. K. Putri, W. J. Ong, W. S. Chang, S. P. Chai, Heteroatom doped graphene in photocatalysis: a review, Applied Surface Science, 358 (2015) 2-14. https://doi.org/10.1016/j.apsusc.2015.08.177

[73]   Y. Boyjoo, H. Sun, J. Liu, V. K. Pareek, S. Wang, A review on photocatalysis for
       air treatment: from catalyst development to reactor design, Chemical Engineering
       Journal, 310 (2017) 537-59. https://doi.org/10.1016/j.cej.2016.06.090

[74]   J. Byrne, P. Dunlop, J. Hamilton, P. Fernández-Ibáñez, I. Polo-López, P. Sharma,
       A. Vennard, A review of heterogeneous photocatalysis for water and surface
       disinfection, Molecules, 20 (2015) 5574-615.
       https://doi.org/10.3390/molecules20045574

[75]   M. N. Chong, B. Jin, C. W. Chow, C. Saint, Recent developments in
       photocatalytic water treatment technology: a review, Water research, 44 (2010) 2997-
       3027. https://doi.org/10.1016/j.watres.2010.02.039

[76]   M.A. Fox, M.T. Dulay, Heterogeneous photocatalysis, Chemical reviews, 93(1)
       (1993) 341-57. https://doi.org/10.1021/cr00017a016

[77]   A. Ibhadon, P. Fitzpatrick, Heterogeneous photocatalysis: recent advances and
       applications. Catalysts, 3 (2013) 189-218. https://doi.org/10.3390/catal3010189

[78]   M. R. Hoffmann, S. T. Martin, W. Choi, D. W. Bahnemann, Environmental
       applications of semiconductor photocatalysis, Chemical reviews, 95 (1995) 69-96.
       https://doi.org/10.1021/cr00033a004

[79]   R. K. Nath, M. F. Zain, A. A. Kadhum, Photocatalysis—a novel approach for
       solving various environmental and disinfection problems: a brief review, Journal of
       Applied Sciences Research, 8 (2012) 4147-55.

[80]   T. Ramesh, B. Nayak, A. Amirbahman, C. P. Tripp, S. Mukhopadhyay,
       Application of ultraviolet light assisted titanium dioxide photocatalysis for food safety:
       A review, Innovative food science & emerging technologies, 38 (2016) 105-15.
       https://doi.org/10.1016/j.ifset.2016.09.015

[81]   D. Pathania, G. Sharma, A. Kumar, M. Naushad, S. Kalia, A. Sharma, Z. A.
       ALOthman, Combined sorptional–photocatalytic remediation of dyes by polyaniline
       Zr (IV) selenotungstophosphate nanocomposite, Toxicological & Environmental
       Chemistry, 97 (2015) 526-37. https://doi.org/10.1080/02772248.2015.1050024

[82]   G. Sharma, A. Kumar, M. Naushad, D. Pathania, M. Sillanpää, Polyacrylamide@
       Zr (IV) vanadophosphate nanocomposite: ion exchange properties, antibacterial
       activity, and photocatalytic behavior, Journal of Industrial and Engineering Chemistry,
       33 (2016) 201-8. https://doi.org/10.1016/j.jiec.2015.10.011

[83]   D. Pathania, G. Sharma, R. Thakur, Pectin@ zirconium (IV) silicophosphate
       nanocomposite ion exchanger: photo catalysis, heavy metal separation and

antibacterial activity, Chemical Engineering Journal, 267 (2015) 235-44.
https://doi.org/10.1016/j.cej.2015.01.004

[84]   V. K. Gupta, D. Pathania, N. C. Kothiyal, G. Sharma, Polyaniline zirconium (IV) silicophosphate nanocomposite for remediation of methylene blue dye from waste water, Journal of Molecular Liquids, 190 (2014) 139-45.
https://doi.org/10.1016/j.molliq.2013.10.027

[85]   V. K. Gupta, S. Agarwal, D. Pathania, N. C. Kothiyal, G. Sharma, Use of pectin–thorium (IV) tungstomolybdate nanocomposite for photocatalytic degradation of methylene blue, Carbohydrate polymers, 96 (2013) 277-83.
https://doi.org/10.1016/j.carbpol.2013.03.073

[86]   D. Pathania, G. Sharma, A. Kumar, N. C. Kothiyal, Fabrication of nanocomposite polyaniline zirconium (IV) silicophosphate for photocatalytic and antimicrobial activity, Journal of Alloys and Compounds, 588 (2014) 668-75.
https://doi.org/10.1016/j.jallcom.2013.11.133

[87]   V. K. Gupta, Sharma G, D. Pathania, N. C. Kothiyal, Nanocomposite pectin Zr (IV) selenotungsto phosphate for adsorptional/photocatalytic remediation of methylene blue and malachite green dyes from aqueous system, Journal of Industrial and Engineering Chemistry, 21 (2015) 957-64. https://doi.org/10.1016/j.jiec.2014.05.001

[88]   D. Pathania, G. Sharma, M. Naushad, A. Kumar, Synthesis and characterization of a new nanocomposite cation exchanger polyacrylamide Ce (IV) silico phosphate: photocatalytic and antimicrobial applications, Journal of Industrial and Engineering Chemistry, 20 (2014) 3596-603. https://doi.org/10.1016/j.jiec.2013.12.054

[89]   V. K. Gupta, D. Pathania, P. Singh, B. S. Rathore, P. Chauhan, Cellulose acetate–zirconium (IV) phosphate nano-composite with enhanced photo-catalytic activity, Carbohydrate polymers, 95 (2013) 434-40.
https://doi.org/10.1016/j.carbpol.2013.02.045

[90]   M. Thakur, G. Sharma, T. Ahamad, A. A. Ghfar, D. Pathania, M. Naushad, Efficient photocatalytic degradation of toxic dyes from aqueous environment using gelatin-Zr (IV) phosphate nanocomposite and its antimicrobial activity, Colloids and Surfaces B: Biointerfaces, 157 (2017) 456-63.
https://doi.org/10.1016/j.colsurfb.2017.06.018

[91]   V. K. Gupta, T. A. Saleh, D. Pathania, B. S. Rathore, G. Sharma, A cellulose acetate based nanocomposite for photocatalytic degradation of methylene blue dye under solar light, Ionics, 21 (2015) 1787-93. https://doi.org/10.1007/s11581-014-1323-9

[92] G. Sharma, M. Naushad, D. Pathania, A. Kumar, A multifunctional nanocomposite pectin thorium (IV) tungstomolybdate for heavy metal separation and photo remediation of malachite green, Desalination and Water Treatment, 57 (2016) 19443-55. https://doi.org/10.1080/19443994.2015.1096834

[93] B. S. Rathore, D. Pathania, Styrene–tin (IV) phosphate nanocomposite for photocatalytic degradation of organic dye in presence of visible light, Journal of Alloys and Compounds, 606 (2014) 105-11. https://doi.org/10.1016/j.jallcom.2014.03.160

[94] D. Pathania, M. Thakur, A. K. Mishra, Alginate-Zr (IV) phosphate nanocomposite ion exchanger: Binary separation of heavy metals, photocatalysis and antimicrobial activity, Journal of Alloys and Compounds, 701 (2017) 153-62. https://doi.org/10.1016/j.jallcom.2017.01.112

[95] D. Pathania, M. Thakur, A. Sharma, S. Agarwal, V. K. Gupta, Synthesis of lactic acid–Zr (IV) phosphate nanocomposite ion exchanger for green remediation, Ionics, 23 (2017) 699-706. https://doi.org/10.1007/s11581-016-1858-z

[96] D. Pathania, M. Thakur, G. Sharma, A. K. Mishra, Tin (IV) phosphate/poly (gelatin-cl-alginate) nanocomposite: Photocatalysis and fabrication of potentiometric sensor for Pb (II), Materials Today Communications, 14 (2018) 282-93. https://doi.org/10.1016/j.mtcomm.2018.01.005

[97] D. Pathania, M. Thakur, V. Puri, S. Jasrotia, Fabrication of electrically conductive membrane electrode of gelatin-tin (IV) phosphate nanocomposite for the detection of cobalt (II) ions, Advanced Powder Technology, 29 (2018) 915-24. https://doi.org/10.1016/j.apt.2018.01.009

[98] M. Thakur, D. Pathania, Sol–gel synthesis of gelatin–zirconium (IV) tungstophosphatenano composite ion exchanger and application for the estimation of Cd (II) ions, Journal of Sol-Gel Science and Technology, (2019) 1-3. https://doi.org/10.1007/s10971-019-04919-2

[99] D. Pathania, S. Agarwal, V. K. Gupta, M. Thakur, N. S. Alharbi, Zirconium (IV) phosphate/poly (gelatin-cl-alginate) nanocomposite as ion exchanger and Al^{3+} potentiometric sensor, International Journal of Electrochemical Science, 13 (2018) 994-1012. https://doi.org/10.20964/2018.01.80

[100] D. Pathania, M. Thakur, S. Jasrotia, S. Agarwal, V. K. Gupta, Gelatin-zirconium dioxide nanocomposite as a Ni (II) selective potentiometric sensor: Heavy Metal Separation and Photocatalysis, International Journal of Electrochemical Science, 12 (2017) 8477-94. https://doi.org/10.20964/2017.09.49

[101] S. Kaushal, R. Badru, P. Singh, S. Kumar, S. K. Mittal, Nanocomposite zirconium phosphoborate ion-exchanger incorporating carbon nanotubes with photocatalytic activity, Separation Science and Technology, 51 (2016) 2896-902. https://doi.org/10.1080/01496395.2016.1234487

Photocatalysis
Materials Research Foundations 100 (2021) 299-331

Materials Research Forum LLC
https://doi.org/10.21741/9781644901359-10

Chapter 10

Structural Modifications of Carbon Nitride for Photocatalytic Applications

Ajay Kumar[1,*], Manisha Chandel[2], Manita Thakur[2]

[1]Department of chemistry, SILB Solan (H.P) India, 173212

[2]Department of Chemistry, IEC University, Baddi, (H.P.), India, 174103

*ajaydogra972@gmail.com

Abstract

The current research on photocatalysis is totally focused on the designing and innovation of various low cost materials. For an efficient photocatalyst, there are some aspects which are to be assessed before practical use, such as optical activity, thermal and chemical stability, easy and availability of raw material, biocompatibility, etc. Fortunately, g-C_3N_4 offers most of these qualities to behave as a star photocatalyst. g-C_3N_4 could be easily prepared from low cost precursor materials such as urea, melamine, cyanimide and dicyandiamide by simple thermal treatment. Furthermore, larger surface area and two-dimensional planar conjugation structure of g-C_3N_4 can provide a large platform for anchoring various substrates. Various researchers have utilized g-C_3N_4 for varieties of applications such as green energy production, energy storage devices, biomedical application, wastewater treatment via photocatalysis and adsorption, photo sensors, etc. Although there are some disadvantages associated with use of g-C_3N_4 when utilized for various applications. To overcome such hitches various structural modifications have been applied to g-C_3N_4. The current chapter summarizes a wide mode of applications of g-C_3N_4 along with various structural modifications which were recently applied to improve the photocatalytic efficacy.

Keywords

g-C_3N_4, Modification, Energy, Wastewater, Photocatalysts

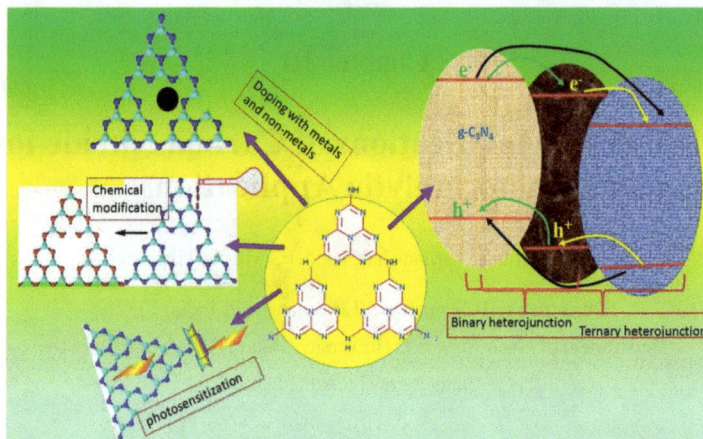

Contents

Materials Research Forum LLC
https://doi.org/10.21741/9781644901359-10

1.   Introduction

The current era is demanding clean energy production and environmental remediation. Various aspects of environment and health related concern have triggered the research towards exploring novel hybrid materials which can be utilized and address these issues without altering the environment quality. Photocatalysis is a greener approach that has proven to be cost effective and highly efficient in the field of clean energy production, wastewater treatment, air pollution remediation, catalytic process, etc.[1–4]. But efficacy and cost effectiveness of photocatalysis process is extensively dependent on the material that is being utilized as "photocatalyst". Now these photocatalysts are the key material as well as center of curiosity and research which attract attention of the worldwide scientist's community. Various metal based and metal free photocatalysts have been explored for the varieties of photocatalytic processes [5]. However, metal based and metal free photocatalysts have their own limits of advantages and disadvantages e.g. Ag, Au, Pt, Ru, etc. based photocatalysts are highly active but their expensive nature limits their exploration for large scale applications [6]. At the same time, metal free photocatalysts such as g-C_3N_4, reduced graphene oxide (RGO), graphene oxide (GO), carbon-quantum dots (CQDTs) and other organic photocatalysts offer an alternate material for being utilized for various photocatalytic applications [7–10]. However chemical, thermal stability and synthesis strategies are various parameters on which the efficacy of a metal free photocatalysts depends.

1.1   g-C_3N_4 as tremendous metal-free photocatalysts

g-C_3N_4 is a Nobel metal-free, stable, nontoxic and inexpensive polymeric photocatalyst [11]. g-C_3N_4 holds many advantages, such as good thermal and chemical stability, metal-free and tunable structure which are the essential requirements. The abundance and low cost of its precursor material such as urea, melamine, cyanimide and dicyandiamide make it approachable for the entire researchers to explore it for high quality research. Furthermore, larger surface area and 2-D planar structure of g-C_3N_4 can provide a large sites for anchoring various substrates [12]. g-C_3N_4 is quite stable up to 600 °C and starts decomposing when temperature is above 700 °C [13]. Due to strong interlayer stacking interaction g-C_3N_4 is almost insoluble in solvents like water, ethanol, methanol, THF, diethyl ether and toluene [14].

Materials Research Forum LLC
https://doi.org/10.21741/9781644901359-10

1.2   Chemical structure of g-C₃N₄

The structure of g-C$_3$N$_4$ consists of tri-s-triazine of units which are interconnected with each other with tertiary amine [15] units as referred with figure 1. It possesses the π-conjugated graphitic planes which are formed due the sp^2 hybridization between the carbon and nitrogen atom. The band gap energy for g-C$_3$N$_4$ is calculated to be ≈2.7 eV which corresponds to the absorption of wavelength ≈460 nm.

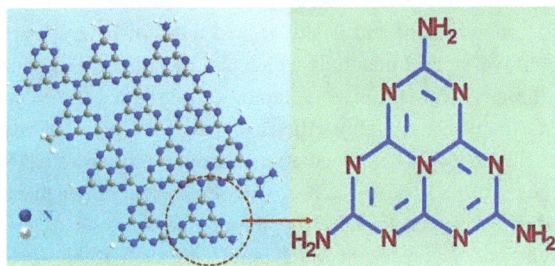

Figure 1: Chemical structure of g-C₃N₄q.

1.3   Band structure of g-C₃N₄

The lone pair of electrons on nitrogen is supposed to be contributing for band structure of g-C$_3$N$_4$. The overlap of 2p orbitals of N together with the overlap carbon 2p orbitals are mainly depot to VB and CB of the g-C$_3$N$_4$ [16].

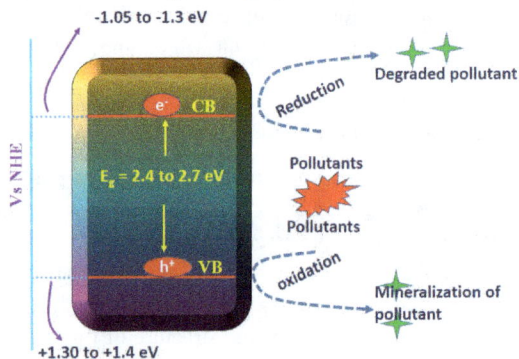

Figure 2: Band structure of g-C₃N₄.

The extra stability for lone pairs of nitrogen has been provided by π-bonding states which are present in conjugation. Therefore, the lone pair electron of nitrogen plays a key role in structure of g-C_3N_4 [17]. Zhang et al. reported that the CB edge and VB edge lies at -1.3 and 1.4 eV (vs NHE) respectively [18] which fulfills the requirements of reduction and oxidation reaction directed by sun light. The band structure of g-C_3N_4 is shown in Fig. 2. The CB & VB edge potential of g-C_3N_4 located at sufficient potentials which enable it for various photo reduction and oxidation reactions.

2.   Synthesis approaches via thermal annealing/Pyrolysis

Thermal annealing is the most common and popular technique used for any modification for the fabrication of template-based and template-free g-C_3N_4. Various precursor materials such as urea, melamine, cyanimide and dicyandiamide etc. are thermally treated at high temperature (400-700 °C) with a heating rate of 2 to 5 °C min^{-1} [19]. Thermal treatment allows the precursor materials to condense into the polymeric form of tri-s-triazine units as shown in fig. 3.

It has been reported that when g-C_3N_4 is prepared in bulk it possesses comparatively low surface area (< 20 m^{2}g^{-1}) [4]. Thus various effective strategies have been employed to exfoliate thin layer of g-C_3N_4 such as oxidation exfoliation [20], ultrasonic liquid exfoliation [21] chemical exfoliation [22] etc. The exfoliated thin sheets of g-C_3N_4 have high surface area as compared to bulk g-C_3N_4. The condensation of precursor materials such as urea, thiourea and melamine have been represented in fig. 4.

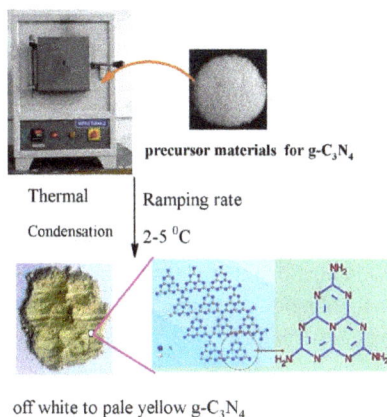

precursor materials for g-C_3N_4

Thermal Condensation | Ramping rate 2-5 °C

off white to pale yellow g-C_3N_4

Figure 3: Method of thermal condensation.

Photocatalysis   Materials Research Forum LLC
Materials Research Foundations **100** (2021) 299-331   https://doi.org/10.21741/9781644901359-10

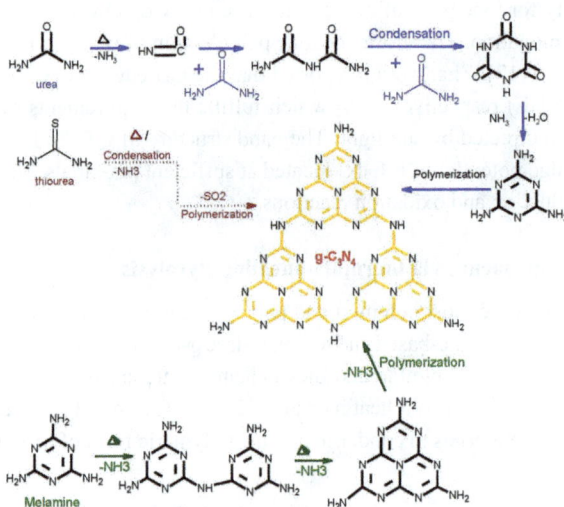

Figure 4. Proposed thermal condensation of various precursor materials to g-C₃N₄.

3.   Strategies followed for the modification of g-C₃N₄

However, g-C₃N₄ has been continuously explored for various photocatalytic applications but there are still some shortcomings and bottlenecks which have to be dealt with, such as rapid charge recombination rate, slightly low absorption in visible spectrum, etc. [23]. To overcome such limitations various modifications have been carried out in g-C₃N₄ to effectively increase the efficiency.

3.1   Doping

Doping is a techniques which focusses on deliberately introducing impurities to the core structure of a photocatalyst is known to be an adequate approach to tune the band gap of g-C₃N₄, which appreciably broadens the light absorption range and reduces the photogenerated charge recombination rate too [24]. Doping has been carried out either by interstitial or substitutional to modify texture and electronic structure of g-C₃N₄ for enhancing the photocatalytic activity [25].

3.2   Metal/non-metal based doping

Doping with metals and non-metals has proven to be an effective approach to increase photo-activity in visible spectrum. Doping not only narrows the band gap but efficiently increases the active sites on the surface of host material. Various metal based doping experiments have been carried out to increase photocatalytic performance of g-C_3N_4. Table 1 & 2 refer to the scientific work carried out in recent years concerning the doping of g-C_3N_4 for enhanced activity.

Table 1*: Metal based doping in g-C₃N₄ along with efficiency.*

Sr.no	Dopant	Performance	Ref
1.	Fe	Apparent quantum efficiency of 78.5% for Photocatalytic water splitting	[26]
2	Co	Exhibits 11 times in rhodamine B degradation than that of pure g-C_3N_4	[27]
3	Ti^{3+}	Exhibits 7.6 times higher degradation efficiency for methylene blue h	[28]
4	Fe	1.7 & 6 times higher efficiency for MB degradation and H_2 production respectively.	[29]
5	Mn	Photocatalytic activity for Cr(VI) reduction and RhB degradation increased to 76.5% & 8.9% respectively	[30]
6	K	Efficiency increased to 5.6 times for photocatalytic hydrogen evolution	[31]
7	Cu	High efficiency for H_2-Evolving Electro catalysis	[32]
8	Ag	Two times higher efficiency for photocatalytic oxidation of methylene blue	[33]
9	Zr	High efficiency for the RhB degradation	[34]
10	Y	Higher degradation efficiency for RhB degradation	[35]
11	Eu(III)	Degradation efficiency of RhB increased to 1.71 fold	[36]
12	Pd	15.3 times higher efficiency toward Hydrogen Evolution Reaction	[37]

Table 2*: Non-metal based doping ing-C₃N₄ for better photocatalytic efficiency.*

Sr.no	Dopant	Performance	Ref
1.	N	4.3 times higher efficiency for enhanced photocatalytic H_2 evolution	[38]
2	N	Enhanced efficiency for tetracycline degradation	[39]
3	S	2.5 fold higher efficiency for photocatalytic CO_2-reduction performance	[40]
4	S	Photoreduction of CO_2 to acetaldehyde and methane increases to 8.3 & 9 times	[41]
5	P	High photocatalytic efficiency for H_2 production	[42]
6	P	High efficiency for photocatalytic oxidation of aromatic alcohols	[43]
7	P	2.9 times higher efficient for photocatalytic activity for H_2 evolution and Rhodamine B degradation	[44]
8	O	5 times higher Photocatalytic CO_2 Reduction activity	[45]
9	O	2.5 times higher H_2 evolution photocatalytic activity	[46]
10	B	2.4 times higher efficiency for photocatalytic H_2 evolution	[47]
11	B	17 times higher photo degradation efficiency for MB dye	[48]
12	C-I	9.8 times higher efficiency for photocatalytic hydrogen evolution	[49]

Photocatalysis   Materials Research Forum LLC
Materials Research Foundations **100** (2021) 299-331   https://doi.org/10.21741/9781644901359-10

3.3 Fabrication of g-C₃N₄ based binary/ternary heterojunctions

$g\text{-}C_3N_4$ based binary/ternary heterojunctions offer propitious and worthwhile strategies to harness solar radiation which sufficiently overcomes the disadvantages associated with other modification strategies. Over the last decades, various $g\text{-}C_3N_4$ based binary/ternary heterojunctions have been successfully fabricated by keeping primary concern such as light response, appropriate lowering in band gap energies, reduced e^-/h^+ pair charge recombination, etc. [50].

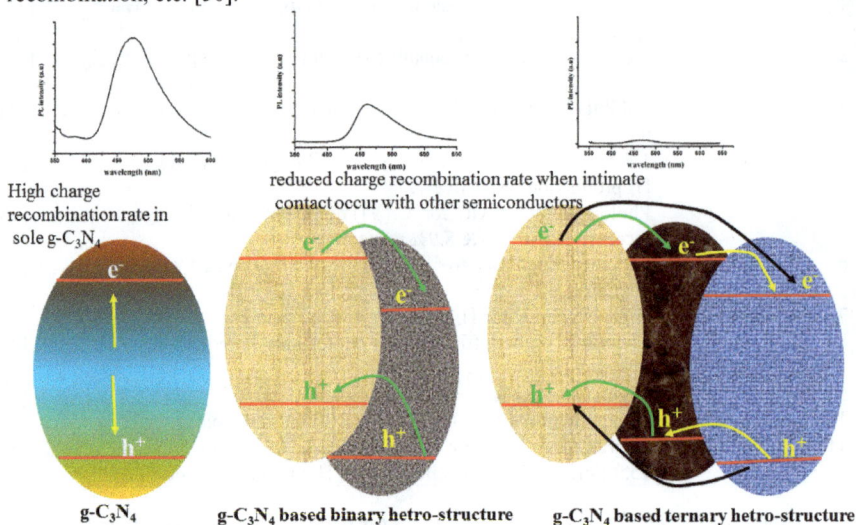

High charge recombination rate in sole g-C₃N₄

reduced charge recombination rate when intimate contact occur with other semiconductors

g-C₃N₄   g-C₃N₄ based binary hetro-structure   g-C₃N₄ based ternary hetro-structure

Figure 5: Band structure for the g-C₃N₄ based binary and ternary hetero-structures.

Fig.5 refers to the band structure and effect on photoluminescence when junction formation takes place between $g\text{-}C_3N_4$ and other semiconductors in various based binary and ternary hetero-structures. However, narrowing of band gap for $g\text{-}C_3N_4$ is not the sole concern to boost its photo activity, no doubt that narrow band gap semiconductors can absorb visible light more efficiently but additionally they suffer from the complications such as high rate of charge recombination, photo corrosion, low quantum yield etc. All these disadvantages are mainly associated with single-component photocatalysts which can be tackled by fabricating the heterojunction of different semiconductors which not only improves the photo-activity but also increase the surface area and life time of photo induced charge carriers [51,52].

Materials Research Forum LLC
https://doi.org/10.21741/9781644901359-10

Table 3: *Binary coupled modification of g-C₃N₄ along with utilization and efficiency.*

Sr no	Binary coupled composite	Enhanced photocatalytic activity evaluated by	Efficiency	Ref
1	WO_3/g-C_3N_4	RhB degradation	91.3% degradation after 150 min	[59]
2	$LaNiO_3$/g-C_3N_4	Tetracycline (TC) degradation.	3.9 and 3.8 times larger than those of sole components respectively	[60]
3	g–C_3N_4/$BiVO_4$	RhB degradation	2.3 times higher than that of the pure g–C_3N_4	[50]
4	(SiC)/g-C_3N_4	Degradation of dyes rhodamine B (RhB) and methyl orange (MO)	1.2 time higher than sole g–C_3N_4	[61]
5	g-C_3N_4/gC_3N_4	Removal of NO at ppb levels	47.6% high as compared to host substrates	[62]
6	$CoTiO_3$/g-C_3N_4	Photocatalytic H_2-evolution	~ 2 times higher than that of the pure g-C_3N_4 under artificial sunlight illumination	[63]
7	$Bi_2O_2CO_3$/g-C_3N_4	Photocatalytic oxidation removal of NO at ppb levels	15.4 times higher rate for NO removal	[64]
8	TiO_2/g-C_3N_4	Potocatalytic reduction of CO_2 and photocatalytic decomposition of N_2O	57% N_2O conversion in 16 h	[65]
9	g-C_3N_4/$Bi_2O_2CO_3$	rhodamine B degradation	3.1 times higher than that of individual C_3N_4	[66]
10	Co_3O_4-g-C_3N_4	degradation of methyl orange	7.5-folds higher than that of pure g-C_3N_4	[67]
11	$BiOCl$–C_3N_4	degradation of methyl orange (MO)	36.2 times higher than sole g-C_3N_4	[68]
12	g-C_3N_4/Bi_2MoO_6	Degradation of Rhodamine B	3 times higher those of pure g-C_3N_4	[69]
13	ZnS/g-C_3N_4	photocatalytic H_2evolution	30 times higher photocatalytic H_2 evolution rate than that of pure g-C_3N_4	[70]
14	α-Fe_2O_3/g-C_3N_4	Photocatalytic CO_2 conversion	2.2 times higher	[56]
15	Ag_2O/g-C_3N_4	Degradation of Rhodamine B (RhB), phenol, imidacloprid, methylene blue (MB) and methyl orange (MO) under visible and NIR light irradiation	RhB degradation is about 26 and 343 times higher than that of pure g-C_3N_4 in visible and NIR respectively	[71]

Table 4: *Ternary coupled modification of g-C$_3$N$_4$ along with utilization and efficiency.*

Sr no	Ternary coupled composite	Enhanced photocatalytic activity evaluated by	Efficiency	Ref
1	g-C$_3$N$_4$/ZnO@α-Fe$_2$O$_3$	Photodegradation of tartrazine	6.8 folds	[72]
2	g-C$_3$N$_4$@Bi/BiOBr	Photo degradation of tetracycline and rhodamine B	Highest efficiency 98%.	[73]
3	C$_3$N$_4$/ZnS/SnS$_2$	Photodegradation of methylene blue (MB)	8.74 times higher	[74]
4	BiOCl/ CdS/g-C$_3$N$_4$	Degradation of rhodamine B (RhB)	12 times higher compared to bared g-C$_3$N$_4$	[75]
5	g-C$_3$N$_4$/Fe$_3$O$_4$/MnWO$_4$	Rhodamine B, methylene blue, methyl orange, and fuchsine removal	7, 10, 25, and 31 time as compared to g-C$_3$N$_4$ respectively	[76]
6	g-C$_3$N$_4$/Ag$_3$VO$_4$/AgBr	RhB degradation	116 times higher	[77]
7	g-C$_3$N$_4$/Ag$_3$PO$_4$/Ag$_2$MoO$_4$	Solar oxygen evolution from water splitting	2.2 times higher	[78]
8	CdS/g-C$_3$N$_4$/CuS	photocatalytic H$_2$-production activity	5 times higher	[79]
9	g-C$_3$N$_4$/Ag/MoS$_2$	Organic pollutanr degradation and production of H$_2$	9.43 & 8.78 time higher for RhB degradation and H$_2$ production respectively	[80]
10	g-C$_3$N$_4$/CeO$_2$/ZnO	Methylene blue (MB) degradation.	11.46 times higher	[81]
11	ZnO-Ag$_2$O/pg-C$_3$N$_4$	Degradation of ciprofloxacin	3.83 times higher	[82]
12	C$_3$N$_4$/Fe$_3$O$_4$/BiOI	Degradation of RhB	21-fold higher	[83]
13	g-C$_3$N$_4$/Fe$_3$O$_4$/Ag$_2$CrO$_4$	Degradation of rhodamine B	6.3 folds higher	[84]
14	g-C$_3$N$_4$/Fe$_3$O$_4$/AgI	Degradation of RhB	8.7 times higher	[85]
15	g-C$_3$N$_4$/Fe$_3$O$_4$/Ag$_3$VO$_4$	Degradation of RhB	14 fold higher	[86]

Various strategies has been approached for fabrication of g-C$_3$N$_4$ based binary/ternary heterojunctions such as hydrothermal/solvothermal, electrodeposition process, chemical bath deposition, sol-gel process, chemical precipitation, self-assembly method etc. [53]. Over the last decades, g-C$_3$N$_4$ based binary/ternary heterojunction are being extensively used in the various fields such as photodegradation and mineralization of noxious pollutants [54], water splitting into H$_2$ [55] CO$_2$ conversion into fuels [56], photo-treatment of some harmful metals Cr(VI), As(III), U(VI) [57,58] etc. Table 3 and 4 referred

Photocatalysis   Materials Research Forum LLC
Materials Research Foundations **100** (2021) 299-331   https://doi.org/10.21741/9781644901359-10

previously fabricated binary and ternary coupled heterojunction utilized in the various filed to evaluate their photoactive potential.

Very recently Kumar et al., 2018, fabricated various g-C$_3$N$_4$ based nano composite such as biochar supported g-C$_3$N$_4$/FeVO$_4$ [54], coal char/g-C$_3$N$_4$/RGO [87] for various environmental applications etc. fig. 6 a & b. referred the XRD and photoluminescence spectra for the g-C$_3$N$_4$/FeVO$_4$. The intimate contact between both of semiconductors and formation of hetro-junction has been shown by XRD result. Furthermore, the PL spectra (referred in fig. 6 b) indicated the reduced charge recombination rate confirmed by the absence of any sharp peak in the composite part.

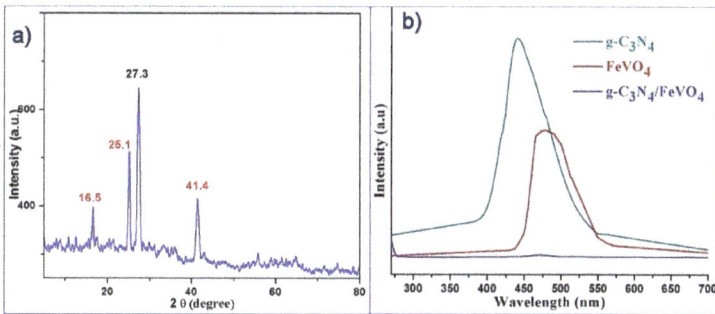

Figure 6: (a) XRD, and (b) Photoluminescence of g-C$_3$N$_4$/FeVO$_4$ (Reproduced with permission from [54] copyright Elsevier).

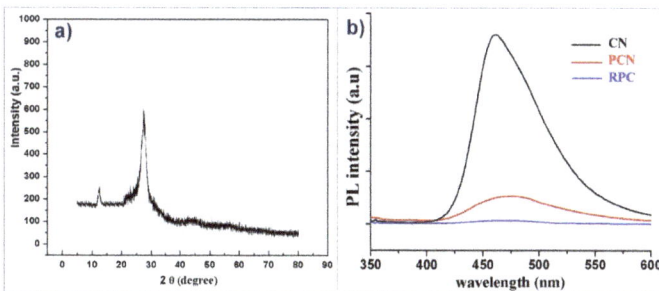

Figure 7: (a) XRD, and (b) Photoluminescence of RPC (Reproduced with permission from [87] copyright RSC).

Fig. 7 a & b refers to the XRD spectra of RPC (coal char/g-C_3N_4/RGO) as fabricated by Kumar et al. The XRD confirms the junction formation between the reduced graphene oxide and g-C_3N_4 on the coal char surface. The increased intensity at 27.2° for (002) graphitic plane observed due to increased staking of conjugation present in aromatic system of RGO and g-C_3N_4 [88]. The PL spectrum of RPC shows absence of sharp peak owing to successful formation of heterojunction between RPC & g-C_3N_4.

3.4   Coupling with metal-free substrates

Introduction of metal free substrate is a novel strategy employed by various researchers owing to their cost effective, ecofriendly nature and abundance of precursor materials. Modification of g-C_3N_4 with carbonaceous materials based motif has proven to be an efficient approach to alter its optical properties [89]. However g-C_3N_4 possesses relatively negative conduction band potential that can able to reduce many pollutants, such as Cr (VI), CO_2, and organic dyes but shows lower oxidation and mineralization activities [90,91].

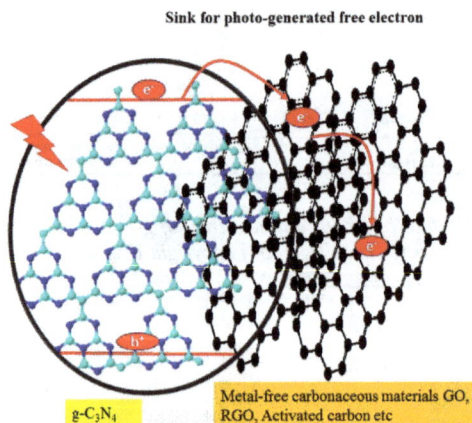

Figure 8: Carbonaceous material as electron sink.

Incorporation of metal free components such as GO, RGO, polyaniline and activated carbon may enhance the conductivity, strength and surface area of as fabricated g-C_3N_4 based photo catalysts [92].

Recently Chen et al., 2016 [93], has synthesized g-C_3N_4/activated carbon composite (g-C_3N_4/AC) by in-situ thermal condensation method. The activity of fabricated composite was evaluated by photodegrading phenol under the visible light radiation. A complete removal of phenol was achieved after 160 min of photodegrading experiment. Fig. 8 shows the route for charge transfer from g-C_3N_4 surface to the carbonaceous material such as RGO, GO etc. under photoexcitation. Some recent metal-free modification that was done on g-C_3N_4 by various researchers has been referred in table 5.

Table 5: Coupling with Metal-free component modified g-C_3N_4 along with utilization and efficiency.

Sr. no	Metal-free photocatalysts	Enhanced photocatalytic activity evaluated by	Efficiency	ref
1	g-C_3N_4/rGO	Degradation of rhodamine B (RhB) and 4-nitrophenol under visible light irradiation	3.0 and 2.7 times higher respectively	[88]
2	g-C_3N_4/rGO	Photocatalytic degradation of rhodamine B	2.6 times that of pristine g-C_3N_4	[94]
3	2D RGO/pCN	Photocatalytic reduction of carbon dioxide to methane	5.4 folds higher	[95]
4	rGO/ g-C_3N_4	Photocatalytic H2-evolution performance	-------	[96]
5	PANI/g- g-C_3N_4	Degradation of methylene blue	5.1-fold higher than that of pure g-C3N4	[97]
6	g-PAN/g-C_3N_4	Photocatalytic H2-evolution	3.8 times over pristine g-C3N4	[98]
7	g-C_3N_4/Biochar	Removal of cationic dye methylene blue (MB)	--------	[99]
8	g-C_3N_4/GO	Degradation of MO and MB and photo reduction of bromate	--------	[100]
9	SiO_2/g-C_3N_4	Degradation of rhodamine B	3.5 times higher	[101]
10	Red phosphor/ g-C_3N_4	Photocatalytic activity for H_2 production and CO_2 conversion	10 timers higher	[102]
11	C-Dots/ g-C_3N_4	Photocatalytic hydrogen H_2 production and RhB degradation	3 times higher	[103]
12	CNNS/CQDs	Photocatalytic generation of hydrogen	3 times higher	[104]

3.5   Chemical modification

Chemical modification of g-C_3N_4 surface is another technique to improve its photocatalytic activity. Introduction of some groups on the surface of g-C_3N_4 increase the active sites which further improves the interaction with the employed substrate. Dong G et al. 2016 [105] use perylene imides modified g-C_3N_4 for the removal of noxious NO and they found this composite 5 times efficient for the photo generation of H_2O_2 as compared to sole g-C_3N_4.

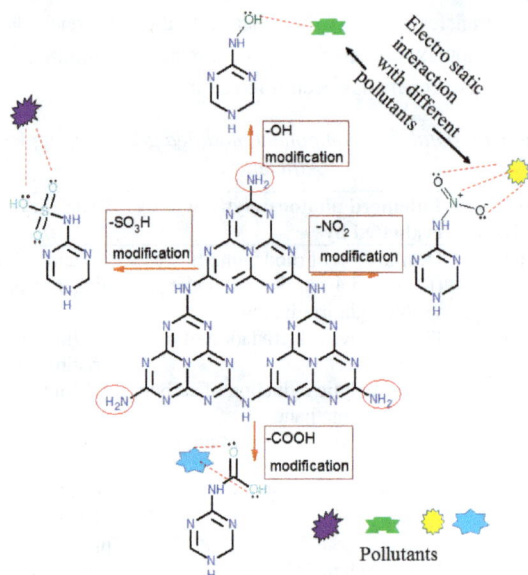

Figure 9: Possible Chemical modification for g-C₃N₄.

Further, Wu X et al. [106] fabricated oxygen-containing groups-modified g-C_3N_4 and the efficiency is evaluated by photocatalytic H_2-evolution performance. Dramatically, they found the efficiency was increased to a factor of 7 as compared to sole g-C_3N_4. Furthermore, the surface decoration with some functional groups such as –OH, -COOH, -SO_3H, -NO_3 etc. may become effective strategies to improve adsorption and photocatalytic activity of pristine g-C_3N_4. Because such groups not only alter the overall charge but also increase the static interaction between the surface and pollutants. Li et al. 2014 has fabricated hydroxyl modified Al_2O_3/g-C_3N_4.

The surface of g-C_3N_4 was modified by dipping in ammonia solution for 5h. The findings has revealed that the –OH modified g-C_3N_4 based composite is 2.4 times more efficient than unmodified, for the photo-degradation of RhB. Fig. 9 refers to the different possible chemical modification on the surface of g-C_3N_4. However, a very limited research has been carried out for the chemical modification of g-C_3N_4 but results depicts that such modification may route for new findings.

3.6   Dye Sensitization of g-C_3N_4

Sensitization is another popular technique to improve the photo-activity of wide band gap semiconductors. Sensitization may be carried by natural or artificial dyes. A sensitizer has ability to convert long wavelength of light to energy for the photocatalytic operation. Various dyes have been used as sensitizer to improve charge transfer mechanism and harvest more amount of solar spectrum. The charge transfer mechanism in dye sensitization is based on position of highest occupied molecular orbital (HOMO) and lowest unoccupied molecular orbital (LUMO) of the dye. When dye sensitized C_3N_4 is subjected to irradiation, the dye gets excited and transfers its charge to CB and VB of C_3N_4. As a consequence, absorption intensity of solar spectrum gets boosted. The sensitized g-C_3N_4 has attracted much attention for DSSC application. Wang et al. 2013 [107] has fabricated g-C_3N_4 nano sheets sensitized with dye Erythrosin B and achieved enhanced activity for the H_2 evolution due to sufficient charge transfer from Erythrosin B to g-C_3N_4 surface.

The exploration of natural sensitizer to sensitize g-C_3N_4 is a greener approach because it implies the use of pigments such as anthocyanin, chlorophyll various flavonoids etc. as referred in fig. 10. These pigments are abundantly found in various plant leaves, fruits and flower and can be extracted by means of simple laboratory scale extraction procedure without involving any harsh chemical use. The modification of g-C_3N_4 with natural sensitizers may be an alternate method that can replace the use of other conventional synthetic sensitizers. Natural sensitizers are basically pigments which consist of auxochrome and chromophores. When such modification is applied to g-C_3N_4, they provided large conjugation and various functionalities which not only inhibit charge recombination but also boost solar light harvesting properties. The main advantages of such photosensitizer are that they enable the large band gap semiconductors to absorb light in the visible region. Buddee et al., 2014, have studied curcumin sensitized TiO_2 nanoparticles and found it highly effective for the cationic azo dye MB and CV photocatalytic removal [108]. Similarly Zyoud et al., 2011, fabricated Anthocyanin sensitize TiO_2 for the photodegradation of methyl orange [109].

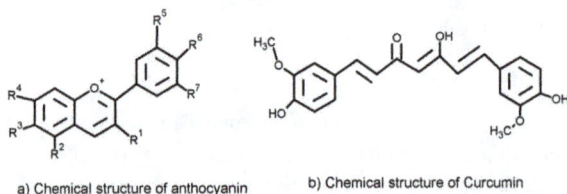

a) Chemical structure of anthocyanin

b) Chemical structure of Curcumin

Figure 10: Chemical structure of natural pigments anthocyanin and curcumin.

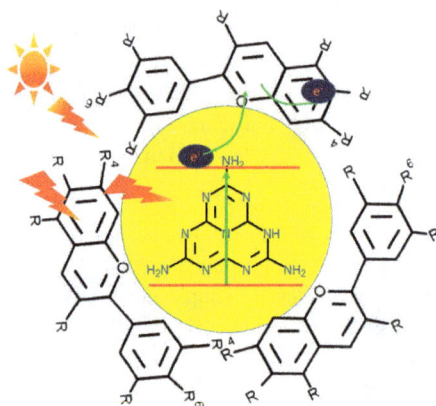

Figure 11: Hypothetical charge transfer between g-C₃N₄ and natural dye molecule.

Fig. 11 refers to the pictorial charge transfer between dye molecule and $g\text{-}C_3N_4$. However very limited work has been done for the sensitization of $g\text{-}C_3N_4$ with natural dye/pigment molecules but this area required more research to explore the use of natural sensitizer in the photocatalytic processes. Table 6 refers to some of previously reported sensitization on $g\text{-}C_3N_4$.

***Table 6**: Dye-sensitized modified g-C₃N₄ along with utilization and efficiency.*

Sr no	Composite	Dye used as sensitizer	Enhanced photocatalytic activity evaluated by	Ref
1	MgPc/Pt/mpg- g-C₃N₄	Magnesium phthalocyanine	Photocatalytic hydrogen evolution	[110]
2	Cu-Cu2O/g- g-C₃N₄	Erythrosin B	Hydrogen evolution	[106]
3	Xanthene Dye-Sensitized Carbon Nitride	Fluorescein, dibromofluorescein, eosin Y, and erythrosine B	-------	[111]
4	g-C₃N₄	Eosin Y	Hydrogen evolution from aqueous triethanolamine	[112]
5	mpg- g-C₃N₄	Eosin Y	Photocatalytic activity for H₂ evolution	[113]

3.7 Shape specific modification

It has been well understood that the shape of a photocatalysts demonstrate a key parameter towards the photocatalytic performance. 2-D nano-sheets are the common morphological shapes of g-C₃N₄ which were fabricated by many researchers [114,115]. Moreover, nanorods [116], nano spheres [117] and flower [118] like shape controlled morphologies also has been achieved for the g-C₃N₄.

Bai et al. 2013 [119] prepared nanorods and compared the photocatalytic properties with nanoplates of C₃N₄ as synthesized by reflux method. They found that photocurrent for nanorods is approximately 2 times higher and more efficient toward the degradation of methylene blue as compared to nanoplates.

Meng et al., 2019 [120], has demonstrated the fabrication of double shell hollow g-C₃N₄.nanosphere. They had found that as fabricated nanospheres possesses outstanding optical and photoelectrical properties. The nanospheres have been prepared by thermal annealing using hallow SiO₂ as template. The activity of nanospheres has been utilized for the CO₂ conversion to formic acid.

Zheng et al., 2014 [121], have demonstrated the fabrication of helical g-C₃N₄ by nanocasting method by using chiral mesoporous silica (CMS) template. They reported that the light absorption and band gap energies are sufficiently altered for the helical structure of g-C₃N₄ as compared to bulk g-C₃N₄. They had evaluated the photocatalytic activities of helical g-C₃N₄.by hydrogen evolution, water oxidation, and CO₂-to-CO conversion. They

Materials Research Forum LLC

https://doi.org/10.21741/9781644901359-10

found that hydrogen evolution rate is 7 folds high for helical g-C_3N_4 as compared to bulk g-C_3N_4.

3.8   Polymerization of g-C_3N_4

Polymerization of g-C_3N_4 is an important modification approach that has been adopted for the g-C_3N_4 which not only increases the specific surface area but also narrows the band gap energies [122]. Ding et al., 2018 [123], have demonstrated the fabrication of polymeric g-C_3N_4 by the thermal condensation of nitric acid pretreated melamine as the precursor material. The polymeric C_3N_4 was utilized for the photocatalytic oxidation of benzyl alcohol and they reported 68.3% conversion with 100% selectivity as compared to pure g-C_3N_4.

3.9   Persulphate activation of g-C_3N_4

The addition of some precursor materials such as H_2O_2, $S_2O_8^{2-}$, O_3, etc. when added with photocatalysts can sufficiently trigger the generation of highly reactive radicals such as OH^* and O_2^{*-}. Activation of g-C_3N_4 by such agents offers a cost effective and highly efficient method for the degradation and mineralization of pollutants. Liu et al., 2017 [124], have studied the degradation of Bisphenol-A by Persulphate/g-C_3N_4 and reported approximately 100% removal at a concentration of 5 mM of Persulphate. Hu et al. 2016 [125] have studied g-C_3N_4/Fe(III)/persulfate system for the photodegradation of phenol and found 240-folds high efficiency. The mechanism supposed to occur during the persulphate activation in the presence of g-C_3N_4 is shown in scheme 1. The photo excited electrons react with persulphate to produce sulfate ion and sulfate radicals which further react with water molecules present in the reaction system to generate OH^* radical. The OH^* radical once formed, undergoes series of photochemical reactions and destroys the targeted pollutant molecules.

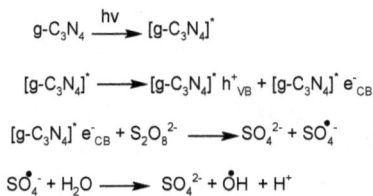

$$g\text{-}C_3N_4 \xrightarrow{h\nu} [g\text{-}C_3N_4]^*$$

$$[g\text{-}C_3N_4]^* \longrightarrow [g\text{-}C_3N_4]^* \, h^+_{VB} + [g\text{-}C_3N_4]^* \, e^-_{CB}$$

$$[g\text{-}C_3N_4]^* \, e^-_{CB} + S_2O_8^{2-} \longrightarrow SO_4^{2-} + SO_4^{\bullet-}$$

$$SO_4^{\bullet-} + H_2O \longrightarrow SO_4^{2-} + \overset{\bullet}{O}H + H^+$$

Scheme 1: *Interaction of g-C_3N_4 with persulphate.*

Photocatalysis   Materials Research Forum LLC
Materials Research Foundations **100** (2021) 299-331   https://doi.org/10.21741/9781644901359-10

4.   Further possible modification perspectives

Various modification strategies have demonstrated that photo-activity of g-C$_3$N$_4$ has been greatly improved. g-C$_3$N$_4$ has shown enhanced efficacy for various applications such as photocatalytic hydrogen evolution, pollutant degradation and mineralization etc.

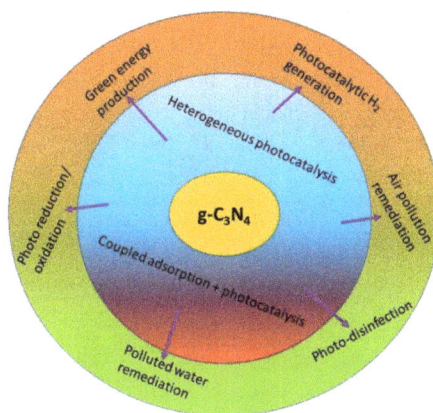

Figure 12: Various photocatalytic operations carried by g-C$_3$N$_4$.

But still there are some unexplored research fields for which g-C$_3$N$_4$ has seldom been explored such as biomedical application, catalytic cracking application for petroleum industries, etc. Cost effectiveness and abundance of its cheaper precursor material can make it as demanding material if we explore it for further modification by keeping the field specific objective in the mind. The wide mode of photocatalytic operations which has been performed by g-C$_3$N$_4$ has been shown in Fig. 12.

Conclusion

The chapter summarizes the various strategies adopted in the present decade for the modification of g-C$_3$N$_4$ to boost its photocatalytic activity. Doping, coupling with binary/ternary semiconductor system, shape specific modification, polymerization, etc. are some approaches that have been explored to augment the photocatalytic potential of g-C$_3$N$_4$. Moreover, all the approaches which are discussed for modification have proven to significantly enhance the activity of g-C$_3$N$_4$ for various photocatalytic targets such as pollutant mineralization, water splitting, CO$_2$ conversion, fuel energy application, etc. g-C$_3$N$_4$ possesses high potential for its exploration in various research fields by undergoing many modifications to its 2-D motif.

References

[1]	A. Kumar, G. Sharma, M. Naushad, A.H. Al-Muhtaseb, A. García-Peñas, G.T. Mola, C. Si, F.J. Stadler, Bio-inspired and biomaterials-based hybrid photocatalysts for environmental detoxification: A review, Chem. Eng. J. 382 (2020) 122937. https://doi.org/10.1016/j.cej.2019.122937

[2]	A. Kumar, G. Sharma, M. Naushad, A.H. Al-Muhtaseb, A. Kumar, I. Hira, T. Ahamad, A.A. Ghfar, F.J. Stadler, Visible photodegradation of ibuprofen and 2,4-D in simulated waste water using sustainable metal free-hybrids based on carbon nitride and biochar, J. Environ. Manage. 231 (2019) 1164–1175. https://doi.org/10.1016/j.jenvman.2018.11.015

[3]	A. Kumar, S.K. Sharma, G. Sharma, M. Naushad, F.J. Stadler, CeO$_2$/g-C$_3$N$_4$/V$_2$O$_5$ ternary nano hetero-structures decorated with CQDs for enhanced photo-reduction capabilities under different light sources: Dual Z-scheme mechanism, J. Alloys Compd. 838 (2020) 155692. https://doi.org/10.1016/j.jallcom.2020.155692

[4]	A. Kumar, G. Sharma, A. Kumari, C. Guo, M. Naushad, D.-V.N. Vo, J. Iqbal, F.J. Stadler, Construction of dual Z-scheme g-C$_3$N$_4$/Bi$_4$Ti$_3$O$_{12}$/Bi$_4$O$_5$I$_2$ heterojunction for visible and solar powered coupled photocatalytic antibiotic degradation and hydrogen production: Boosting via I$^-$/I$_3^-$ and Bi^{3+}/Bi^{5+} redox mediators, Appl. Catal. B Environ. (2020) 119808. https://doi.org/10.1016/j.apcatb.2020.119808

[5]	A. Kumar, S.K. Sharma, G. Sharma, C. Guo, D.V.N. Vo, J. Iqbal, M. Naushad, F.J. Stadler, Silicate glass matrix@Cu$_2$O/Cu$_2$V$_2$O$_7$ p-n heterojunction for enhanced visible light photo-degradation of sulfamethoxazole: High charge separation and interfacial transfer, J. Hazard. Mater. 402 (2021) 123790. https://doi.org/10.1016/j.jhazmat.2020.123790

[6]	R. Ravindranath, P. Roy, A.P. Periasamy, H.T. Chang, Effects of deposited ions on the photocatalytic activity of TiO$_2$-Au nanospheres, RSC Adv. 4 (2014) 57290–57296. https://doi.org/10.1039/C4RA10192H

[7]	D. Ma, J. Wu, M. Gao, Y. Xin, T. Ma, Y. Sun, Fabrication of Z-scheme g-C$_3$N$_4$/RGO/Bi$_2$WO$_6$ photocatalyst with enhanced visible-light photocatalytic activity, Chem. Eng. J. 290 (2016) 136–146. https://doi.org/10.1016/j.cej.2016.01.031

[8]	L.K. Putri, L.L. Tan, W.J. Ong, W.S. Chang, S.P. Chai, Graphene oxide: Exploiting its unique properties toward visible-light-driven photocatalysis, Appl. Mater. Today. 4 (2016) 9–16. https://doi.org/10.1016/j.apmt.2016.04.001

Photocatalysis   Materials Research Forum LLC
Materials Research Foundations **100** (2021) 299-331   https://doi.org/10.21741/9781644901359-10

[9]   T. Wang, X. Liu, C. Ma, Z. Zhu, Y. Liu, Z. Liu, M. Wei, X. Zhao, H. Dong, P. Huo, C. Li, Y. Yan, Bamboo prepared carbon quantum dots (CQDs) for enhancing $Bi_3Ti_4O_{12}$ nanosheets photocatalytic activity, J. Alloys Compd. 752 (2018) 106–114. https://doi.org/10.1016/j.jallcom.2018.04.085

[10]   R. Saravanan, E. Sacari, F. Gracia, M.M. Khan, E. Mosquera, V.K. Gupta, Conducting PANI stimulated ZnO system for visible light photocatalytic degradation of coloured dyes, J. Mol. Liq. 221 (2016) 1029–1033. https://doi.org/10.1016/j.molliq.2016.06.074

[11]   W. Zhao, Y. Guo, S. Wang, H. He, C. Sun, S. Yang, A novel ternary plasmonic photocatalyst: Ultrathin g-C3N4 nanosheet hybrided by $Ag/AgVO_3$ nanoribbons with enhanced visible-light photocatalytic performance, Appl. Catal. B Environ. 165 (2015) 335–343. https://doi.org/10.1016/j.apcatb.2014.10.016

[12]   Y. Gong, C. Fu, G. Zhang, H. Zhou, Y. Kuang, Three-dimensional Porous C_3N_4 Nanosheets@Reduced Graphene Oxide Network as Sulfur Hosts for High Performance Lithium-Sulfur Batteries, Electrochim. Acta. 256 (2017) 1–9. https://doi.org/10.1016/j.electacta.2017.10.032

[13]   J. Yang, X. Wu, X. Li, Y. Liu, M. Gao, X. Liu, L. Kong, S. Yang, Synthesis and characterization of nitrogen-rich carbon nitride nanobelts by pyrolysis of melamine, Appl. Phys. A Mater. Sci. Process. 105 (2011) 161–166. https://doi.org/10.1007/s00339-011-6471-4

[14]   E.G. Gillan, Synthesis of nitrogen-rich carbon nitride networks from an energetic molecular azide precursor, Chem. Mater. 12 (2000) 3906–3912. https://doi.org/10.1021/cm000570y

[15]   J.P. Zou, L.C. Wang, J. Luo, Y.C. Nie, Q.J. Xing, X.B. Luo, H.M. Du, S.L. Luo, S.L. Suib, Synthesis and efficient visible light photocatalytic H_2 evolution of a metal-free g-C3N4/graphene quantum dots hybrid photocatalyst, Appl. Catal. B Environ. 193 (2016) 103–109. https://doi.org/10.1016/j.apcatb.2016.04.017

[16]   X. Ma, Y. Lv, J. Xu, Y. Liu, R. Zhang, Y. Zhu, A strategy of enhancing the photoactivity of g-C3N4 via doping of nonmetal elements: A first-principles study, J. Phys. Chem. C. 116 (2012) 23485–23493. https://doi.org/10.1021/jp308334x

[17]   G. Fanchini, A. Tagliaferro, N.M.J. Conway, C. Godet, Role of lone-pair interactions and local disorder in determining the interdependency of optical constants of (formula presented) thin films, Phys. Rev. B - Condens. Matter Mater. Phys. 66 (2002) 1–9. https://doi.org/10.1103/PhysRevB.66.195415

[18]   J. Zhang, X. Chen, K. Takanabe, K. Maeda, K. Domen, J.D. Epping, X. Fu, M. Antonieta, X. Wang, Synthesis of a carbon nitride structure for visible-light catalysis by copolymerization, Angew. Chemie - Int. Ed. 49 (2010) 441–444. https://doi.org/10.1002/anie.200903886

[19]   Q. Wei, X. Yan, Z. Kang, Z. Zhang, S. Cao, Y. Liu, Y. Zhang, Carbon Quantum Dots Decorated C_3N_4/TiO_2 Heterostructure Nanorod Arrays for Enhanced Photoelectrochemical Performance , J. Electrochem. Soc. 164 (2017) H515–H520. https://doi.org/10.1149/2.1281707jes

[20]   P. Niu, L. Zhang, G. Liu, H.M. Cheng, Graphene-like carbon nitride nanosheets for improved photocatalytic activities, Adv. Funct. Mater. 22 (2012) 4763–4770. https://doi.org/10.1002/adfm.201200922

[21]   S. Yang, Y. Gong, J. Zhang, L. Zhan, L. Ma, Z. Fang, R. Vajtai, X. Wang, P.M. Ajayan, Exfoliated graphitic carbon nitride nanosheets as efficient catalysts for hydrogen evolution under visible light, Adv. Mater. 25 (2013) 2452–2456. https://doi.org/10.1002/adma.201204453

[22]   J. Feng, T. Chen, S. Liu, Q. Zhou, Y. Ren, Y. Lv, Z. Fan, Improvement of $g-C_3N_4$ photocatalytic properties using the Hummers method, J. Colloid Interface Sci. 479 (2016) 1–6. https://doi.org/10.1016/j.jcis.2016.06.040

[23]   M. Faisal, A.A. Ismail, F.A. Harraz, S.A. Al-Sayari, A.M. El-Toni, A.E. Al-Salami, M.S. Al-Assiri, Fabrication of highly efficient TiO_2/C_3N_4 visible light driven photocatalysts with enhanced photocatalytic activity, J. Mol. Struct. 1173 (2018) 428–438. https://doi.org/10.1016/j.molstruc.2018.07.014

[24]   M.R.D. Khaki, M.S. Shafeeyan, A.A.A. Raman, W.M.A.W. Daud, Application of doped photocatalysts for organic pollutant degradation - A review, J. Environ. Manage. 198 (2017) 78–94. https://doi.org/10.1016/j.jenvman.2017.04.099

[25]   Y.P. Zhu, T.Z. Ren, Z.Y. Yuan, Mesoporous Phosphorus-Doped $g-C_3N_4$ Nanostructured Flowers with Superior Photocatalytic Hydrogen Evolution Performance, ACS Appl. Mater. Interfaces. 7 (2015) 16850–16856. https://doi.org/10.1021/acsami.5b04947

[26]   Z. Li, C. Kong, G. Lu, Visible Photocatalytic Water Splitting and Photocatalytic Two-Electron Oxygen Formation over Cu- and Fe-Doped $g-C_3N_4$, J. Phys. Chem. C. 120 (2016) 56–63. https://doi.org/10.1021/acs.jpcc.5b09469

[27] J. Mu, J. Li, X. Zhao, E.C. Yang, X.J. Zhao, Cobalt-doped graphitic carbon nitride with enhanced peroxidase-like activity for wastewater treatment, RSC Adv. 6 (2016) 35568–35576. https://doi.org/10.1039/C6RA02911F

[28] K. Li, S. Gao, Q. Wang, H. Xu, Z. Wang, B. Huang, Y. Dai, J. Lu, In-situ-reduced synthesis of Ti^{3+} self-doped TiO_2/g-C_3N_4 heterojunctions with high photocatalytic performance under LED light irradiation, ACS Appl. Mater. Interfaces. 7 (2015) 9023–9030. https://doi.org/10.1021/am508505n

[29] J. Gao, Y. Wang, S. Zhou, W. Lin, Y. Kong, A Facile One-Step Synthesis of Fe-Doped g-C_3N_4 Nanosheets and Their Improved Visible-Light Photocatalytic Performance, ChemCatChem. 9 (2017) 1708–1715. https://doi.org/10.1002/cctc.201700492

[30] J.C. Wang, C.X. Cui, Y. Li, L. Liu, Y.P. Zhang, W. Shi, Porous Mn doped g-C_3N_4 photocatalysts for enhanced synergetic degradation under visible-light illumination, J. Hazard. Mater. 339 (2017) 43–53. https://doi.org/10.1016/j.jhazmat.2017.06.011

[31] Y. Wang, S. Zhao, Y. Zhang, J. Fang, Y. Zhou, S. Yuan, C. Zhang, W. Chen, One-pot synthesis of K-doped g-C_3N_4 nanosheets with enhanced photocatalytic hydrogen production under visible-light irradiation, Appl. Surf. Sci. 440 (2018) 258–265. https://doi.org/10.1016/j.apsusc.2018.01.091

[32] X. Zou, R. Silva, A. Goswami, T. Asefa, Cu-doped carbon nitride: Bio-inspired synthesis of H_2 -evolving electrocatalysts using graphitic carbon nitride (g-C_3N_4) as a host material, Appl. Surf. Sci. 357 (2015) 221–228. https://doi.org/10.1016/j.apsusc.2015.08.197

[33] M. Faisal, A.A. Ismail, F.A. Harraz, S.A. Al-Sayari, A.M. El-Toni, M.S. Al-Assiri, Synthesis of highly dispersed silver doped g-C_3N_4 nanocomposites with enhanced visible-light photocatalytic activity, Mater. Des. 98 (2016) 223–230. https://doi.org/10.1016/j.matdes.2016.03.019

[34] Y. Wang, Y. Wang, Y. Li, H. Shi, Y. Xu, H. Qin, X. Li, Y. Zuo, S. Kang, L. Cui, Simple synthesis of Zr-doped graphitic carbon nitride towards enhanced photocatalytic performance under simulated solar light irradiation, Catal. Commun. 72 (2015) 24–28. https://doi.org/10.1016/j.catcom.2015.08.022

[35] Y. Wang, Y. Li, X. Bai, Q. Cai, C. Liu, Y. Zuo, S. Kang, L. Cui, Facile synthesis of Y-doped graphitic carbon nitride with enhanced photocatalytic performance, Catal. Commun. 84 (2016) 179–182. https://doi.org/10.1016/j.catcom.2016.06.020

[36]   M. Wang, P. Guo, Y. Zhang, C. Lv, T. Liu, T. Chai, Y. Xie, Y. Wang, T. Zhu, Synthesis of hollow lantern-like Eu(III)-doped g-C_3N_4 with enhanced visible light photocatalytic perfomance for organic degradation, J. Hazard. Mater. 349 (2018) 224–233. https://doi.org/10.1016/j.jhazmat.2018.01.058

[37]   N. Wang, J. Wang, J. Hu, X. Lu, J. Sun, F. Shi, Z.H. Liu, Z. Lei, R. Jiang, Design of Palladium-Doped g-C_3N_4 for Enhanced Photocatalytic Activity toward Hydrogen Evolution Reaction, ACS Appl. Energy Mater. 1 (2018) 2866–2873. https://doi.org/10.1021/acsaem.8b00526

[38]   Y. Zhou, L. Zhang, W. Huang, Q. Kong, X. Fan, M. Wang, J. Shi, N-doped graphitic carbon-incorporated g-C_3N_4 for remarkably enhanced photocatalytic H_2 evolution under visible light, Carbon N. Y. 99 (2016) 111–117. https://doi.org/10.1016/j.carbon.2015.12.008

[39]   L. Jiang, X. Yuan, G. Zeng, J. Liang, Z. Wu, H. Yu, D. Mo, H. Wang, Z. Xiao, C. Zhou, Nitrogen self-doped g-C_3N_4 nanosheets with tunable band structures for enhanced photocatalytic tetracycline degradation, J. Colloid Interface Sci. 536 (2019) 17–29. https://doi.org/10.1016/j.jcis.2018.10.033

[40]   K. Wang, Q. Li, B. Liu, B. Cheng, W. Ho, J. Yu, Sulfur-doped g-C_3N_4 with enhanced photocatalytic CO_2-reduction performance, Appl. Catal. B Environ. 176–177 (2015) 44–52. https://doi.org/10.1016/j.apcatb.2015.03.045

[41]   N.D. Shcherban, S.M. Filonenko, M.L. Ovcharov, A.M. Mishura, M.A. Skoryk, A. Aho, D.Y. Murzin, Simple method for preparing of sulfur-doped graphitic carbon nitride with superior activity in CO_2 photoreduction, ChemistrySelect. 1 (2016) 4987–4993. https://doi.org/10.1002/slct.201601283

[42]   S. Guo, Y. Tang, Y. Xie, C. Tian, Q. Feng, W. Zhou, B. Jiang, P-doped tubular g-C_3N_4 with surface carbon defects: Universal synthesis and enhanced visible-light photocatalytic hydrogen production, Appl. Catal. B Environ. 218 (2017) 664–671. https://doi.org/10.1016/j.apcatb.2017.07.022

[43]   M. Bellardita, E.I. García-López, G. Marcì, I. Krivtsov, J.R. García, L. Palmisano, Selective photocatalytic oxidation of aromatic alcohols in water by using P-doped g-C_3N_4, Appl. Catal. B Environ. 220 (2018) 222–233. https://doi.org/10.1016/j.apcatb.2017.08.033

[44]   Y. Zhou, L. Zhang, J. Liu, X. Fan, B. Wang, M. Wang, W. Ren, J. Wang, M. Li, J. Shi, Brand new P-doped g-C_3N_4: Enhanced photocatalytic activity for H_2 evolution and Rhodamine B degradation under visible light, J. Mater. Chem. A. 3 (2015) 3862–3867. https://doi.org/10.1039/C4TA05292G

Materials Research Forum LLC
https://doi.org/10.21741/9781644901359-10

[45]   J. Fu, B. Zhu, C. Jiang, B. Cheng, W. You, J. Yu, Hierarchical Porous O-Doped g-C_3N_4 with Enhanced Photocatalytic CO_2 Reduction Activity, Small. 13 (2017). https://doi.org/10.1002/smll.201603938

[46]   J. Li, B. Shen, Z. Hong, B. Lin, B. Gao, Y. Chen, A facile approach to synthesize novel oxygen-doped g-C_3N_4 with superior visible-light photoreactivity, Chem. Commun. 48 (2012) 12017–12019. https://doi.org/10.1039/c2cc35862j

[47]   P. Chen, P. Xing, Z. Chen, H. Lin, Y. He, Rapid and energy-efficient preparation of boron doped g-C_3N_4 with excellent performance in photocatalytic H_2-evolution, Int. J. Hydrogen Energy. 43 (2018) 19984–19989. https://doi.org/10.1016/j.ijhydene.2018.09.078

[48]   Q. Yan, G.F. Huang, D.F. Li, M. Zhang, A.L. Pan, W.Q. Huang, Facile synthesis and superior photocatalytic and electrocatalytic performances of porous B-doped g-C_3N_4 nanosheets, J. Mater. Sci. Technol. 34 (2018) 2515–2520. https://doi.org/10.1016/j.jmst.2017.06.018

[49]   C. Yang, W. Teng, Y. Song, Y. Cui, C-I codoped porous g-C_3N_4 for superior photocatalytic hydrogen evolution, Cuihua Xuebao/Chinese J. Catal. 39 (2018) 1615–1624. https://doi.org/10.1016/S1872-2067(18)63131-6

[50]   Y. Ji, J. Cao, L. Jiang, Y. Zhang, Z. Yi, g-C_3N_4/$BiVO_4$ composites with enhanced and stable visible light photocatalytic activity, J. Alloys Compd. 590 (2014) 9–14. https://doi.org/10.1016/j.jallcom.2013.12.050

[51]   A. Kumar, A. Kumar, G. Sharma, A.H. Al-Muhtaseb, M. Naushad, A.A. Ghfar, F.J. Stadler, Quaternary magnetic BiOCl/g-C_3N_4/Cu_2O/Fe_3O_4 nano-junction for visible light and solar powered degradation of sulfamethoxazole from aqueous environment, Chem. Eng. J. 334 (2018) 462–478. https://doi.org/10.1016/j.cej.2017.10.049

[52]   A. Kumar, A. Kumar, G. Sharma, A.H. Al-Muhtaseb, M. Naushad, A.A. Ghfar, C. Guo, F.J. Stadler, Biochar-templated g-C_3N_4/$Bi_2O_2CO_3$/$CoFe_2O_4$ nano-assembly for visible and solar assisted photo-degradation of paraquat, nitrophenol reduction and CO_2 conversion, Chem. Eng. J. 339 (2018) 393–410. https://doi.org/10.1016/j.cej.2018.01.105

[53]   J.S. Jang, H.G. Kim, J.S. Lee, Heterojunction semiconductors: A strategy to develop efficient photocatalytic materials for visible light water splitting, in: Catal. Today, Elsevier, 2012: pp. 270–277. https://doi.org/10.1016/j.cattod.2011.07.008

[54]   A. Kumar, A. Kumar, G. Sharma, M. Naushad, F.J. Stadler, A.A. Ghfar, P. Dhiman, R. V. Saini, Sustainable nano-hybrids of magnetic biochar supported g-

C_3N_4/FeVO$_4$ for solar powered degradation of noxious pollutants- Synergism of adsorption, photocatalysis & photo-ozonation, J. Clean. Prod. 165 (2017) 431–451. https://doi.org/10.1016/j.jclepro.2017.07.117

[55] E. Liu, J. Chen, Y. Ma, J. Feng, J. Jia, J. Fan, X. Hu, Fabrication of 2D SnS$_2$/g-C$_3$N$_4$ heterojunction with enhanced H$_2$ evolution during photocatalytic water splitting, J. Colloid Interface Sci. 524 (2018) 313–324. https://doi.org/10.1016/j.jcis.2018.04.038

[56] Z. Jiang, W. Wan, H. Li, S. Yuan, H. Zhao, P.K. Wong, A Hierarchical Z-Scheme α-Fe$_2$O$_3$/g-C$_3$N$_4$ Hybrid for Enhanced Photocatalytic CO$_2$ Reduction, Adv. Mater. 30 (2018). https://doi.org/10.1002/adma.201706108

[57] Y. Zhang, Q. Zhang, Q. Shi, Z. Cai, Z. Yang, Acid-treated g-C$_3$N$_4$ with improved photocatalytic performance in the reduction of aqueous Cr(VI) under visible-light, Sep. Purif. Technol. 142 (2015) 251–257. https://doi.org/10.1016/j.seppur.2014.12.041

[58] X.H. Jiang, Q.J. Xing, X.B. Luo, F. Li, J.P. Zou, S.S. Liu, X. Li, X.K. Wang, Simultaneous photoreduction of Uranium(VI) and photooxidation of Arsenic(III) in aqueous solution over g-C$_3$N$_4$/TiO$_2$ heterostructured catalysts under simulated sunlight irradiation, Appl. Catal. B Environ. 228 (2018) 29–38. https://doi.org/10.1016/j.apcatb.2018.01.062

[59] F. Chang, J. Zheng, F. Wu, X. Wang, B. Deng, Binary composites WO$_3$/g-C$_3$N$_4$ in porous morphology: Facile construction, characterization, and reinforced visible light photocatalytic activity, Colloids Surfaces A Physicochem. Eng. Asp. 563 (2019) 11–21. https://doi.org/10.1016/j.colsurfa.2018.11.058

[60] X. Zhou, Y. Chen, C. Li, L. Zhang, X. Zhang, X. Ning, L. Zhan, J. Luo, Construction of LaNiO$_3$ nanoparticles modified g-C$_3$N$_4$ nanosheets for enhancing visible light photocatalytic activity towards tetracycline degradation, Sep. Purif. Technol. 211 (2019) 179–188. https://doi.org/10.1016/j.seppur.2018.09.075

[61] F. Chang, J. Zheng, X. Wang, Q. Xu, B. Deng, X. Hu, X. Liu, Heterojuncted non-metal binary composites silicon carbide/g-C$_3$N$_4$ with enhanced photocatalytic performance, Mater. Sci. Semicond. Process. 75 (2018) 183–192. https://doi.org/10.1016/j.mssp.2017.11.043

[62] F. Dong, Z. Zhao, T. Xiong, Z. Ni, W. Zhang, Y. Sun, W.K. Ho, In situ construction of g-C$_3$N$_4$/g-C$_3$N$_4$ metal-free heterojunction for enhanced visible-light photocatalysis, ACS Appl. Mater. Interfaces. 5 (2013) 11392–11401. https://doi.org/10.1021/am403653a

[63]   R. Ye, H. Fang, Y.Z. Zheng, N. Li, Y. Wang, X. Tao, Fabrication of CoTiO$_3$/g-C$_3$N$_4$ Hybrid Photocatalysts with Enhanced H$_2$ Evolution: Z-Scheme Photocatalytic Mechanism Insight, ACS Appl. Mater. Interfaces. 8 (2016) 13879–13889. https://doi.org/10.1021/acsami.6b01850

[64]   Z. Wang, Y. Huang, W. Ho, J. Cao, Z. Shen, S.C. Lee, Fabrication of Bi$_2$O$_2$CO$_3$/g-C$_3$N$_4$ heterojunctions for efficiently photocatalytic NO in air removal: In-situ self-sacrificial synthesis, characterizations and mechanistic study, Appl. Catal. B Environ. 199 (2016) 123–133. https://doi.org/10.1016/j.apcatb.2016.06.027

[65]   M. Reli, P. Huo, M. Šihor, N. Ambrožová, I. Troppová, L. Matějová, J. Lang, L. Svoboda, P. Kuśtrowski, M. Ritz, P. Praus, K. Kočí, Novel TiO$_2$/C$_3$N$_4$ Photocatalysts for Photocatalytic Reduction of CO$_2$ and for Photocatalytic Decomposition of N$_2$O, J. Phys. Chem. A. 120 (2016) 8564–8573. https://doi.org/10.1021/acs.jpca.6b07236

[66]   N. Tian, H. Huang, Y. Guo, Y. He, Y. Zhang, A g-C$_3$N$_4$/Bi$_2$O$_2$CO$_3$ composite with high visible-light-driven photocatalytic activity for rhodamine B degradation, Appl. Surf. Sci. 322 (2014) 249–254. https://doi.org/10.1016/j.apsusc.2014.10.071

[67]   C. Han, L. Ge, C. Chen, Y. Li, X. Xiao, Y. Zhang, L. Guo, Novel visible light induced Co$_3$O$_4$-g-C$_3$N$_4$ heterojunction photocatalysts for efficient degradation of methyl orange, Appl. Catal. B Environ. 147 (2014) 546–553. https://doi.org/10.1016/j.apcatb.2013.09.038

[68]   X. jing Wang, Q. Wang, F. tang Li, W. yan Yang, Y. Zhao, Y. juan Hao, S. jun Liu, Novel BiOCl-C$_3$N$_4$ heterojunction photocatalysts: In situ preparation via an ionic-liquid-assisted solvent-thermal route and their visible-light photocatalytic activities, Chem. Eng. J. 234 (2013) 361–371. https://doi.org/10.1016/j.cej.2013.08.112

[69]   H. Li, J. Liu, W. Hou, N. Du, R. Zhang, X. Tao, Synthesis and characterization of g-C$_3$N$_4$/Bi$_2$MoO$_6$ heterojunctions with enhanced visible light photocatalytic activity, Appl. Catal. B Environ. 160–161 (2014) 89–97. https://doi.org/10.1016/j.apcatb.2014.05.019

[70]   X. Hao, J. Zhou, Z. Cui, Y. Wang, Y. Wang, Z. Zou, Zn-vacancy mediated electron-hole separation in ZnS/g-C$_3$N$_4$ heterojunction for efficient visible-light photocatalytic hydrogen production, Appl. Catal. B Environ. 229 (2018) 41–51. https://doi.org/10.1016/j.apcatb.2018.02.006

[71]   S. Liang, D. Zhang, X. Pu, X. Yao, R. Han, J. Yin, X. Ren, A novel Ag$_2$O/g-C$_3$N$_4$ p-n heterojunction photocatalysts with enhanced visible and near-infrared light activity, Sep. Purif. Technol. 210 (2019) 786–797. https://doi.org/10.1016/j.seppur.2018.09.008

[72]   S. Balu, S. Velmurugan, S. Palanisamy, S.W. Chen, V. Velusamy, T.C.K. Yang, E.S.I. El-Shafey, Synthesis of α-Fe$_2$O$_3$ decorated g-C$_3$N$_4$/ZnO ternary Z-scheme photocatalyst for degradation of tartrazine dye in aqueous media, J. Taiwan Inst. Chem. Eng. 99 (2019) 258–267. https://doi.org/10.1016/j.jtice.2019.03.011

[73]   H. Liu, H. Zhou, X. Liu, H. Li, C. Ren, X. Li, W. Li, Z. Lian, M. Zhang, Engineering design of hierarchical g-C$_3$N$_4$@Bi/BiOBr ternary heterojunction with Z-scheme system for efficient visible-light photocatalytic performance, J. Alloys Compd. 798 (2019) 741–749. https://doi.org/10.1016/j.jallcom.2019.05.303

[74]   K. Dai, J. Lv, J. Zhang, C. Liang, G. Zhu, Band structure engineering design of g-C$_3$N$_4$/ZnS/SnS$_2$ ternary heterojunction visible-light photocatalyst with ZnS as electron transport buffer material, J. Alloys Compd. 778 (2019) 215–223. https://doi.org/10.1016/j.jallcom.2018.11.127

[75]   S. Bellamkonda, G. Ranga Rao, Nanojunction-mediated visible light photocatalytic enhancement in heterostructured ternary BiOCl/CdS/g-C$_3$N$_4$ nanocomposites, Catal. Today. 321–322 (2019) 18–25. https://doi.org/10.1016/j.cattod.2018.03.025

[76]   M. Mousavi, A. Habibi-Yangjeh, D. Seifzadeh, Novel ternary g-C$_3$N$_4$/Fe$_3$O$_4$/MnWO$_4$ nanocomposites: Synthesis, characterization, and visible-light photocatalytic performance for environmental purposes, J. Mater. Sci. Technol. 34 (2018) 1638–1651. https://doi.org/10.1016/j.jmst.2018.05.004

[77]   J. Barzegar, A. Habibi-Yangjeh, A. Akhundi, S. Vadivel, Novel ternary g-C$_3$N$_4$/Ag$_3$VO$_4$/AgBr nanocomposites with excellent visible-light-driven photocatalytic performance for environmental applications, Solid State Sci. 78 (2018) 133–143. https://doi.org/10.1016/j.solidstatesciences.2018.03.001

[78]   W. Liu, J. Shen, X. Yang, Q. Liu, H. Tang, Dual Z-scheme g-C$_3$N$_4$/Ag$_3$PO$_4$/Ag$_2$MoO$_4$ ternary composite photocatalyst for solar oxygen evolution from water splitting, Appl. Surf. Sci. 456 (2018) 369–378. https://doi.org/10.1016/j.apsusc.2018.06.156

[79]   F. Cheng, H. Yin, Q. Xiang, Low-temperature solid-state preparation of ternary CdS/g-C$_3$N$_4$/CuS nanocomposites for enhanced visible-light photocatalytic H$_2$ - production activity, Appl. Surf. Sci. 391 (2017) 432–439. https://doi.org/10.1016/j.apsusc.2016.06.169

[80]   D. Lu, H. Wang, X. Zhao, K.K. Kondamareddy, J. Ding, C. Li, P. Fang, Highly efficient visible-light-induced photoactivity of Z-scheme g-C$_3$N$_4$/Ag/MoS$_2$ ternary photocatalysts for organic pollutant degradation and production of hydrogen, ACS

Sustain. Chem. Eng. 5 (2017) 1436–1445.
https://doi.org/10.1021/acssuschemeng.6b02010

[81]   Y. Yuan, G.F. Huang, W.Y. Hu, D.N. Xiong, B.X. Zhou, S. Chang, W.Q. Huang, Construction of g-C_3N_4/CeO_2/ZnO ternary photocatalysts with enhanced photocatalytic performance, J. Phys. Chem. Solids. 106 (2017) 1–9. https://doi.org/10.1016/j.jpcs.2017.02.015

[82]   X. Rong, F. Qiu, Z. Jiang, J. Rong, J. Pan, T. Zhang, D. Yang, Preparation of ternary combined ZnO-Ag_2O/porous g-C_3N_4 composite photocatalyst and enhanced visible-light photocatalytic activity for degradation of ciprofloxacin, Chem. Eng. Res. Des. 111 (2016) 253–261. https://doi.org/10.1016/j.cherd.2016.05.010

[83]   M. Mousavi, A. Habibi-Yangjeh, Magnetically separable ternary g-C_3N_4/Fe_3O_4/BiOI nanocomposites: Novel visible-light-driven photocatalysts based on graphitic carbon nitride, J. Colloid Interface Sci. 465 (2016) 83–92. https://doi.org/10.1016/j.jcis.2015.11.057

[84]   A. Habibi-Yangjeh, A. Akhundi, Novel ternary g-C_3N_4/Fe_3O_4/Ag_2CrO_4 nanocomposites: Magnetically separable and visible-light-driven photocatalysts for degradation of water pollutants, J. Mol. Catal. A Chem. 415 (2016) 122–130. https://doi.org/10.1016/j.molcata.2016.01.032

[85]   A. Akhundi, A. Habibi-Yangjeh, Ternary magnetic g-C_3N_4/Fe_3O_4/AgI nanocomposites: Novel recyclable photocatalysts with enhanced activity in degradation of different pollutants under visible light, Mater. Chem. Phys. 174 (2016) 59–69. https://doi.org/10.1016/j.matchemphys.2016.02.052

[86]   M. Mousavi, A. Habibi-Yangjeh, Ternary g-C_3N_4/Fe_3O_4/Ag_3VO_4 nanocomposites: Novel magnetically separable visible-light-driven photocatalysts for efficiently degradation of dye pollutants, Mater. Chem. Phys. 163 (2015) 421–430. https://doi.org/10.1016/j.matchemphys.2015.07.061

[87]   A. Kumar, A. Kumar, G. Sharma, M. Naushad, R.C. Veses, A.A. Ghfar, F.J. Stadler, M.R. Khan, Solar-driven photodegradation of 17-β-estradiol and ciprofloxacin from waste water and CO_2 conversion using sustainable coal-char/polymeric-g-C_3N_4/RGO metal-free nano-hybrids, New J. Chem. 41 (2017) 10208–10224. https://doi.org/10.1039/C7NJ01580A

[88]   Y. Li, H. Zhang, P. Liu, D. Wang, Y. Li, H. Zhao, Cross-Linked g-C_3N_4/rGO Nanocomposites with Tunable Band Structure and Enhanced Visible Light Photocatalytic Activity, Small. 9 (2013). https://doi.org/10.1002/smll.201203135

[89]   J. Wan, C. Pu, R. Wang, E. Liu, X. Du, X. Bai, J. Fan, X. Hu, A facile dissolution strategy facilitated by H_2SO_4 to fabricate a 2D metal-free g-C_3N_4/rGO heterojunction for efficient photocatalytic H_2 production, Int. J. Hydrogen Energy. 43 (2018) 7007–7019. https://doi.org/10.1016/j.ijhydene.2018.02.134

[90]   H. Wang, Y. Liang, L. Liu, J. Hu, W. Cui, Highly ordered TiO_2 nanotube arrays wrapped with g-C_3N_4 nanoparticles for efficient charge separation and increased photoelectrocatalytic degradation of phenol, J. Hazard. Mater. 344 (2018) 369–380. https://doi.org/10.1016/j.jhazmat.2017.10.044

[91]   F. Hussin, H.O. Lintang, L. Yuliati, Enhanced Activity of C_3N_4 with Addition of ZnO for Photocatalytic Removal of Phenol under Visible Light 4G-PHOTOCAT View project vapochromic chemosensor of VOCs View project, Artic. Malaysian J. Anal. Sci. (2016). https://doi.org/10.17576/mjas-2016-2001-11

[92]   A. Kumar, A. Kumari, G. Sharma, B. Du, M. Naushad, F.J. Stadler, Carbon quantum dots and reduced graphene oxide modified self-assembled S@C_3N_4/B@C_3N_4 metal-free nano-photocatalyst for high performance degradation of chloramphenicol, J. Mol. Liq. 300 (2020) 112356. https://doi.org/10.1016/j.molliq.2019.112356

[93]   L. Chen, Y. Man, Z. Chen, Y. Zhang, Ag/g-C_3N_4 layered composites with enhanced visible light photocatalytic performance, Mater. Res. Express. 3 (2016) 115003. https://doi.org/10.1088/2053-1591/3/11/115003

[94]   B. Yuan, J. Wei, T. Hu, H. Yao, Z. Jiang, Z. Fang, Z. Chu, Simple synthesis of g-C_3N_4/rGO hybrid catalyst for the photocatalytic degradation of rhodamine B, Cuihua Xuebao/Chinese J. Catal. 36 (2015) 1009–1016. https://doi.org/10.1016/S1872-2067(15)60844-0

[95]   W.J. Ong, L.L. Tan, S.P. Chai, S.T. Yong, A.R. Mohamed, Surface charge modification via protonation of graphitic carbon nitride (g-C_3N_4) for electrostatic self-assembly construction of 2D/2D reduced graphene oxide (rGO)/g-C_3N_4 nanostructures toward enhanced photocatalytic reduction of carbon dioxide to methane, Nano Energy. 13 (2015) 757–770. https://doi.org/10.1016/j.nanoen.2015.03.014

[96]   Q. Sun, P. Wang, H. Yu, X. Wang, In situ hydrothermal synthesis and enhanced photocatalytic H_2-evolution performance of suspended rGO/g-C_3N_4 photocatalysts, J. Mol. Catal. A Chem. 424 (2016) 369–376. https://doi.org/10.1016/j.molcata.2016.09.015

[97]   L. Ge, C. Han, J. Liu, In situ synthesis and enhanced visible light photocatalytic activities of novel PANI-g-C_3N_4 composite photocatalysts, J. Mater. Chem. 22 (2012) 11843–11850. https://doi.org/10.1039/c2jm16241e

[98]   F. He, G. Chen, Y. Yu, S. Hao, Y. Zhou, Y. Zheng, Facile approach to synthesize g-PAN/g-C$_3$N$_4$ composites with enhanced photocatalytic H$_2$ evolution activity, ACS Appl. Mater. Interfaces. 6 (2014) 7171–7179. https://doi.org/10.1021/am500198y

[99]   L. Pi, R. Jiang, W. Zhou, H. Zhu, W. Xiao, D. Wang, X. Mao, g-C$_3$N$_4$ Modified biochar as an adsorptive and photocatalytic material for decontamination of aqueous organic pollutants, in: Appl. Surf. Sci., Elsevier B.V., 2015: pp. 231–239. https://doi.org/10.1016/j.apsusc.2015.08.176

[100] L. Tang, C. tao Jia, Y. cheng Xue, L. Li, A. qi Wang, G. Xu, N. Liu, M. hong Wu, Fabrication of compressible and recyclable macroscopic g-C$_3$N$_4$/GO aerogel hybrids for visible-light harvesting: A promising strategy for water remediation, Appl. Catal. B Environ. 219 (2017) 241–248. https://doi.org/10.1016/j.apcatb.2017.07.053

[101] X. Wang, S. Wang, W. Hu, J. Cai, L. Zhang, L. Dong, L. Zhao, Y. He, Synthesis and photocatalytic activity of SiO$_2$/g-C$_3$N$_4$ composite photocatalyst, Mater. Lett. 115 (2014) 53–56. https://doi.org/10.1016/j.matlet.2013.10.016

[102] M. Shen, L. Zhang, J. Shi, Converting CO$_2$ into fuels by graphitic carbon nitride-based photocatalysts, Nanotechnology. 29 (2018) 412001. https://doi.org/10.1088/1361-6528/aad4c8

[103] S. Fang, Y. Xia, K. Lv, Q. Li, J. Sun, M. Li, Effect of carbon-dots modification on the structure and photocatalytic activity of g-C$_3$N$_4$, Appl. Catal. B Environ. 185 (2016) 225–232. https://doi.org/10.1016/j.apcatb.2015.12.025

[104] K. Li, F.Y. Su, W. De Zhang, Modification of g-C$_3$N$_4$ nanosheets by carbon quantum dots for highly efficient photocatalytic generation of hydrogen, Appl. Surf. Sci. 375 (2016) 110–117. https://doi.org/10.1016/j.apsusc.2016.03.025

[105] G. Dong, L. Yang, F. Wang, L. Zang, C. Wang, Removal of Nitric Oxide through Visible Light Photocatalysis by g-C$_3$N$_4$ Modified with Perylene Imides, ACS Catal. 6 (2016) 6511–6519. https://doi.org/10.1021/acscatal.6b01657

[106] P. Zhang, T. Wang, H. Zeng, Design of Cu-Cu$_2$O/g-C$_3$N$_4$ nanocomponent photocatalysts for hydrogen evolution under visible light irradiation using water-soluble Erythrosin B dye sensitization, Appl. Surf. Sci. 391 (2017) 404–414. https://doi.org/10.1016/j.apsusc.2016.05.162

[107] Y. Wang, J. Hong, W. Zhang, R. Xu, Carbon nitride nanosheets for photocatalytic hydrogen evolution: Remarkably enhanced activity by dye sensitization, Catal. Sci. Technol. 3 (2013) 1703–1711. https://doi.org/10.1039/c3cy20836b

Materials Research Forum LLC

https://doi.org/10.21741/9781644901359-10

[108] S. Buddee, S. Wongnawa, P. Sriprang, C. Sriwong, Curcumin-sensitized TiO_2 for enhanced photodegradation of dyes under visible light, J. Nanoparticle Res. 16 (2014) 1–21. https://doi.org/10.1007/s11051-014-2336-z

[109] A. Zyoud, N. Zaatar, I. Saadeddin, M.H. Helal, G. Campet, M. Hakim, D. Park, H.S. Hilal, Alternative natural dyes in water purification: Anthocyanin as TiO_2-sensitizer in methyl orange photo-degradation, Solid State Sci. 13 (2011) 1268–1275. https://doi.org/10.1016/j.solidstatesciences.2011.03.020

[110] K. Takanabe, K. Kamata, X. Wang, M. Antonietti, J. Kubota, K. Domen, Photocatalytic hydrogen evolution on dye-sensitized mesoporous carbon nitride photocatalyst with magnesium phthalocyanine, Phys. Chem. Chem. Phys. 12 (2010) 13020–13025. https://doi.org/10.1039/c0cp00611d

[111] H. Zhang, S. Li, R. Lu, A. Yu, Time-Resolved Study on Xanthene Dye-Sensitized Carbon Nitride Photocatalytic Systems, ACS Appl. Mater. Interfaces. 7 (2015) 21868–21874. https://doi.org/10.1021/acsami.5b06309

[112] J. Xu, Y. Li, S. Peng, G. Lu, S. Li, Eosin Y-sensitized graphitic carbon nitride fabricated by heating urea for visible light photocatalytic hydrogen evolution: The effect of the pyrolysis temperature of urea, Phys. Chem. Chem. Phys. 15 (2013) 7657–7665. https://doi.org/10.1039/c3cp44687e

[113] S. Min, G. Lu, Enhanced electron transfer from the excited eosin y to mpg-C_3N_4 for highly efficient hydrogen evolution under 550 nm irradiation, J. Phys. Chem. C. 116 (2012) 19644–19652. https://doi.org/10.1021/jp304022f

[114] J. Fu, Q. Xu, J. Low, C. Jiang, J. Yu, Ultrathin 2D/2D WO_3/g-C_3N_4 step-scheme H_2-production photocatalyst, Appl. Catal. B Environ. 243 (2019) 556–565. https://doi.org/10.1016/j.apcatb.2018.11.011

[115] X. She, L. Liu, H. Ji, Z. Mo, Y. Li, L. Huang, D. Du, H. Xu, H. Li, Template-free synthesis of 2D porous ultrathin nonmetal-doped g-C_3N_4 nanosheets with highly efficient photocatalytic H_2 evolution from water under visible light, Appl. Catal. B Environ. 187 (2016) 144–153. https://doi.org/10.1016/j.apcatb.2015.12.046

[116] B. Tahir, M. Tahir, N.A.S. Amin, Photo-induced CO_2 reduction by CH_4/H_2O to fuels over Cu-modified g-C_3N_4 nanorods under simulated solar energy, Appl. Surf. Sci. 419 (2017) 875–885. https://doi.org/10.1016/j.apsusc.2017.05.117

[117] D.A. Giannakoudakis, N.A. Travlou, J. Secor, T.J. Bandosz, Oxidized g-C_3N_4 Nanospheres as Catalytically Photoactive Linkers in MOF/g-C_3N_4 Composite of

Hierarchical Pore Structure, Small. 13 (2017) 1601758.
https://doi.org/10.1002/smll.201601758

[118] J. Kavil, S. Pilathottathil, M.S. Thayyil, P. Periyat, Development of 2D nano heterostructures based on g-C_3N_4 and flower shaped MoS_2 as electrode in symmetric supercapacitor device, Nano-Structures and Nano-Objects. 18 (2019) 100317. https://doi.org/10.1016/j.nanoso.2019.100317

[119] X. Bai, L. Wang, R. Zong, Y. Zhu, Photocatalytic activity enhanced via g-C_3N_4 nanoplates to nanorods, J. Phys. Chem. C. 117 (2013) 9952–9961. https://doi.org/10.1021/jp402062d

[120] J. Meng, Y. Tian, C. Li, X. Lin, Z. Wang, L. Sun, Y. Zhou, J. Li, N. Yang, Y. Zong, F. Li, Y. Cao, H. Song, A thiophene-modified doubleshell hollow g-C_3N_4 nanosphere boosts NADH regeneration: Via synergistic enhancement of charge excitation and separation, Catal. Sci. Technol. 9 (2019) 1911–1921. https://doi.org/10.1039/C9CY00180H

[121] Y. Zheng, L. Lin, X. Ye, F. Guo, X. Wang, Helical Graphitic Carbon Nitrides with Photocatalytic and Optical Activities, Angew. Chemie. 126 (2014) 12120–12124. https://doi.org/10.1002/ange.201407319

[122] C.Q. Xu, K. Li, W. De Zhang, Enhancing visible light photocatalytic activity of nitrogen-deficient g-C_3N_4 via thermal polymerization of acetic acid-treated melamine, J. Colloid Interface Sci. 495 (2017) 27–36. https://doi.org/10.1016/j.jcis.2017.01.111

[123] J. Ding, W. Xu, H. Wan, D. Yuan, C. Chen, L.W.-A.C.B., undefined 2018, Nitrogen vacancy engineered graphitic C_3N_4-based polymers for photocatalytic oxidation of aromatic alcohols to aldehydes, Elsevier. (n.d.). (accessed August 28, 2020). https://doi.org/10.1016/j.apcatb.2017.09.048

[124] B. Liu, M. Qiao, Y. Wang, L. Wang, Y. Gong, T. Guo, X. Zhao, Persulfate enhanced photocatalytic degradation of bisphenol A by g-C_3N_4 nanosheets under visible light irradiation, Chemosphere. 189 (2017) 115–122. https://doi.org/10.1016/j.chemosphere.2017.08.169

[125] J.Y. Hu, K. Tian, H. Jiang, Improvement of phenol photodegradation efficiency by a combined g-C_3N_4/Fe(III)/persulfate system, Chemosphere. 148 (2016) 34–40. https://doi.org/10.1016/j.chemosphere.2016.01.002

Keyword Index

About the Editors

Dr. Gaurav Sharma research activity started in 2009 at Shoolini University (India) as master of philosophy student, and then, He continued his research work as PhD student with the preparation and characterization of diverse multifunctional nanomaterials, and their composites, specially focused on potential applications in environmental remediation (as photocatalysts and adsorbents). For four years He worked as assistant professor in the School of chemistry at Shoolini University (India), where He carried out diverse research lines, interrelated to each other based on synthesis and characterization of nanocomposites, hydrogels, bi and trimetallic nanoparticles, ion exchangers, adsorbents and photocatalysts etc. Moreover, He performed and taught different courses as nanochemistry, polymer chemistry, spectroscopy and natural products, among others. On the other hand, He supervised 3 PhD, 5 Master of Philosophy, and more than 25 Master and Bachelors students. He established collaborative research with various professors in countries as Finland, Saudi Arabia, China, Spain and South Africa. In this context, He was invited as visiting research professor from University of KwaZuklu-Natal (South Africa) in 2017 and 2019. In 2017, He joined as postdoctoral fellow at college of materials science and engineering, Shenzhen University. He got a project from China postdoctoral science foundation in 2018. The outcome of his research work was depicted in more than 140 publications, in various journals such as Renewable and Sustainable Energy Reviews, Chemical Engineering Journal, Journal of Cleaner Production, Carbohydrate Polymers, ACS Applied Materials and Interfaces, Journal of Hazardous Materials, Applied Catalysis B, and International Journal of Biological Macromolecules etc, 9 book chapters and 6 edited books. He is also serving as Director, International Research Centre of Nanotechnology for Himalayan Sustainability (IRCNHS), Shoolini University, India.

He is Highly Cited Researcher -2020 Crossfield (Web of Science); and also Ranked among World top 2% Scientists (Current year 2019 category) as per Stanford.

His h-index is 50, citations: 5682 (web of science); Google Scholar: h-index is 51, citations: 6068. He is Associate Editor of International Journal of Environmental Science and Technology (Springer) IF: 2,031. Editorial Board member of Current Organic Chemistry (IF:1.9), Current analytical Chemistry (IF:1.3), Materials-MDPI (IF:3.057) Innovations in Corrosion and Materials Science, Journal of Nanostructure in Chemistry (IF:4.0), Nanotechnology for Environmental Engineering (Springer), Letters in Applied

NanoBioScience etc, and Academic Editor of Journal of Nanomaterials (IF: 1.980) Advances in Polymer Technology (IF: 1.539).

Dr. Amit Kumar started his research career with Ph.D in Chemistry at Himachal Pradesh University, Shimla, India in 2008. He was awarded with the prestigious predoctoral scholarship of the Council of Scientific and Industrial Research (Government of India) in 2007. After his PhD, he worked as assistant professor at Shoolini University (India) for 3.5 years, where he was involved in research and teaching at undergraduate, postgraduate and doctoral level (2014-2017). He has experience of supervising three Ph.D students (all graduated). Furthermore, he has supervised 10 Masters students. He has worked as post doctoral fellow in the College of Materials Science and Engineering of Shenzhen University (2017-2019). He is involved in the field of optical designing of semiconductor heterojunctions as catalysts for environmental remediation and energy production. Currently He is a senior researcher at College of Materials Science and Engineering of Shenzhen University, PR China (01.01.2020 onwards). In addition he holds position of Co-director, International Research Centre of Nanotechnology for Himalayan Sustainability (IRCNHS), Shoolini University, India. He is also a Visiting Professor at School of technology, Glocal University, India. Moreover, he is a visiting faculty at Department of Chemistry and Physics from University of KwaZulu Natal, South Africa.

He has diverse research collaborations in Instituto de Catálisis y Petroleoquímica (Spain), King Saud University (Saudi Arabia), Chinese Research Academy of Environmental Sciences (China) and Universidade Federal do Rio Grande do Sul (Brazil). He is also ranked among World top 2% Scientists (Current year 2019 category) as per Stanford University rankings, 2020.

His scientific production encompasses 110 papers (WoS) including Applied Catalysis B: Environmental, Chemical Engineering Journal, ACS Applied Materials and Interfaces (I.F 8.75), Journal of Hazardous Materials, Journal of Cleaner Production etc. He has received h-index 47 and more than 5000 citations.